Engineering Systems for Safety

Related Titles

Constituents of Modern System-safety Thinking
Proceedings of the Thirteenth Safety-critical Systems Symposium, Southampton, UK, 2005
Redmill and Anderson (Eds)
1-85233-952-7

Developments in Risk-based Approaches to Safety
Proceedings of the Fourteenth Safety-critical Systems Symposium, Bristol, UK, 2006
Redmill and Anderson (Eds)
1-84628-333-7

The Safety of Systems
Proceedings of the Fifteenth Safety-critical Systems Symposium, Bristol, UK, 2007
Redmill and Anderson (Eds)
978-1-84628-805-0

Improvements in System Safety
Proceedings of the Sixteenth Safety-critical Systems Symposium, Bristol, UK, 2008
Redmill and Anderson (Eds)
978-1-84800-099-5

Safety-Critical Systems: Problems, Process and Practice
Proceedings of the Seventeenth Safety-critical Systems Symposium, Brighton, UK, 2009
Dale and Anderson (Eds)
978-1-84882-348-8

Making Systems Safer
Proceedings of the Eighteenth Safety-critical Systems Symposium, Bristol, UK, 2010
Dale and Anderson (Eds)
978-1-84996-085-4

Advances in Systems Safety
Proceedings of the Nineteenth Safety-critical Systems Symposium, Southampton, UK, 2011
Dale and Anderson (Eds)
978-0-85729-132-5

Achieving Systems Safety
Proceedings of the Twentieth Safety-critical Systems Symposium, Bristol, UK, 2012
Dale and Anderson (Eds)
978-1-4471-2493-1

Assuring the Safety of Systems
Proceedings of the Twenty-first Safety-critical Systems Symposium, Bristol, UK, 2013
Dale and Anderson (Eds)
978-1481018647

Addressing Systems Safety Challenges
Proceedings of the Twenty-second Safety-critical Systems Symposium, Brighton, UK, 2014
Dale and Anderson (Eds)
978-1491263648

Mike Parsons • Tom Anderson
Editors

Engineering Systems for Safety

Proceedings of the Twenty-third
Safety-critical Systems Symposium,
Bristol, UK, 3rd-5th February 2015

**Safety-Critical
Systems Club**

The publication of these proceedings is
sponsored by BAE Systems plc

Editors

Mike Parsons
CGI UK Ltd
Kings Place
90 York Way
London
N1 9AG
United Kingdom

Tom Anderson
Centre for Software Reliability
Newcastle University
Newcastle upon Tyne
NE1 7RU
United Kingdom

ISBN 978-1505689082

© Safety-Critical Systems Club 2015. All Rights Reserved

Individual chapters © as shown on respective first pages

Preface

This volume contains the papers presented at the twenty-third Safety-critical Systems Symposium (SSS'15). This year's authors have, as usual, produced interesting and informative material covering many topics that are of concern to the safety-critical systems community; we are grateful to them for their contributions.

The first day focuses on Sector Safety Engineering, and Tools and Techniques. The first two papers are concerned with the marine sector, the third with rail. The keynote paper is given by Nancy Leveson and compares the techniques of SAE ARP 4761 with STPA. This is followed by papers on how to apply an evidence-based methodology to safety engineering and formal methods applicability to ISO 26262. Approaches to safety design are discussed in the last paper of the day.

The two themes of the second day are Cases and Arguments, and Risks. The first papers look at construction of safety arguments and security cases. The next two papers examine the underlying arguments of two existing safety standards, DO-178C and IEC 61508. The keynote address by Peter Bernard Ladkin takes an interesting view on risks and how we respond to them, drawing on topical examples. The next two papers are about living and working with risk – one on board a yacht, the other in aerospace manufacturing. The final paper of the day is concerned with the uncertainty of demonstrating requirements.

New Methods, and People and Skills are the themes of the final day. The morning's papers consider new approaches to safety problems: one by placing data at the heart of the assessment, another examining the safety issues of swarms. The keynote address, given by Tim Kelly and Chris Megone, covers the ethics of acceptable safety, looking at issues safety practitioners have to address daily. The final papers concern people issues within an organisation.

We are grateful to our sponsors for their valuable support and to the exhibitors at the Symposium's tools and services fair for their participation. And we thank Joan Atkinson and her team at Newcastle for laying the event's foundation through their exemplary planning and organisation.

MP & TA
December 2014

A message from the sponsors

BAE Systems is pleased to support the publication of these proceedings. We recognise the benefit of the Safety-Critical Systems Club in promoting safety engineering in the UK and value the opportunities provided for continued professional development and the recognition and sharing of good practice. The safety of our employees, those using our products and the general public is critical to our business and is recognised as an important social responsibility.

The Safety-Critical Systems Club

organiser of the

Safety-critical Systems Symposium

Safety-critical systems and the accidents that don't happen

When an aircraft crashes, it makes headlines. That hundreds of thousands of flights each week do not crash is accepted as routine. Airliners, air traffic control systems, railway signalling, car braking systems, defence systems, nuclear power stations and medical equipment (increasingly including home medical electronics) are some of the complex, largely digital, systems in use, on which life and property depend. That these safety-critical systems do work well is because of the expertise and diligence of professional systems safety engineers, regulators and other practitioners who work to minimise both the likelihood that accidents will occur, and the consequences of those that do. Their efforts prevent untold deaths every year. The Safety-Critical Systems Club (SCSC) has been actively engaged for nearly twenty-five years to help to ensure that this continues to be the case.

What is the Safety-Critical Systems Club?

The SCSC is the UK's professional network and community for sharing knowledge about safety-critical systems. It brings together engineers and specialists from a range of disciplines working on safety-critical systems in a wide variety of industries, academics researching the arena of safety-critical systems, providers of the tools and services that are needed to develop the systems, and the regulators who oversee safety. It provides, through publications, seminars, workshops, tutorials, a web site and, most importantly, at the annual Safety-critical Systems Symposium, opportunities for them to network and benefit from each other's experience in working hard at the accidents that don't happen. It focuses on current and emerging practices in safety engineering, software engineering, and product and process safety standards.

What does the SCSC do?

The SCSC maintains a website (scsc.org.uk), which includes directories of tools and services that assist in the development of safety-critical systems. It publishes a regular newsletter, Safety Systems, three times a year. It organises seminars, workshops and training on general matters or specific subjects of current concern, which are prepared and led by world experts. Since 1993 it has organised the annual Safety-critical Systems Symposium (SSS) where leaders in different aspects of safety, from different industries, including regulators and academics, meet to exchange information and experience, with the papers published in a proceedings volume. From time to time, the SCSC supports relevant initiatives, such as the current Data Safety Initiative that is addressing concerns raised about data in safety-related systems. The SCSC carries out all these activities to support its mission:

> ... *to raise awareness and facilitate technology transfer in the field of safety-critical systems* ...

History

The SCSC began its work in 1991, supported by the Department of Trade and Industry and the Engineering and Physical Sciences Research Council. The Club has been self-sufficient since 1994, but enjoys the active support of the Health and Safety Executive, the Institution of Engineering and Technology, and BCS, The Chartered Institute for IT; all are represented on the SCSC Steering Group.

Membership

Membership may be either corporate or individual. Individual membership, which costs £95 a year, entitles the member to Safety Systems three times a year, other mailings, and discounted entry to seminars, workshops and the annual Symposium. Frequently individual membership is paid by the employer.

Corporate membership is for organisations that would like several employees to take advantage of the benefits of SCSC programmes. The amount charged is tailored to the needs of the organisation.

For more information about membership, or to join, please call Joan Atkinson on +44 191 221 2222 or email Joan.Atkinson@ncl.ac.uk

Contents

Subsea Safety Shutdown Architectures: Present and Future
Eberechi Weli, Adrian Allan 1

Are modern safety systems leading to deficiencies in post-accident control measures?
Martin Toland 23

Formal Modelling of Railway Safety and Capacity
Alexei Iliasov and Alexander Romanovsky 39

Keynote Address
A Comparison of SAE ARP 4761 and STPA Safety Assessment Processes
Nancy Leveson et al 55

Can Evidence-Based Software Engineering Contribute To Safer Software?
K.R.Wallace 79

Applicability of Formal Methods for Safety-Critical Systems in the Context of ISO 26262
Susanne Kandl et al 95

Functional Safety by Design – Magic or Logic?
Derek Fowler 117

Controlled Expression for Assurance Case Development
Katrina Attwood and Tim Kelly 143

Systematically Self-Reflecting Safety-Arguments: Introduced, Illustrated and Commended
Stephen E. Paynter 167

A Case Study of Security Case Development
Benjamin D. Rodes et al 187

Explicate '78: Uncovering the Implicit Assurance Case in DO–178C
C. Michael Holloway 205

Using a Goal-Based Approach to Improve the IEC 61508-3 Software Safety Standard
Thor Myklebust et al 227

Keynote Address
Risks People Take and Games People Play
Peter Bernard Ladkin 245

Risk Tolerance: A tale of professional yachtsman and meddling bastards
Les Chambers 265

Assessing the Safety Risk of Collaborative Automation within the UK Aerospace Manufacturing Industry
Amira Hamilton and Phil Webb 285

Uncertainty in Demonstrating Requirements
Clive Lee 301

Copernic Safety
José Miguel Faria 321

The Data Elephant
Paul Hampton and Mike Parsons 335

Approximate verification of swarm-based systems: a vision and preliminary results
Benjamin Herd et al 361

Demonstrating Compliance in the Arctic
Nick Golledge 379

Keynote Address
The Ethics of Acceptable Safety
Tim Kelly and Chris Megone, with Ibrahim Habli, Mark Nicholson, Kevin Macnish and Andrew Rae 393

Combining Organisational and Safety Culture Models
Elizabeth Jacob 409

Author Index 427

Subsea Safety Shutdown Architectures: Present and Future

Eberechi Weli, Adrian Allan

> Aker Solutions
>
> Aberdeen, UK

Abstract *This paper provides a general overview of the subsea architecture for safe system shutdowns as currently defined in today's generic subsea systems application requirements. Subsequently, this paper identifies emerging requirements and associated characteristic complexities in the design and use of the key features of emerging subsea shutdown systems. This paper critically examines the objectives and benefits of the shift towards the selective high integrity controlled or commanded shut-down patterns. Using two similar projects with comparable application parameters for safe shutdown, we submit a platform that can be used when differentiating between the pre-requisites and solutions for subsea control systems; now and in the future*

1 Introduction

Subsea safety systems are currently undergoing a significant step change in the Oil & Gas industry.

Historically, it has typically been "good enough" to isolate production flow from a subsea well by removing power to the field, resulting in valves to spring-return to their failsafe position, but as this paper proffers, there has recently been a trend towards requirements for more controlled, targeted shutdowns becoming the norm. This in turn leads to the safety system having to be more flexible and focussed, thus the architecture becomes more complex.

Risk reduction is a primary concern for the industry, with recent incidents highlighting the potential impact of failure (e.g. Deepwater Horizon) in terms of humans, environment, assets and economics. Considering that the number of easy to exploit fields are dwindling, meaning exploration moves to harsher environments in deep water and arctic areas, safety not only continues to be a key consideration in the design and to exploitation of future fields but it is also developing and adapting suit the emerging needs. In turn, this has seen associated technology develop and mature resulting in requirements for subsea safety systems to become more refined and onerous. As will be described by this paper, this includes a shift from simple hardwired safety systems to more complex, communication based systems with an increased number of valves (final elements) requiring individual, sequenced control and monitoring of subsea instrumentation (also requiring communication protocols suitable for safety related systems).

This paper provides an overview of a typical subsea process, including the role of a subsea safety shutdown system. Typical present-day shutdown safety system architectures are described followed by the developments being seen in the emerging requirements which is used to demonstrate, via a case study, the fundamental changes required to the subsea safety shutdown system architecture.

2 Subsea Safety Shutdown (Overview)

Control and monitoring of subsea wells requires a number of components which typically include:

Hydraulic Power Unit (HPU)
Electrical Power Unit (EPU), supplied by an Uninterruptable Power Supply (UPS)
Master Control Station (MCS)

- Which contains a Subsea Power and Communications Unit (SPCU), or Subsea Control Unit (SCU)

The system is monitored and, in some design applications, controlled from an Operator Supervisory System (DCS) located on a host, or topside facility[1].

Key interfaces from subsea to topside that provide a complete subsea control system, and are relevant to the safety shutdown functionality as depicted in Figure 1, include:

Subsea manifold
Subsea Control Modules (SCM)
Subsea valves under control within the Xmas Trees (XMT).

---

[1] On an offshore installation, topside systems or facilities are those located on a platform or process/production unit above sea level. Figure 1 shows topside and subsea located control systems

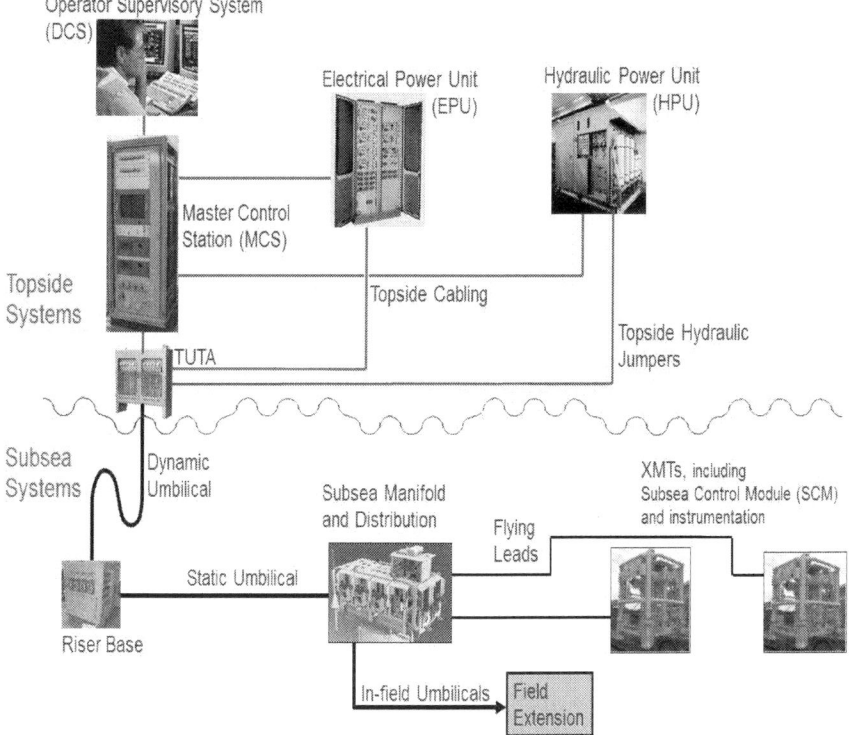

Fig. 1. Subsea Control System Overview

The Subsea Manifold is made up of pipes and valves designed to transfer the process fluid from one or more wellheads to a common subsea pipeline.

The Subsea Control System provides a means of controlling and monitoring the subsea production systems located on the seabed. This includes the following:

Operator control of tree, including choke and manifold valves used for the control of production from the subsea wells.

- Includes the necessary interlocks to avoid inadvertent, or out of sequence actions, and ensure that choke operation [1] does not reduce hydraulic pressure sufficiently to cause valve "dropout".

Monitor and display inferred position of actuated XMT spring return valves and actual position of XMT mounted choke valves.

Monitor and display processed data from all subsea instrumentation.

---

[1] Choke operations allow throttling control of the production flow from the XMT to the pipeline. The operation maintains required production characteristics downstream of the production wing valve (PWV) by controlling flow, pressure or temperature of a process fluid. Figure 5 presents PWV and choke locations on a XMT.

Monitor and display control system equipment housekeeping parameters
Provide necessary operator station displays, to permit full control and monitoring of the subsea wells.
Generate the required shutdowns from safety shutdown inputs.
Set parameters, enunciate, log time, tag, group and suppress alarms
Permit low and high level access for operator/engineer reconfiguration as required
Perform system fault diagnosis

In addition to the above, risk reduction (i.e. protection against environmental, safety or asset damage) is provided by the safety shutdown capabilities of the Subsea Control System. This is generally expected to provide the means for a safe shutdown on failures of the subsea production or production control subsystems, such as overpressures, loss of containment, leakage or ingress of seawater.

A further need for system shutdown can originate from loss of communications with the dedicated subsea production control systems tasked with safety shutdown related functions. This includes diagnostics and monitoring, particularly for remotely dependent control systems.

The HPU supplies correct and stable clean hydraulic fluid pressure to the subsea system. The HPU typically can provide high-pressure (HP) and low-pressure (LP) outputs to the systems subsea using HP/LP pumps, pressure regulators, pressure safety valves, filters and accumulators. During activation of a safety shutdown (abandon vessel or platform scenario), the HPU depressurises and vents hydraulic power from subsea, depending on application requirements.

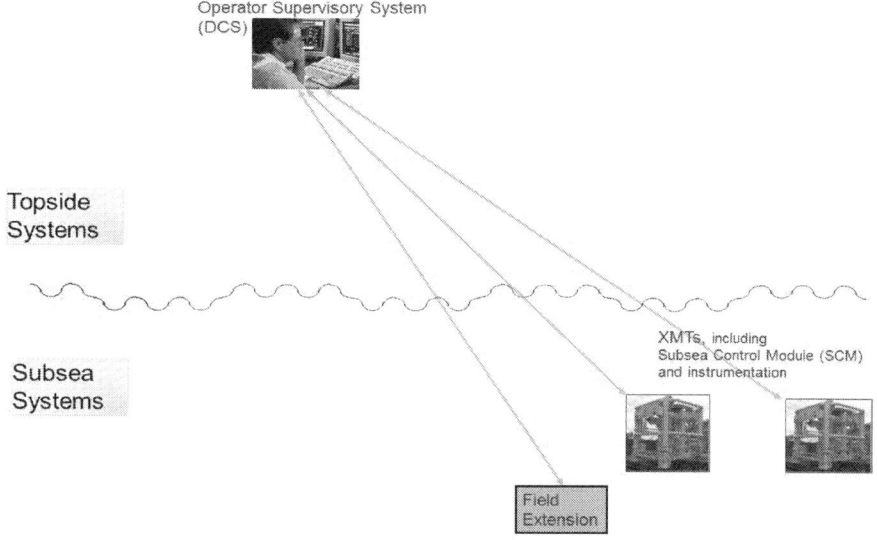

Fig. 2. Operator Supervisory System

The SPCU has a hardwired ESD signal interface, which turns off power to subsea equipment. Control and monitoring functions in the SPCU are routed to the Operator Supervisory System. Monitoring of status of all data communication components in the subsea control system is achieved through the Operator Supervisory System. Additionally a subsea control system network administration function is available for diagnostic and maintenance on the topside facility.

The Subsea Control Modules are the heart of the subsea control and monitoring system which includes the electronics and hydraulics required to action commands sent from the Operator Supervisory System. The retrievable housing for the subsea control module primarily consists of:

Subsea Electronic Modules (SEM)
Electro-hydraulic directional control valves (DCV) and other valves
Feed through connectors (electric and hydraulic)
Control Module Base
Control Module Housing
Lock/unlock mechanism for running tool (installation)
Internal sensors and transmitters
Filters/strainers
Accumulators
Pressure intensifiers (optional)
Pressure reducers (optional)
Chemical injection regulation valves

The Subsea Tree (or XMT) system provides subsea monitoring and controls production from a subsea well. The subsea valves under control include, but are not limited to:

Downhole Safety Valve (DHSV) / Surface Controlled Subsea Safety Valve (SCSSV)
Production Master Valve (PMV)
Production Wing Valve (PWV)
Annulus Master Valve (AMV)
Annulus Wing Valve (AWV)
Crossover Injection Valve (XOV)
Methanol/Chemical Injection Valve (MIV/CIV)
Scale Inhibitor Injection Valve
Corrosion Inhibitor Injection Valve
Production Choke valve (PCV)
Injection Choke Valve (ICV)
Manifold Valves
Chemical Injection Control Valve

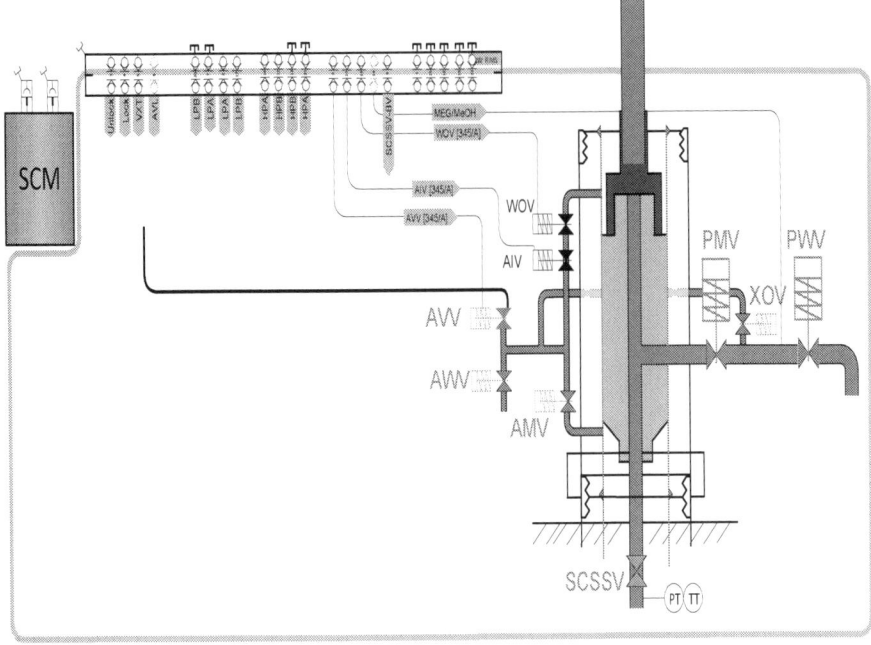

Fig. 3. XMT Schematic, showing selected valves

Some of these valves are considered primary to safety shutdowns depending on the specific operational or application requirements. Subsea valves that typically feature in standard safety shutdowns (i.e. achieving well and associated systems safe condition in the event of abnormalities such as failures, leakages or overpressure) irrespective of application characteristics are the SCSSV, PMV and PWV.

For this paper, the subsea system includes all elements involved in subsea production from the well head and the Surface Controlled Subsea Safety Valve (SCSSV) to riser emergency shutdown valves (ESDV), including the production control systems (topside and subsea). The safety shutdown is the remotely controlled or manual action taken by the dedicated subsea system to prevent or minimise asset damage, loss of production, environmental degradation or human safety.

2.1 Requirements and Architectures

The generally accepted operating modes of the safety shutdown systems as highlighted in (Curran 2014) are:

(a) detection of an abnormal condition potentially leading to an undesirable event by automatic monitoring,
(b) automatic protective action if manually activated by personnel who observes or is alerted to an unsafe condition following detection,
(c) continuous protection by support systems that minimise the effects of escaping hydrocarbons or other undesirable events.

Most subsea applications inherently come with a great level of unpredictability. This inadvertently impacts on the technology developed to meet the existing application requirements and results in substantial technological developments to meet varying applications needs. Subsequently, standardisation of safety shutdown requirements and systems continues to evolve. In this section, we introduce some key requirements that directly influence the design of the subsea safety shutdown systems and form part of the final safety systems validation.

Several publications related to subsea control and safety systems unintentionally focus on production loss and asset damage. This is somewhat influenced by the HSE publication (HSE 2005) on Safety Instrumented Systems for pipeline riser overpressure protection which advocates the use of fully rated pipelines to reduce or prevent major safety hazards with less reliance on the safety shutdown system. The same publication provides a striking note of caution that a safety shutdown system should be implemented where self-acting full flow mechanical relief is impracticable and it is uneconomical to fully rate the pipeline and riser to maximum pressure (e.g. where the pipeline is so long that rating it for maximum pressure is feasible but renders the project uneconomic), or in cases where it is not possible to fully rate the pipeline and riser.

This implies that specific application engineering assessments supplemented by cost-benefit and risk reduction studies are the ideal drivers for subsea systems design. That being the case, environmental and safety integrity driven subsea safety shutdowns may become the norm as we delve further into high pressure, high temperature developments.

With this in mind, we can clearly and concisely state the high level safety architectural requirements with the primary objective of reducing or preventing major safety hazards. Subsequently, other supporting subsea safety shutdown systems requirements to achieve and maintain the primary objectives are described in the following paragraphs.

High Level Safety Requirements (Primary Objectives for Subsea Shutdown)
Pipeline rupture is a major safety hazard if it occurs near people. However this could also result in major environmental and economic losses. (HSE 1996) stipulates that the operation of the subsea pipeline must be within the safe operating limits and a safety shutdown system is required to restrict operations in the event of any abnormal operating conditions or faults giving rise to a potential for overpressure.

The requirement for controlling the subsea loss of containment/leakage resulting in hydrocarbon release can also find its origins in (HSE 1995) which requires the use of safety critical systems to prevent the uncontrolled release of flammable or explosive substances.

High Level Supplementary Requirements (Secondary Objectives for Subsea Shutdown)
With the primary safety shutdown requirements established, a summary of the requirements to achieve and maintain the systems are comprehensively discussed below and follows the subsea control and shutdown systems design overview as illustrated previously.

One of the objectives of the safety shutdown is to isolate and sectionalise subsea production subsystems in a fast and reliable manner without any unintended and unknown system failures. This objective brings with it the need for systems integrity in the design and implementation of safety shutdown systems. Systems critical to safety shutdown must be application specific and designed/developed to determined safety integrity levels. In so doing, this generally provides confidence in the safety shutdown strategy where the shutdown logic reflects the overall installation arrangement, system design and operational characteristics. The high level safety requirements typically consist of:

Specific application characteristics, such as reservoir chemical and flow characteristics
The process safety time, including the required safety shutdown response time
Safety shutdown independence and segregation requirements
Application specific systems expected in achieving the fail-safe shutdown
Potential interactions (especially for existing host facilities with an existing safety case)
Operating modes for the safe shutdown systems - start, reset/re-start and abnormal/degraded conditions.
Considerations for maintenance, testing, replacements/retrievals, modifications and obsolescence management

Operational maintenance and testing philosophies that ensure adequate availability of the safety shutdown system are comparatively not emphasised in the safety shutdown requirements (i.e. ensuring safety integrity of the shutdown system is not compromised through its application life).

2.2 Safety Shutdown Hierarchies

A typical subsea safety shutdown system will have a number of safety functions which are generally structured in a defined hierarchy with escalating priority, where superior levels can initiate lower levels, but not the other way round.

Figure 4 below shows the simplified flow of a typical hierarchy, ranging from a higher priority ESD-1 down to the lower priority ESD-2, followed by PSD. Terminology may vary across projects but the diagram below is sufficient for illustrative purpose in this paper. Figure 5 presents an example of where the hierarchies are implemented in a subsea safety shutdown and control design.

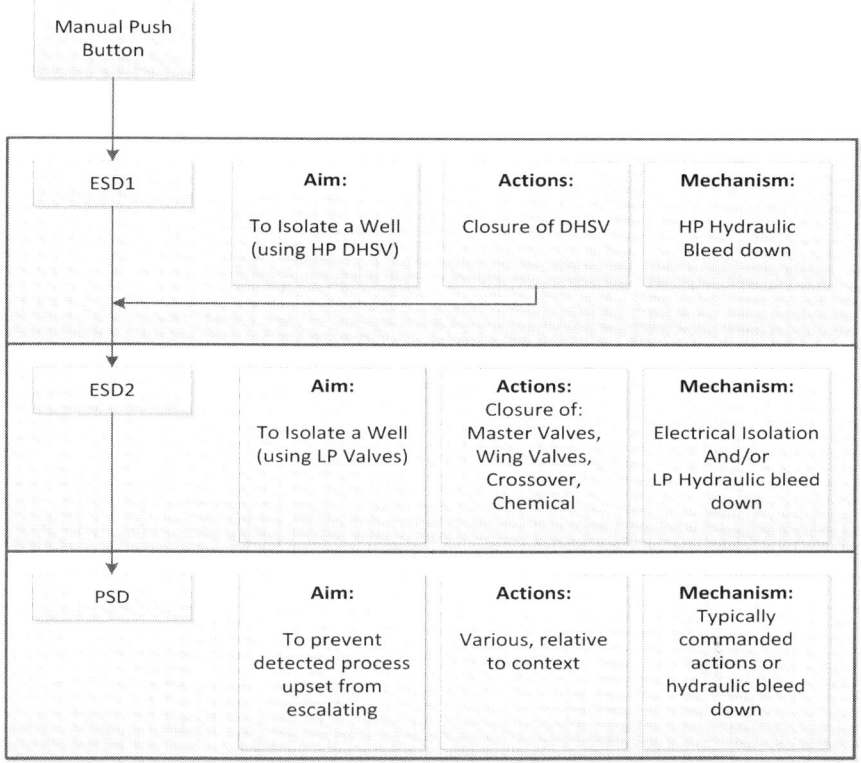

Figure 4. Typical ESD Hierarchy (simplified)

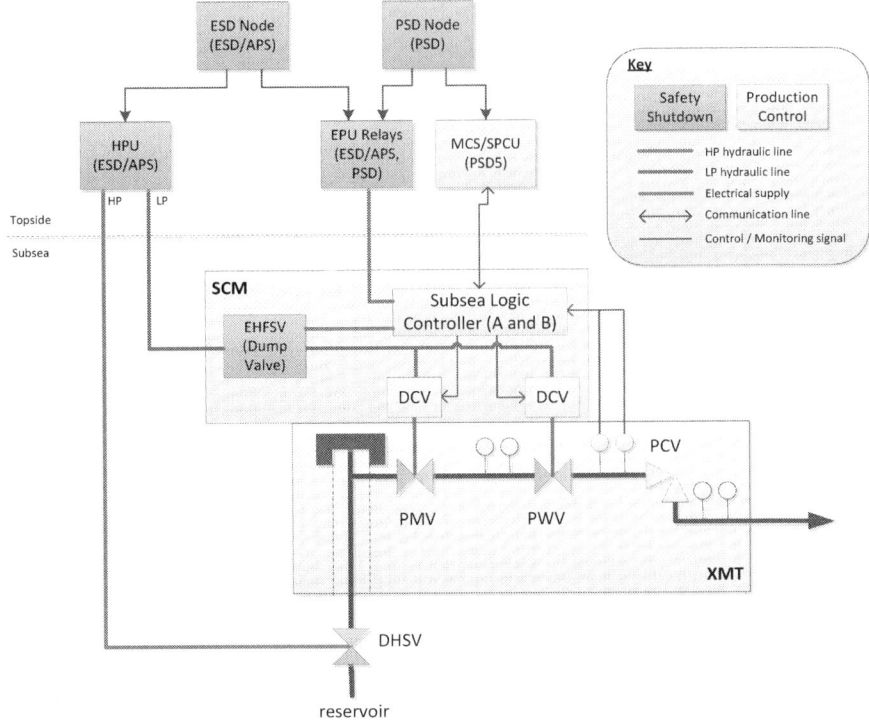

Figure 5. An example of a design application of the shutdown hierarchies

2.3 Subsea Fail-Safe Philosophy

The fail-safe requirement is established in all subsea operational principles and a constitutional element of most subsea development contracts. As applied in the subsea industry and presented in the industry accepted Subsea Engineering Handbook (Bai & Bai 2010), subsea production control and safety shutdown systems currently rely on the hydraulic and electrical fail-safe design philosophy. These electro-hydraulic systems have been in existence for decades as depicted in (Locheed & Phillips 1979). A more recent example of this dependency on electrical designs includes the first all-electric subsea system (Gerardin et al 2009).

In order to achieve this, all subsea valves related to safety shutdown functions are designed to return to a safe position on loss of hydraulic pressure or electrical power. For systems using the combined electro-hydraulic design, the hydraulic supply in each SCM on a XMT is designed to depressurise the Low Pressure (LP) circuit and isolate the umbilical through an electrically held discharge/depressurise valve. To support these safety functions, the communications subsystems of the subsea control and shutdown system are designed with message prioritisation functions. The prioritisation is in line with the fail-safe design and follows the shutdown hierarchies of the subsea application or operations. Thus, the combination of fail-safe design for safety shutdown and subsea control hierarchy takes the order (as described in the previous section):

ESD and PSD functions (generally fail-safe and first line of defence in the event of abnormal and safety-related hazards being detected)
Subsea production control functions (maintenance, testing, shutdown – where shutdown is of less priority to the ESD/PSD function).

The above subsea control and shutdown functions are influenced by the fail-safe philosophy which ensures that in all subsea operations, and solutions using control systems, all system failures or abnormalities result in a safe condition

2.4 Subsea Safety Shutdown Requirements

By introducing the general safety shutdown requirements and systems in the preceding sections, we now have a platform to comprehensively establish and introduce the primary safety shutdown functional requirements that is of utmost concern to the subsea system development team. These are presented in no systematic order.

The safe state can generally be achieved by closure of the wing valves, crossover valves and, depending on application requirements, closure of the annulus valves (as indicated in Figure 5). The conventional subsea safety shutdown is reliant on:

PWV closed or (PMV and XOV) closed
AWV closed or (AMV and XOV) closed
DHSV closed

However, this paper submits that this base case is shifting towards additional shutdown requirements and prompts the need for design changes and implementation. The emerging requirements are generally influenced by the increasing activities in subsea safety assessments. These are further driven by the potential economic benefits derived from minimising the impact of shutdowns (i.e. selected shutdowns with quicker resets and restarts) and derivation of specific application requirements for subsea safety shutdowns.

There follows a brief re-introduction of the current conventional shutdown requirements prior to presenting the potential emerging safety shutdown requirements.

Protection of the XMT and associated systems
The rationale behind this shutdown requirement is generally due to overpressure situations downstream of the production choke or abnormalities detected within

the subsea manifold and XMT. The safe shutdown prevents damage to the XMT from potential blowouts, and other effects of overpressure leading to significant loss of production (individual well isolation depending on design), asset damage and loss of environmental integrity. This is achieved by isolating the flow of hydrocarbons to the subsea production systems. This particular shutdown function provides for protection of the well, well-base, well casings, individual XMT, the Subsea Manifold, umbilicals and associated piping. In order to meet this safety shutdown requirement, the safety shutdown systems implementation considers:

Isolation of individual XMT on overpressure detection downstream of the production choke (subsea autonomous shutdown)
Isolation of individual (or several) XMTs on activation of the designated topside Emergency Shutdown (ESD)
Isolation of individual XMT by topside activation of the designated Process Shutdown (PSD)

Protection of the flowline and riser overpressure
The protection of assets downstream of the manifold such as risers and subsea flowlines is essentially achieved by isolating the flow of hydrocarbons to the flowline on detection of overpressure usually by a high integrity pressure sensor. The protection of the riser or pipelines downstream of the riser base reduces potential consequences of failure subsea to the topside facilities (personnel safety hazards). There is usually some emphasis on system solutions to this shutdown requirement. The pipeline can have a design safety factor, or can be fully rated, thereby reducing the risk of overpressure. A fully rated pipeline is one designed to withstand maximum pressures i.e. designed against application specific overpressure or loss of containment risks. An under-rated pipeline or riser system may compromise the overall ALARP solution for topside systems (such as ventilation, deluge release[1] on gas detection and fire and gas detection and suppression) that are reliant in the overall operational risk reduction. Cost-benefit assessments, and in some cases field application unknowns, may have an impact on decisions to go with a robust pipeline design without the additional integrity of an overpressure detection and safe shutdown requirement. In this paper, we assume the decision on safe shutdown requirement takes priority over the pipeline design.

Similarly, considerations for the flowline protection safety shutdown design and implementation are influenced by:

Isolation of individual XMT on overpressure detection downstream of the production choke (subsea autonomous shutdown)
Isolation of the Manifold on detection of overpressure downstream of the Manifold (subsea autonomous shutdown)
Isolation of individual (or several) XMTs by topside ESD
Isolation of individual (or several) XMTs by topside PSD

---

[1] Deluge release equipment is part of the topside fire and gas safety systems installed to prevent escalation of radiated heat or fire.

3 Emergency Safety Shutdown – Present/Conventional

The primary purpose of the ESD is to prevent escalation of abnormal conditions into a major safety hazard and to restrict the period and expansion of any such safety hazards. As we narrow down focus to the ESD system, it is worth noting that the subsea ESD operational concepts are derived from the need to mitigate against hydrocarbon releases through leakages and overpressure protection. (Bratland 1995) presents an improved understanding of emergency shutdown systems whilst (Shanks et al 2012) offers a broad description of the safety-critical systems involved in emergency shutdowns. The ESD operational requirements (introduced in Section 2.4) generally offer the basis for a detailed assessment and comparison between subsea shutdown systems in subsequent sections of this paper. However, for the purpose of this paper we shall narrow down the ESD requirements and design assumptions to:

Isolation of an individual well including DHSV/SCSSV
Isolation of an individual excluding DHSV/SCSSV

The isolation of an individual well from the manifold/flowline by activation of topside ESD logic solver [1] and closure of well valves (relevant XMT valves including DHSV) is also introduced in the ESD hierarchies and relates to ESD-1.

ESD-2 is the isolation of one well from manifold/flow-line by activation of topside ESD logic solver and closure of well valves (relevant XMT-valves excluding DHSV).

Other possible ESD functional requirements not included in this paper may include the isolation of one well (annulus & process flow) from manifold utility/service lines by activation of topside ESD logic solver and closure of relevant XMT valves.

The typical ESD System is topside initiated by cutting power to dedicated topside relays that de-energise the XMT Electrically Held Failsafe Valve (EHFSV) and venting of the hydraulic system.

3.1 Interfaces

In general, the ESD system for an offshore installation will interface with fire and gas detection, emergency depressurisation and flare vent, topside process safety, alarms and communications, human machine interface, security, emergency power and lighting, drilling hoisting, ballast water and positioning, heating ventilation and air conditioning (HVAC) and other similar safety systems.

---

[1] The Subsea ESD logic solvers are typically electrical, electronic or programmable electronic based equipment in the form of relays, trip amplifiers or programmable logic controllers (PLC). IEC 61511 (2004) Part 1 defines a logic solver as "*the portion of either a basic process control system or safety instrumented system that performs one or more logic functions*".

This demonstrates that the supplementary requirements (presented in Section 2.4) such as segregation and independence are reasonably realised.

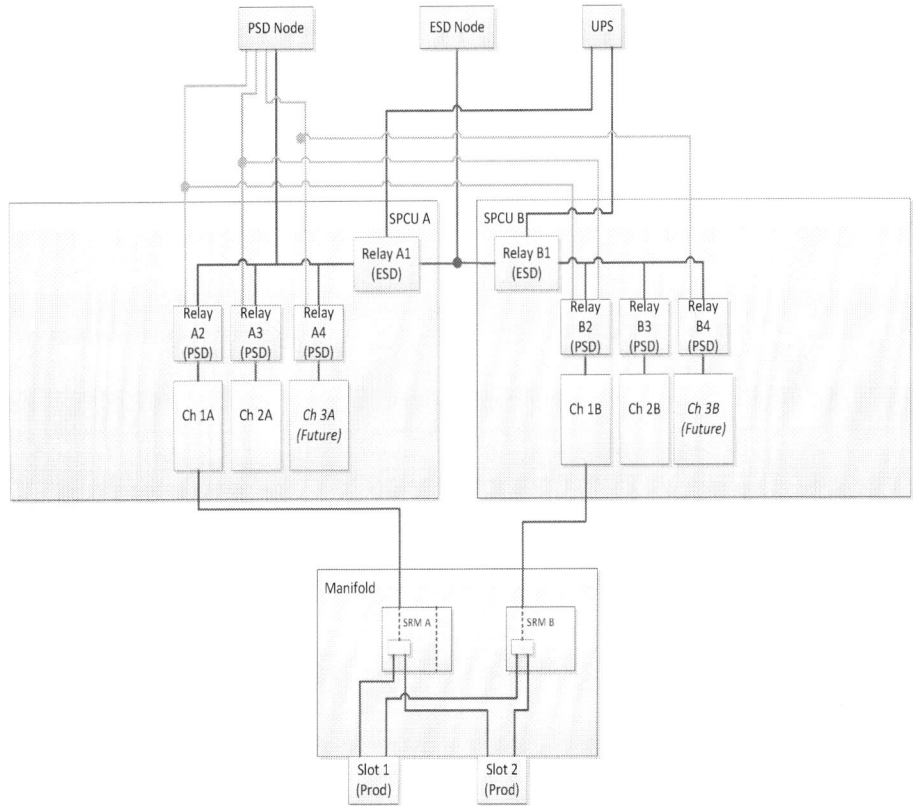

Fig. 6. ESD Solution using electrical isolation

4 Emerging Safety Shutdown Requirements and Challenges

The opinion that subsea system designs and field developments were restricted to a limited number of wells due to a lack of confidence in manifold systems, flow-line connections, productions riser capacities and difficulties with down-hole and equipment maintenance as presented in (Barker & Lim 1986) is now a distant memory.

Current offerings in the subsea systems field are synonymous with more audacious expansions by technologically savvy engineering institutions and organisations undertaking the most complex of subsea production and systems development. These complexities are usually associated with increased level of risks (with environmental, safety and production loss consequences) some of which are unknown, others not yet realised. The emerging roles for subsea trees as presented in (Fenton 2009) offers an initial introduction into potentially significant influences of subsea trees and subsea systems on the performance of subsea production systems.

With the expanding subsea activities around the world and the growing safety, environmental and asset damage risks that are to be managed, advanced technology and associated applications have also escalated. To bring it all back to the failsafe principles, what is currently on offer, where there is a genuine gap to fill etc., it is important to highlight the emerging safety shutdown requirements and assess how these apply to Emergency Shutdown applications.

There are several requirements that support safety shutdown claims, however we have chosen to bring to the fore those that are challenging emergency shutdown designs and applications. These are the selective sequential control for safety shutdown systems, segregation of shutdown and control systems (i.e. elimination of shared elements as far as is reasonably practicable) and subsea safety monitoring. These design and implementation challenges are presented in subsequent sections.

In addition, other requirements intensifying the conditions upon which the integrity of safe shutdowns now depend include testability, process shutdown or response times, resetting and restarting in the event of shutdown actions, additional safety function capabilities in different operating modes.

The ability to provide the required in-service testing, proof testing (test intervals) for all critical safety function lines and elements directly impacts the integrity of the safety shutdown functions.

As systems become more defined and subsea operators look to optimise risk reduction, the maximum response time of the safe shutdown function is gradually reducing. Consequently, the total reaction time for each safety function (i.e. subsea safe state) is further plummeting despite the cumulative number of safe functions to be achieved. This includes durations for monitoring, selection or voting and activation.

4.1 Sequential Actions

A growing expectation due to technological advancements in control and communication systems is that a number of subsea valves can be selectively commanded for a shutdown; this is now integral to safety shutdown strategy.

With the conventional ESD systems still highly dependent on hardwired (electro-hydraulic) designs, modularity was always the significant challenge (and configurability of the shutdown in the sequences required within a hardwired control system). International safety standards generally consider modularity and configurability as beneficial to safety system designs as it provides the following essential features to system fault tolerance and safety shutdown claims:

Ability to maintain system functionality despite an internal fault
Testability of multi-functional safety-critical elements (unit to integrated system)
Removes the existing restrictions on mode selection and operating parameters/modes.

4.2 Segregation

One noteworthy and potentially non-compliant feature of most existing ESD systems is that well safety functions (PSD and ESD) and control functions are implemented by the use of a control system for safety and non-safety functions (no segregation) and shared final elements (in this case, subsea valves). Independence between PSD and ESD is not ensured and furthermore, elements of the safety shutdown functions are not independent of well control used during normal operations. An example of this is the subsea valves; however, common subsea valves and associated hydraulic components are considered acceptable.

This feature is essential to the avoidance and control of systematic failures and invariably, the subsea safety shutdown strategy. There are two main reasons for segregation in safety system design practice and as stipulated in a growing number of safety shutdown requirements, the development of safety systems must consider:

The use of physical separation (i.e. segregation) of "channels" can provide defence against a wide range of common mode failures in redundant systems.
Safety functions must be separate from the control functions operated by a control system

Clear segregation to meet this requirement can be achieved at different levels of the system architecture. SW design with segregated functions undertaken by different functional blocks (high complexity – most difficult to prove), electronic/control safety system (medium complexity) and overall safety system (physical separation can be easily proved, however is associated with scale or size challenges)

4.3 Monitoring – Subsea Safe State

Some ambiguity still exists in monitoring of functions and the level of integrity required for monitoring designated "safety functions" subsea. If the fail-safe design philosophy is fully adopted, we find there are conflicts with the more detailed requirements and design considerations for ESD systems. A note in IEC 61784 Part 3 states *"The resulting safety integrity claim of a system depends on the implementation of the selected functional safety communications profile within this system*". This statement removes all ambiguity that currently exists between proposed design requirements and the standards/guides on monitoring.

The importance of monitoring of safety functions is now made more pronounced by the Deepwater Horizon accident investigations (DHSG 2011). The subsea shutdown requirements now consider the monitoring of safety shutdown functions to have similar integrity to the monitored functions. (Altamiranda 2007) proposes intelligent supervision, fault detection and diagnosis framework for subsea control systems. This piece of work provides a good level of understanding of subsea detection and diagnostics; however it doesn't address the integrity requirements for specific subsea control and shutdown applications.

In a communications based ESD system, the safety-critical data communications should now look to use an established safety comms protocol robust enough for use in safety system monitoring. This is even more pertinent as use of the communications protocol for triggering shutdowns relies heavily on the primary safety-critical elements for status monitoring. The integrity of the monitoring/comms is a primary requirement for claiming diagnostics especially for claims on dangerous failure detection i.e. reduction in dormant undetected failures.

For the developing ESD systems to meet the growing need for safe monitoring of subsea systems, multi-function safety monitoring is a feature that must be implemented. Table A in (IEC 61508 2010a) shows the importance of monitoring/data comms to the maximum diagnostic coverage creditable. This is also considerably addressed in Annex A of (IEC 61508 2010b). The tricky part of fully complying with this requirement is how to factor in monitoring and detection of localised failures to the design.

5 Case Study

In the case of isolating an individual well, the assumption that the XMT valves and down-hole safety valve are sufficient to provide a safe state in the event of shutdown is increasingly becoming considered a flawed concept. This is due to the increasingly complex parameters encountered in new subsea exploits, such as:

subsea depth
environmental conditions
high pressure, high temperature

To illustrate the differences between conventional and emerging ESD requirements, let us consider two projects with similar application characteristics. We will assume the shutdown requirement is to isolate one production well.

Project A (shown in Table 1 below) is derived from safety shutdown functional requirements for a conventional subsea development project, as introduced in Section 2.

Project B (shown in Table 2 below) is based on emerging requirements as introduced in Section 4. The emerging requirements are a result of subsea developments heading further into complex application environments. This brings with it the potential for additional safety functions via the closure of an increasing number of subsea valves, commanded sequentially and providing the same goal of safe state under all operating scenarios.

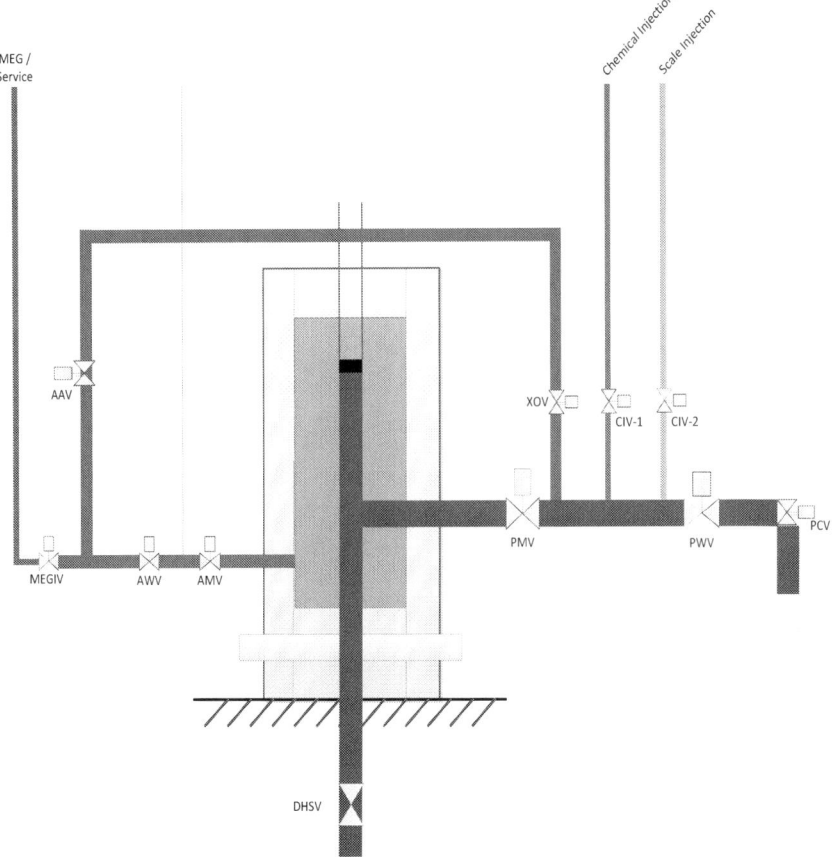

Fig. 7. Simplified XMT (showing potential additional safety shutdown valves)

Table 1. Project A Requirements (Conventional)

Project A (Conventional)	Safety Shutdown Function (ESD)	
	ESD-01	ESD-02
Definition	Isolate one subsea well (including DHSV)	Isolate one Subsea well (XMT valves only, without DHSV)
Functional Boundaries	Isolate one well from manifold, flowline by activation through topside ESD logic solver and closure of well valves (relevant XMT-valves including DHSV).	Isolate one well from manifold, flowline by activation through topside ESD logic solver and closure of well valves (relevant XMT-valves excluding DHSV).
Subsystems	ESD node, SPCU ESD relays, topside HPU ESD valves, subsea valves	ESD node, SPCU ESD relay, topside HPU ESD valves, Subsea valves.
Safe State	[PWV] or [PMV & XOV] & [AWV] or [AMV & XOV] & [DHSV] closed	[PWV] or [PMV & XOV] & [AWV] or [AMV & XOV]

Table 2. Project B Requirements (Emerging)

Project B (Emerging)	Safety Shutdown Function (ESD)	
	ESD-01	ESD-02
Definition	Isolate one subsea well (including DHSV)	Isolate one Subsea well (XMT valves only, without DHSV)
Functional Boundaries	Isolate one well from manifold, flowline by activation through topside ESD logic solver and closure of well valves (relevant XT-valves including DHSV).	Isolate production and chemical injection lines from production wells, excluding closure of DHSV, upon ESD activation.
Subsystems	ESD Node, ESD (Topside) Logic Solver, HPU ESD valves, Subsea Logic Solver, Subsea Solenoid valves, subsea valves	ESD Node, ESD (Topside) Logic Solver, Subsea Logic Solver, Subsea Solenoid valves subsea valves.
Safe State	Combinations of either: [PWV & AMV & CIV] or [PWV & AWV & CIV] & DHSV closed OR [PMV & AMV] or [PMV & AWV & AAV & XOV closed] or [PMV & AWV & AAV & XOV & CIV] & DHSV closed	Combinations of either: [PWV & AMV & CIV] or [PWV & AWV & CIV] closed OR [PMV & AMV] or [PMV & AWV & AAV & XOV closed] or [PMV & AWV & AAV & XOV & CIV] closed
Additional Requirements	Selective sequential shutdown of subsea valves in the combinations shown above.	Selective sequential shutdown of subsea valves in the combinations shown above

5.1 Proposed Safety Design Solution

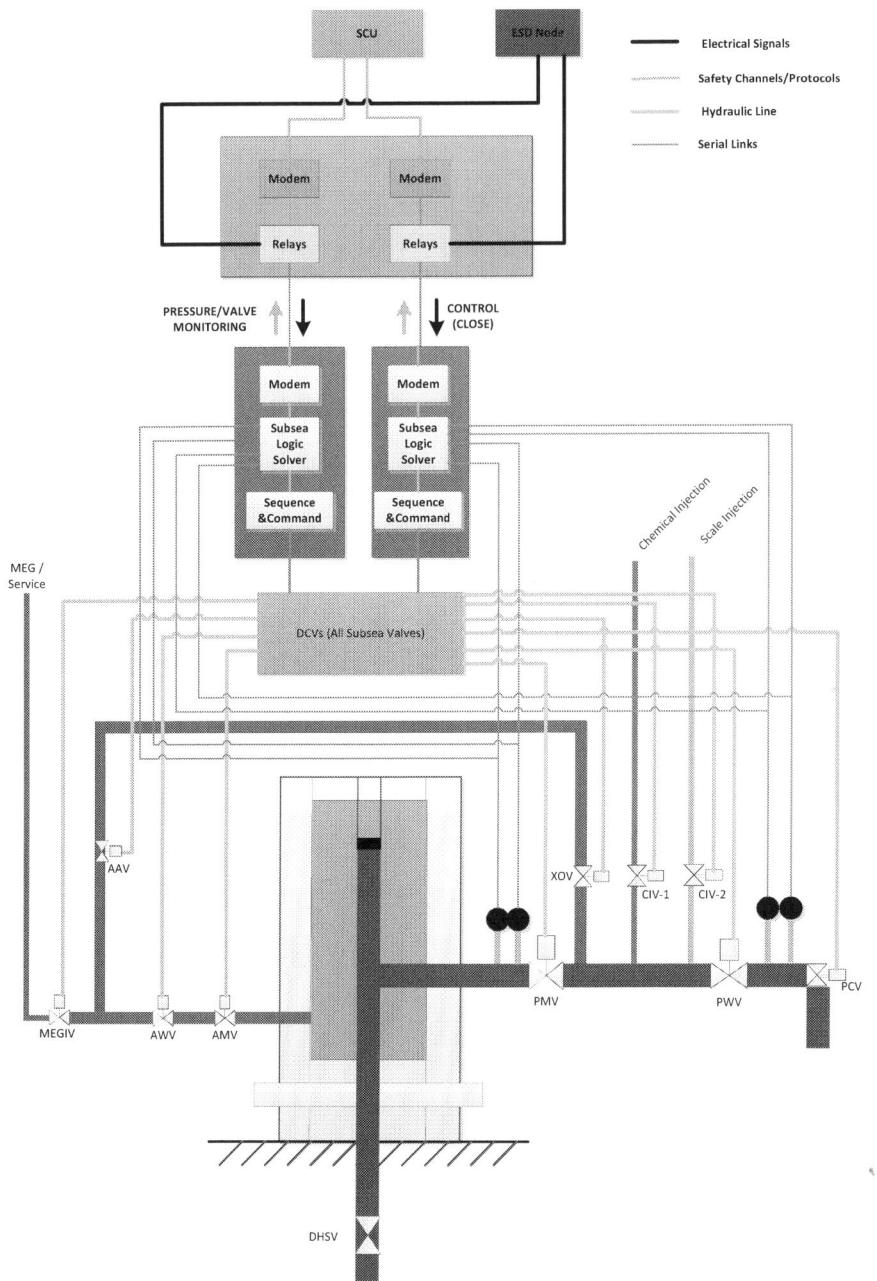

Fig. 8. Proposed design to meet emerging requirements

Future designs to meet the growing complexity of subsea safety shutdown functions will, as a necessity, have:

High integrity subsea status monitoring
Use of safety communications link to topside
Quick action selective commands to subsea as a paramount feature
A good level of assurance of the state of the subsea valves and condition of critical assets following a safe shutdown action

The design proposed to meet these emerging changes to conventional applications can be subsea initiated or topside controlled. The topside controlled safety shutdown system is presented in Figure 8 above.

The proposed solution provides for the correct combination and sequence of XMT valves being closed where sequences are considered primary to the safety shutdown function (i.e. achieving the safe state).

6 Conclusions

The limitations of orthodox ESD hardwired design which was suitable for conventional safety shutdown requirements includes the inability to shutdown individual production wells via ESD isolation of subsea supplies. This means that a shutdown request on one well of a template will result in all wells of that type on that template being shutdown.

As operational characteristics become more complex the emerging requirements suggest that a greater number of tree valves than normally encountered will become commonplace. This will have significant impact on the demand for advanced control systems with corresponding safety integrity. In addition, long term maintenance of the integrity of these advanced systems also suggests an increase in the need for robust maintainability and testability as primary requirements for safety shutdowns.

The benefits of the proposed system, and similar, specifically for ESD requirements where the subsea valves required for a safe condition are increased beyond the norm and selected sequential commands are paramount to obtaining a safe state, include:

1. Assured commanded controlled safety shutdowns
2. Optimised high integrity system depressurisation and downstream well control
3. Greater choice of valves to shut down as part of overall safety function resulting in reduced asset damage i.e. greater valve control based on high integrity status monitoring.
4. Optimised shutdowns – less susceptibility to response constraints

5. Greater control of recovery and resetting of the subsea systems resulting in increased safety and availability.

References

Curran J.C. (2014). Safety Systems for Subsea Applications. Offshore Technology Conference, 2014. OTC-25366-MS

HSE (2014). Safety instrumented systems for the overpressure protection of pipeline risers. SPC/TECH/OSD/31. http://www.hse.gov.uk/foi/internalops/hid_circs/technical_osd/spc_tech_osd_31.htm. Accessed December 1, 2014.

HSE (1996). Pipelines Safety Regulations. HSE Books

HSE (1995). Offshore Installations (Prevention of Fire and Explosion, and Emergency Response) Regulations. HSE Books.

Bratland O (1995). Emergency Shutdown Systems: Improved understanding of design requirements. 27th Annual Offshore Technology Conference. OTC-7718.

Shanks E, Pfeifer W, Savage S, Jain A (2012). Enhanced Subsea Safety critical Systems. Offshore Technology Conference. OTC-23480.

Lim J, Barker G (1986). Deepwater Multiwell Subsea Production System. Offshore Technology Conference. OTC-5092.

Fenton S (2009). Emerging roles for subsea trees: portals for subsea system functionality. Offshore Technology Conference. OTC-20108.

Bai Y, Bai Q (2010). Subsea Engineering Handbook. Elsevier Inc.

Locheed E, Phillips R (1979). A high integrity electrohydraulic subsea production control system. Offshore Technology Conference. OTC-3357.

DHSG (2011). Final Report on the Investigation of the Macondo Well Blowout. Deepwater Horizon Study Group, Centre for Catastrophic Risk Management, University of California, Berkeley.

Altamiranda, E (2007). Intelligent Supervision and Integrated Fault Detection and Diagnosis for subse control systems.

Gerardin P, Mackenzie R, van den Akker, J (2009). The first all-electric subsea system on stream: development, operational feedback and benefits for future applications. Offshore Europe Oil & Gas Conference, 2009. SPE 124290

IEC 61784 (2010). Industrial communication networks – Profiles – Part 3: Functional safety fieldbuses – general rules and profile definitions.

IEC 61508 (2010a). Functional safety of electrical/electronic/programmable electronic safety-related systems - Part 1: General requirements.

IEC 61508 (2010b). Functional safety of electrical/electronic/programmable electronic safety-related systems – Part 7: Overview of techniques and measures.

IEC 61511 (2004). Functional safety – Safety instrumented systems for the process industry sector – Part 1: Framework, definitions, system, hardware and software requirements

Are modern safety systems leading to deficiencies in post-accident control measures?

Martin Toland

STS Defence

Gosport, UK

Abstract *Modern advances in technology and highly coordinated safety systems have successfully reduced the frequency of major accidents in the marine industry. Has this decrease in frequency come at a cost? The myth of an "unsinkable" ship went down with the Titanic and yet accidents resulting in loss of life still occur. On occasion the control measures put in place to mitigate against disaster are not sufficient due to reliance on human factors or technology which may not be available after the accident. This paper discusses the issues caused by the focus on pre-accident mitigation and scenarios deemed too implausible by hazard identification techniques.*

1 Introduction

There are many hierarchies of risk control in place and followed as best practice across a number of industries. All of these hierarchies abide by the principle that, where possible, a hazard should be designed out to reduce the risk associated with a given task or operation. This has had a positive effect within the marine industry achieving a reduction of accident frequencies and severities, but with an unforeseen side effect.

By making the likelihood of an accident significantly lower the time, effort and ultimately money spent on control measures for that accident may be reduced, while still satisfying an As Low As Reasonably Practicable (ALARP) argument.

This paper will explore examples of accidents at sea, showing how the scenario was attempted to be prevented and how effective, or otherwise, the post-accident control measures were in reducing the accident severity. The major elements to be explored are the unforeseen side effects of the human factors and safety culture; declared the cause of most accidents at sea in the 2013 Safety and Shipping Review (Allianz 2013).

2 Lessons from the past

A number of historical incidents have shaped the approach to maritime safety we have today. Undoubtedly the most famous of these is the loss of RMS Titanic in 1912. The sinking of a high profile vessel and the deaths of 1517 people, including many celebrities of the day transformed the approach to provision of life saving equipment on-board passenger ships and saw design improvements relating to watertight integrity and ultimately survivability of a craft until rescue was available. The International Convention for the Safety of Life at Sea (SOLAS) which is used today as the benchmark for maritime safety standards was born out of the events in the Atlantic that night. For example until that incident the provision of life boats was determined by ship size not number of passengers.

Further changes to vessel design for Roll On Roll Off ships and procedural controls to ensure watertight integrity control measures were improved after the 1987 capsize of the MV Herald of Free Enterprise shortly after sailing from Zeebrugge. This disaster that was accompanied by the loss of 193 passengers and crew, followed the failure by crew to shut the bow doors, resulting in water ingress and the subsequent free surface effect (Department of Transport 1987).

Further changes in vessel design standards occurred in 1989 following the inquiry into the fire on board the Tor Scandinavia passenger ferry resulting in recognition that greater provision for firefighting alarms and equipment was necessary. Although the ship was saved, two people died in the fire and had it not been for the actions of individuals this total would have been far higher after the failure of passenger alarm systems (Steele and Steele 1996).

Over 100 years since the Titanic was declared by some, although notably not the designers at Harland and Wolf, to be "unsinkable", similar complacency is being shown by reliance on design to ensure safety while controls put in place post-incident are often found to be lacking when thoroughly exercised or called upon for real. With accident rates still averaging 9 losses a month globally (Allianz, 2013) the importance of controlling these risks remains high.

3 Regulation, design and emergency provision

It is clear that significant events, especially in this age of transparency and media coverage, are always followed by extensive inquiries and analysis to identify failings and root causes. Each subsequent tightening of design rules and modifications to procedures makes an accident less likely and improves safety, but as the perceived likelihood of an incident is reduced the potential for complacency grows. This section discusses how the lessons of accidents are translated into formal requirements by regulatory bodies.

Much has been written on the subject of ship classification and the rigours that are applied to ship design so this section will briefly outline the links between regulatory bodies, classification standards and the manner in which ships are operated and equipped in order to maximise the safety of the vessel and those onboard.

3.1 Regulation

Regulation at sea is laid down by the International Maritime Organisation (IMO). They have established best practice and protocols to be followed. In addition to this nations can set their own rules for their own waters or vessels operating under their flag provided the local rule conforms to the conventions of the IMO and is an enhancement, not relaxation of the rules.

Although they are the de facto arbiter of standards the IMO do not have any enforcement capability. This responsibility is passed to the Flag Sate Authority who will enforce the IMO rules and any of their own. Some countries have adopted significantly higher standards of safety, environmental protection and welfare which need to be met for a vessel to fly the flag of that country and be registered therein. Other nations have lower standards merely adhering to IMO minimum requirement and are often dubbed "flags of convenience" giving opportunity to circumvent improvements in best practice and thereby reduce the level of safety achieved; the second highest country for registered tonnage is Liberia with 99.10 million tonnes (Allianz, 2012), but its safety standards are below that of less popular flag states. In an attempt to deliver uniform implementation of the rules the IMO introduced an amendment to SOLAS in 1994 called the International Safety Management Code, or ISM code; although local variation still applies the ISM code reduces this variability of standards (Trafford 2009).

3.2 Design and classification

The classification societies set the technical standards and specifications that must be achieved in order for a vessel to be certified as conforming to classification standards. This certification is not a warranty or guarantee of safety or that a vessel is seaworthy or fit for sea; it only certifies that a ship conforms to the standards developed and published by the society issuing the classification.

Design and classification also considers the future operating environment of the ship, for example any vessel wishing to use the Panama Canal must conform to Panamax (or new Panamax) criteria for size and tonnage, similar restrictions apply for the Suez Canal.

3.3 Emergency provision

As previously stated the SOLAS regulations detailing the Life Saving Apparatus (LSA) to be a carried by ships are a result of the 1912 Titanic disaster. Although the rules have been revised over the years the fundamental principle still remains that the ship in question should have sufficient provision of LSA appropriate to the vessel's size, number of passengers and task. The scope of SOLAS has also increased to include the firefighting arrangements onboard, first aid, radio communications etc. as well as details on technical specifications for vessels and their management and operation. Further details on vessel management and operation are described in the ISM Code issued by the IMO in response to the loss of the Herald of Free Enterprise.

In addition to stating the equipment to be carried there are standards for the performance of LSA given in SOLAS. Additional product specification and quality checks are afforded through the work of other organisations. For example the European Union has detailed the Marine Equipment Directive (MED) to give details on product requirements and manufacture of that product.

While the requirements for emergency provision are clearly defined it must be noted that these measures are reactive and are present to deal with the situations caused by failures of design or procedure that have led to an accident. This emergency provision will only reduce an accident severity post initiating event, not reduce the frequency of the event occurring.

4 Pre vs. Post Mitigation

The regulatory regime outlined in section 3 provides the basis for well-designed ships with systems that attempt to prevent failures through effective systems and controls. Secondarily it details measures to be taken to alleviate the severity of an accident where it cannot be prevented completely. The engineering response to a failure is to identify the probable causes and then design features to reduce the frequency of such an event. These design features typically form the barriers on the left hand side of a Bow Tie Diagram (Figure 1).

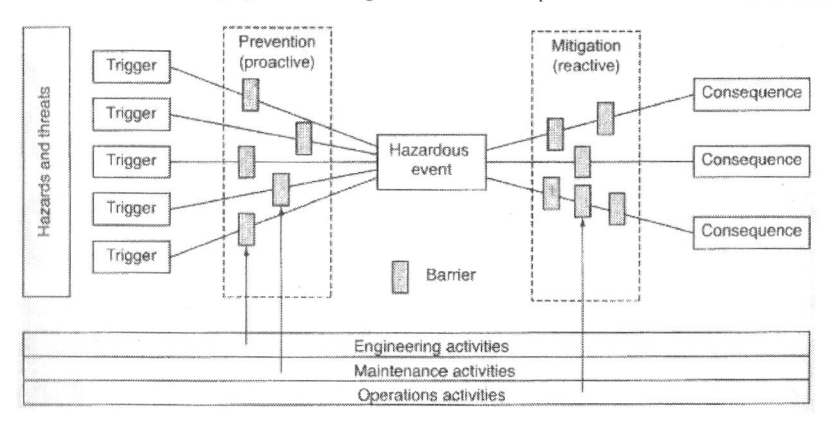

Fig 1. Bow Tie Diagram

In the context of marine accidents the threat would come from fire, collision or grounding while the control measures are typically designed features to meet regulation or best practice e.g. watertight integrity, propulsion and manoeuvrability. The recovery measures would include the LSA or firefighting equipment coupled with procedures for their use to prevent the consequences of loss of ship, loss of life, loss of cargo etc.

The engineering response of putting additional barriers in place to prevent the initiating event can result in a left loaded Bow Tie Diagram where focus is on prevention to the detriment of recovery. As will also be seen some of the recovery measures may not be available after an incident if there is insufficient resilience in design e.g. LSA being unavailable due to the list of the ship or electrical and hydraulic supplies for ship services being out of action after an incident. The other feature that can occur is that control measures are engineered design solutions while recovery measures are procedural and subject to greater variance because of the actions of the people involved.

5 Failures of designed control features

Globally the marine industry spends significant time and money to ensure that ships are well designed, maintained and operated. Despite this, accidents occur which result in loss of life, vessel and goods. The example of the Estonia (Figure 2) shows adherence to requirements; the provision of indicator lights for the status of the bow doors. However this design feature was insufficient to prevent the sinking of the ship when damage was sustained to the bow.

Fig 2. Estonia

The passenger ferry Estonia was sailing between Tallinn and Stockholm when she sank on 28 September 1994. It was established at the subsequent inquiry that the bow visor (shown raised in Figure 2) that was designed to protect the ship's ramp and access to the car decks had suffered a catastrophic failure. This led to it separating from the hull and pulling the ramp open allowing water to access the car deck (HSVA 2008). Although fitted, neither the door nor ramp sensors indicated as open on the bridge, because the nature of the failure was such that it did not move the sensor from the closed position. This system was a lesson identified after the Herald of Free Enterprise accident yet was not sufficient on this occasion to alert the bridge watch keepers to the failure that had occurred.

The water that flowed onto the car decks caused a free surface effect and the ship to capsize in a manner similar to that of the Herald of Free Enterprise. On this occasion however the depth of water meant the vessel was lost with 852 fatalities. With the ship on its side evacuation was nearly impossible and most of the casualties failed to escape the ship before she sank. Of those that did make it to the upper deck approximately half perished in the cold water of the Baltic Sea (HSVA, 2008).

As with the Costa Concordia this accident shows that LSA can only be used in the right circumstances and that while in many cases the casualty vessel is the safest place to be, early preparations for evacuation are essential in case the situation deteriorates.

A second issue identified was associated with the bow door. Although the Estonia was built to the correct specifications, the bow door was more suitable for coastal work, not the heavier seas experienced out in the open waters of the Baltic (HSVA 2008). Whilst this may not have precluded the Estonia from this particular route it should have seen a review of the maintenance schedules, especially the degree of material fatigue around the bow visor; the focal point for the wave's energy as the ship crossed the rough water. Faith in the design and classification approvals could have led to complacency in this regard. Modern ships will frequently pass to new owners across the course of their life in service and changes in their operating environment must be considered to ensure the applicability of existing checks and controls.

Without doubt the most prominent accident in recent memory is that of the cruise ship Costa Concordia (Figure 3). Unlike many accidents, where a ship will disappear beneath the waves, the wreck of the Costa Concordia has until very recently been sat on the foreshore of Isola del Giglio. Of note in figure 3 are the two life rafts hanging redundantly on the ship's port side unable to reach the water for evacuation and rescue.

Fig 3. Costa Concordia

Based on the account given in Marine Emergencies for Mates and Masters (House 2014) the Costa Concordia was sailing on the 13 January 2012 when she manoeuvred too close to land and tore a 50 metre hole in the ship's port side. The ship soon began to list and lost power before drifting back onto the island where it came to rest. Of the 4252 people on-board just 32 died, but had the ship rolled over instead of resting on the shore this total would have been far higher.

A point of continued debate in the media, and in the courts, is the role of the crew in the evacuation, especially the senior members of the ship's company. It is understandable for passengers to be panicked and afraid, however crew responses should be more proficient especially when looking at the command and control of an evacuation. Crew responses are discussed further in subsequent sections. Following Costa Concordia the cruise industry has already implemented further safety policies as a direct response; particularly with regard to passenger briefings and muster drills (Allianz 2013). Although too late for the 32 that died this self-regulation and drive to greater levels of safety shows the positive stride in safety culture that are being made.

As the example shows once a significant list had developed (regulations allow for up to 20°) the ability of crew to deploy and for passengers to use the life boats was severely limited and evacuation along corridors now became a desperate attempt to climb seeming vertical shafts. Costa Concordia carried the mandatory number of life rafts and lifeboats (125% of the maximum number of personnel on-board), however some of the equipment on her starboard side was out of use by being under the overhang of the ship or submerged while that on the port side could not be lowered to the water because of the list. Sadly this demonstrates that adherence to regulation and class requirements may afford sufficient equipment, but does not provide any guarantee that it will be available post incident.

While trials are conducted to establish evacuation times they are done in harbours and in controlled situations, not in the rapidly evolving scenario of a genuine emergency. In spite of compliant passengers and controlled environments simulated evacuations conducted on-board the Stena Invicta in 1996 saw just 315 of the 800 passengers evacuated within the target time of 30 minutes (Steele and Steele 1996). Now imagine that scenario for a cruise ship, such as that in figure 4, with over 4000 passengers and crew on-board as was the case with the Costa Concordia and it is easy to see how such high loss of life can occur in a short space of time. Current regulations require evacuation within 30 minutes from the order to abandon ship with the ship to have sufficient systems to be able to sustain conditions to continue the evacuation for a further 3 hours (MSC Circular 1214 2007). It is worth considering that the Herald of Free Enterprise, Sewol (discussed in the next section), Estonia and Costa Concordia all sank or capsized in less than three hours.

Fig 4. Cruise ship evacuation exercise

An additional example is that of the Costa Allegra (Figure 5). On 27 February 2012 she suffered a fire in the generator room. In a testament to the fire suppression systems fitted on-board the fire was evacuated and there were no casualties however the ship was left without power (Carnival Corporation and Plc 2012). The ship was taken in tow by a French fishing boat and taken to the Seychelles for repairs.

Fig 5. Costa Allegra under tow

It is once again interesting to see that although the ship dealt with the initial emergency, the follow-on actions to resume operations or evacuate passengers were hampered by the damage caused. In this instance a lack of power led to poor conditions on board. Passengers slept on deck to escape the stifling heat below deck and there was no hot food or sanitary facilities. Additionally there were security concerns because of the threat of pirates in the Indian Ocean.

As shown in other examples the ship carried the right firefighting equipment to successfully deal with a dangerous situation. However this incident shows that post-accident control measures are not always available and the severity of an incident may still be high even though designed counter measures are present. This is particularly significant given the relatively high number of fires on cruise ships; 72 between 1990 and 2011 (www.shipcruise.org, 2014).

It is not the intention of this paper to discredit the design process, technological advances or the adherence to standards. They are crucial to establishing the baseline of a vessel which is safe to function. However they must be considered in conjunction with the likely failures of the designed features and the behaviours of those involved when faced with such a failure, which is something that will be discussed further.

6 Safety culture

It is very difficult to determine the safety culture that was present on-board the ships in question at the times of their accidents. Culture cannot be measured or established by referring to a shipping forecast or the passenger or cargo manifests as one would for physical issues in force at the time of an incident. In order for a culture to be effective it must be accepted by all those on-board regardless of their position or responsibilities (Ritchie 2012). Although anecdotal evidence may exist of crew not following procedures or owners cutting corners or overloading a ship as a matter of routine it is very hard to substantiate or quantify this. There may be signs indicative of problems such as high rates of accidents or lack of audit trail or accountability for decisions which would suggest that a poor safety culture is in place. But these cannot be considered as definitive.

A recent example with clear evidence of a failure in safety culture is the loss of the South Korean ferry the Sewol, in April 2014 while on route from Incheon to Jeju Island. She capsized and sank with the loss of 294 of her passengers and crew, many of the passengers were secondary school students which increased the media attention on this accident.

Fig 6. Sewol

The inquiries into this loss have suggested that a sharp turn to starboard resulted in the cargo shifting and the ship developing a list from the offset loading from which she could not recover. The loading of the vessel with over 3000 tons of cargo was over three times the 987 ton limit, this coupled with a lower than recommended amount of ballast water being carried all contributed to the instability that ultimately led to the ship's sinking.

Parallels may be drawn to the Costa Concordia where the reactions of passengers and information from the crew were confused in the immediate aftermath and led to delays in evacuation. Sadly on this occasion the ship did not come to rest in shallow water and people drowned while trapped inside the hull. This again highlights the importance of preparing for the worst scenario with full evacuation in good time.

Also significant is the failure of those responsible to adhere to loading guidelines and heeding warnings that the ship was overloaded. Much like the sinking of the Estonia the issues of loading were not inherent design issues. The ship was deemed safe by the flag state. It was the subsequent actions by owners and crew that led to an unsafe situation developing; a situation not identified by class or state inspection.

More significant to gauge the safety culture is the presence of a questioning attitude towards safety issues. In the case of the Sewol concerns were raised by a former master of the vessel that she was being routinely overloaded and the ballast was at a dangerously low level to accommodate this extra loading. Every accident is always accompanied by people being wise after the event, but it is clear that in this case concerns were raised, but played down by the operator of the ship. This has highlighted broader issues in the safety culture within the South Korean marine industry.

The South Korean regulator was very closely linked to industry which led to a lack of autonomy or genuine ability to look objectively at an issue without being embroiled in the commercial considerations; factors which should not be unduly influencing safety decisions. It is interesting to note that a very similar issue was experienced when looking at nuclear power in Japan and the shortcomings that led to the meltdown at Fukishima Daiichi. This highlights the broader cultural issues which must be overcome, in this instance the strong hierarchal nature of society in some countries which can lead to a failure to question those in a position of authority and investigate problems to the required level of detail.

Interestingly the system in the United Kingdom (and beyond) of classification operated by Lloyds began in Edward Lloyd's Coffee House where ship owners and insurance brokers would get together to share the risk of an expedition or voyage in return for a share of the profits. To agree an acceptable standard for vessels and judge the associated risks a form of classification was needed and this is the origin of the system in place today. This has, over time, seen the required separation of owners, brokers and classification to allow for an effective and objective system.

The safety culture of the ship in question must also be considered. The standards of safety, watchkeeping and overall adherence to standards will be driven by the senior management team on-board. In the case of the loss of the car carrier Tricolor (Figure 7) a number of issues can be seen. The ship itself was sunk in the English Channel in December 2002 after a collision with the container ship the Kariba. The loss of the ship itself is in many ways unremarkable and resulted in no loss of life, although 3000 cars went down with her.

Fig 7. Tricolor

Of greater interest are the subsequent actions of other ships. The Tricolor was resting on her side in approximately 30 metres of water and was only just submerged; consequently radio warnings were issued, a buoy positioned to mark the wreck and guard ships allocated. Despite this, another ship, the Nicola collided with the wreck the following night. Further buoys and additional guard ships were provided and yet approximately two weeks on from the original incident another ship, the Vicky also hit the wreck (BEAMer 2004).

The point to note here is the failings of bridge teams on the Nicola and the Vicky to see warning signs of the new wreck. Procedures were followed by the authorities, but the failure of the watch keepers to adhere to the standards required by IMO regulations for prevention of collision at sea and the standards detailed in Standards of Training Certification and Watchkeeping (STCW) meant that follow on incidents still occurred.

A further point raised in the initial sinking is the confusion that the use of VHF communication caused between vessels when actions to avoid collision were being discussed. As designs and technological complexity increases the faith placed upon it increases too. In the sinking of the Tricolor discussions took place over VHF to try and avoid a collision. By relying on these communications and not seeing the situation unfolding before them a conversation was had with a different ship and as such the discussed collision avoidance and manoeuvring did not have the desired effect.

The example of the Tricolor shows issues that reflect poorly upon safety culture on board ships. The Nicola and the Vicky, which subsequently hit the wreck, both failed to keep an effective look out as required by the regulations so did not see the situation developing which ultimately led to collision. Instead complacency set in and the Navtex and radio warnings and presence of the guard ships went unheeded resulting in further accidents. On these occasions they were fortunate that the secondary accidents did not result in ships being sunk, lives being lost or significant environmental impact.

The final area of safety culture is that of the behaviour of passengers which will be discussed in the context of human factors.

7 Human Factors

Safety Culture extends beyond the management of a vessel and is crucial in its day to day operation. In the examples shown there are several failures of the crew who should be highly trained and familiar with emergency situations likely to face their ship. As seen by the prosecutions in the cases of both the Costa Concordia and Sewol responsibility is placed upon the crew of a ship to safely evacuate the passengers when the vessel cannot be saved.

On both occasions there are reports of senior members of the ship's company abandoning their posts before evacuation is complete and the communication with the passengers was poor, misleading and often slow in being given. This demonstrates poor leadership and a lack of safety culture where people know and understand their safety responsibilities.

The behaviour of crew should be easier to understand and predict; they have been trained to deal with the accident situations they are facing. The passengers however have been put into an unfamiliar scenario and things are starting to go wrong. Research has been conducted looking at passenger behaviour and some key observations made. As stated in section 5, trials of evacuation procedures are conducted on ships, but these will normally be in controlled circumstances and the passengers are often volunteers who will behave in a compliant fashion.

In the paper Human Factors Management of Passenger Ship Evacuation (Jorgensen and May, 2002) noncompliance effects are discussed. In short these are the tendencies for passengers to stick in family groups, or depart from muster points and take undesignated routes between locations to find friends or loved ones. Passengers may also choose to move away from muster stations following a self-preservation instinct; if mustered in a cruise liner's theatre the perceived safety of the upper deck can seem like a long way away which might lead to passengers moving of their own accord to the upper deck. This may in turn cause them to miss vital safety information or updates from the crew.

It is this tendency of the passengers to display non-compliance effects that mean a ship's company must be well drilled in controlling passengers during an emergency scenario and keeping them calm and well informed. This is something which can only be achieved through training and experience. It is important to understand that design features can be used to control the human factors witnessed. Adequate lighting (including emergency lighting) will enable passengers to see exit routes and keep them calm. Good stairway design and provision of hand rails coordinates the flow of people as they move around the ship reducing choke points. Understanding the likely reaction of individuals is as crucial to the design of layout and facilities as regulation or convention. Crucial too is effective communications, both from crew controlling muster routines and via public address (PA) systems. If passengers are informed and aware they are more likely to act as directed. Without this understanding of human factors it is not possible for the correct features to be designed into a ship or system.

It is human nature to trust a complex system (Lee and See 2004). We assume that the greater the level of engineering involved the lower will be the probability of failure. While this may be true it does not absolve an individual from failing to follow best practice and pacing all their trust in one system. Indeed within the Convention on the International Regulations for Preventing Collision at Sea (IMO 1972), the Highway Code for ships, Rule 2 Responsibility, states:

2. Responsibility
(a) Nothing in these Rules shall exonerate any vessel, or the owner, master or crew thereof, from the consequences of any neglect to comply with these Rules or of the neglect of any precaution which may be required by the ordinary practice of seamen, or by the special circumstances of the case
(b) In construing and complying with these rules due regard shall be had to all dangers of navigation and collision and to any special circumstances,

including the limitations of the vessels involved, which may make a departure from these rules necessary to avoid immediate danger

To paraphrase; while the rules should be adhered to, sticking to the rules like dogma is not sufficient when judgment and common sense dictates otherwise and the situation requires alternative action. When considered in the broader context of engineered solutions and safety it is not sufficient to build in safe guards and controls if they cannot be demonstrated to be effective or work when put into a real world scenario.

8 Conclusions

It is clear that modern regulations detailing safety requirements have seen significant efforts put into the design of ships, systems and emergency equipment and procedures. These enhanced designs have led to systems that have ever higher levels of integrity and resilience and so improving the chances of surviving a major accident at sea. There remains though the possibility that a failure, no matter how remote, could still happen and that when it does the resilience of design is insufficient to deal with the scenario as it develops.

The safety culture in the broader industry and local to the ship are very important in two regards; adherence to rules, regulation and maintenance routines to keep a vessel operating as intended and then the training required to ensure familiarity with emergency procedures and actions.

While there will always be a degree of uncertainty as to how people will react the understanding of human factors will help in coordinating responses. Design features can be used to assist behaviour and passenger actions and good communications will keep passengers informed about the situation and increase the likelihood of them complying with crew instructions.

The key lessons to observe are that compliance with regulation and legislation are no guarantee of safety, systems fail and people panic, while the building in of suitable safeguards can mitigate this situation, but can never truly eliminate risk.

Acknowledgments With thanks to my colleagues at STS Defence and Associate Professor Iain Anderson ME, PhD (Engineering Science) of the University of Auckland for their assistance in reviewing this paper.

References
Allianz (2012) Safety and Shipping 1912-2012 From Titanic to Costa Concordia. Allianz Global Corporate and Specialty
Allianz (2013) Safety and Shipping Review 2013. Allianz Global Corporate and Specialty
BEAMer (2004) Supplementary report to the inquiry into The collision between the car Carrier TRICOLOR and the container vessel KARIBA. Bureau d'enquêtes sur les évènements de mer
Carnival Corporation and Plc (2012) Advisory Notice 02/2012 Costa Allegra. Carnival Corporation and Plc

Department of Transport (1987) MV Herald of Free Enterprise Report of Court No. 8074 Formal Investigation. Her Majesty's Stationery Office
House D (2014) Marine Emergencies for Masters and Mates. Routledge
HSVA (2008) Research Study on the Sinking Sequence and Evacuation of the MV Estonia – Final Report. The Hamburg Ship Model Basin
International Maritime Organisation (1972) Convention on the International Regulations for the Preventing Collision at Sea. International Maritime Organisation
Jorgensen HD, May M (2002) Human factors Management of passenger Ship Evacuation. RINA
Lee JD, See KA (2004) Trust in Automation: Designing for Appropriate Reliance. University of Iowa
Maritime Safety Committee (2007) Maritime Safety Committee Circular 1214. Maritime Safety Committee
Ritchie G (2009) Onboard Safety. Witherby Seamanship International
ShipCruise.org (2014) Cruise Ship Accidents http://www.shipcruise.org/cruise-ship-accidents/ Accessed 20/08/2014
Steele N, Steele J (1996) Burning Ships. Argyll Publishing
Trafford SM (2009) Maritime Safety The Human Factors. Book Guild Ltd

Formal Modelling of Railway Safety and Capacity

Alexei Iliasov and Alexander Romanovsky

CSR, Newcastle University

Newcastle, UK

Abstract *Development of future railway systems requires a rigorous modelling of safety and capacity conducted in an integrated way. Supported by EPSRC and Rail Safety and Standards Board* the SafeCap project laid the foundations for overcoming challenges to railway capacity without undermining rail network safety. The main outcome of the project is the SafeCap Toolset, which relies on a formal Domain Specific Language, safety verification and capacity simulation methodologies. The work was conducted in close cooperation with Siemens Rail Automation and evaluated using the layouts of a number of UK stations. The Toolset is being further actively developed and evaluated in a series of industrial and impact acceleration projects.

1 Introduction

The Rail Technical Strategy 2012 produced by the UK Technical Strategy Leadership Group (TSLG) sets out a number of challenging objectives to be achieved by the UK railway industry in the coming 40 years (TSLG 2012). As part of this work a series of the SafeCap projects have been funded to develop theories, methods and tools that address the challenges of improving railway capacity at the same time providing strong guarantees of system safety. The approach taken by the project is motivated by the needs to deal with the railway systems/networks of growing complexity, to reduce their development time, and to increase the confidence in the products developed.

One of the main decisions we made in this work was to use formal methods to model systems, stations, layouts and control tables, and to formally verify their consistency and safety. Even though the railway industry is the main success story in accepting formal methods, their application is still patchy. One of the barriers is the high cost of training and deployment. To address this issue we developed a graphical Domain Specific Language that has formal semantics and allows us to fully hide formal methods and tools. This is complemented by a high performance

© Alexei Iliasov and Alexander Romanovsky 2015.
Published by the Safety-Critical Systems Club. All Rights Reserved

automated verification back-end capable of verifying large stations automatically. The capacity is evaluated by simulation of the formal models and the results are shown in terms of the Domain Specific Language.

In this work we are relying on the extensive experience we gained in the FP7 DEPLOY Integrated Project on developing the Rodin toolset supporting the Event-B method and in deploying it in Bosch, SAP, Siemens Transportation Systems, and other companies (Romanovsky and Thomas 2013). This experience has helped us to develop an extensible SafeCap Eclipse-based environment supporting the work of railway signalling engineers in designing stations/junctions that are safe and have the improved capacity.

2 Problems and objectives

Let us discuss the principal problems and objectives of railway signalling verification. Within the hierarchy defined in (Fokkink and Hollingshead 1998) we focus exclusively on the middle interlocking layer leaving out details of the lower layer of physical equipment functioning and the upper layer of railway logics and exploitation largely out of the view. We identify five kinds of railway safety verification concerns. The first three are the fundamental safety properties that may be reasoned about at a specification level abstracting away from minute details of physical track topology and the setting in which the track is laid. The remaining two require consideration of concrete track geometry, topography, train exploitation characteristics and prospective service requirements.

A schema must be free from collisions. A collision happens when two trains occupy the same part of a track. Reasoning about collisions must take into an account concrete topology, requires an explicit train notion, the definition of laws of train movement and assumptions about train driver (either human or automatic) behaviour. Note that if train drivers choose to ignore whatever means of indication of track occupation states are available to them (e.g., track side signals) there is nothing preventing two trains from colliding. Hence, the absence of collisions is ensured by demonstrating the compatibility of specific topology, signalling and certain driving rules.

The basic safety mechanism is that of route locking and holding (see Figure 1 below). A train is given permission to enter an area of a railway once there is a continuous and safe path through this area assigned exclusively to this train. Such a path is normally called a *route* and is delineated by *signals* - either physical trackside signals with lamps or conceptual signals displayed to a driver via a computer screen. Two-aspect signals (red/green or stop/proceed) are positioned at the maximum braking distance from each other and this defines the smallest train separation. 3- and 4- and higher aspect signalling allows trains to come closer by advising drivers on the safe speed and the extent of free track available in front.

Fig.1. A part of mid-size junction topology and an excerpt from its control table. Route `S21_S22` consists of three train detection circuits: `BH`, `XA` and `BB1`. A train detection circuit is a part of a railway (a sub-graph in an abstract topology) with some equipment capable of reporting the presence or absence of a train in this part. To avoid derailment, the movement of point `P100` requires that circuit `XA` is clear. This, in its turn, requires that routes `S53_S22` and `S21_S22` are not set. To control speed on a curved track, the layout uses fixed speed limits (circles with numbers) and approach speed control. One example of the latter is a control table mandating that a train travelling over the route `S71_S23` occupies `BR3` for at least 15 seconds.

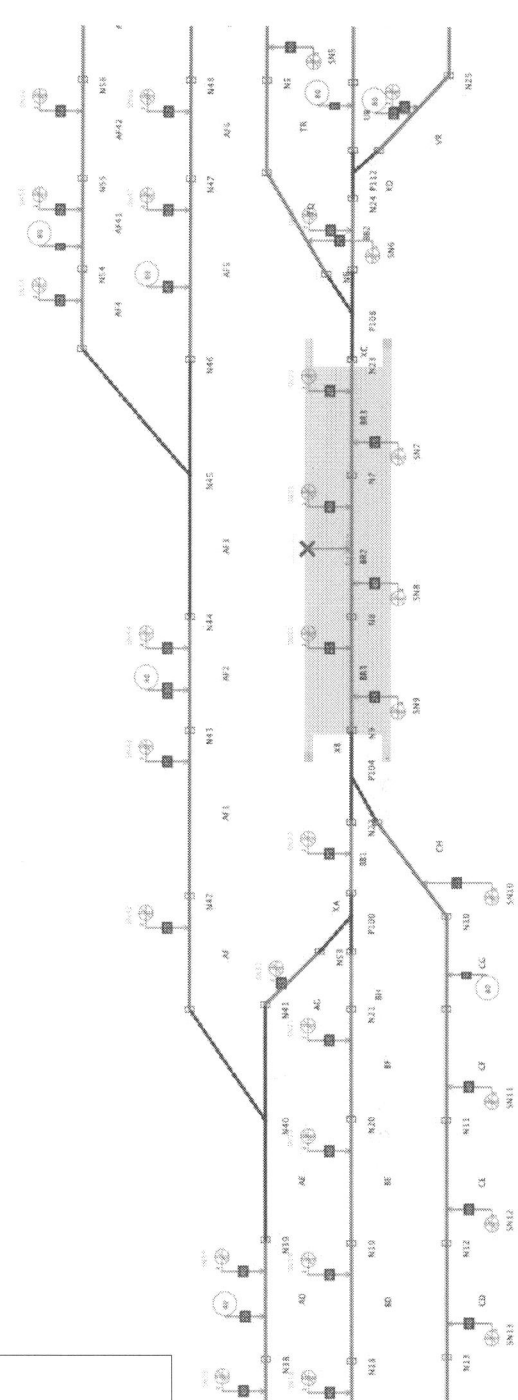

```
   point P100 : clear(XA)
te SN39_SN42 : clear(AE, AF) ∧ reverse(P102)
 route S23_S22 : clear(BH, XA, BB1) ∧ clear(AE, AG) ∧ reverse(P102)
 route S22_S81 : clear(XB, BR1, BR2, BR3, XC) ∧ normal(P104)
         ... : ...
```

When a route is locked, all the movable equipment such as *points* or level crossings must be set and detected in a position that would let a train safely travel on its desired route. They must remain locked in such a state until the train passage is positively confirmed.

A schema must be free from derailments. A derailment may happen when a train moves over a point that is not set in any specific direction and thus may move under a train. To avoid this, a point must be positively confirmed to be *locked* before a train may travel over it. In a control table one writes a condition defining when a point reconfiguration may happen.

Another reason for derailment is driving a train through a curve at an unsafe speed. As a train goes over a curve, the combination of gravitational, centripetal and centrifugal forces exerts a rolling force on train carriages and a substantial lateral force on rails. This effect can be mitigated by track canting although no single canting is a perfect fit for all train types. Hence, enforcing a safe speed limit before a train enters a curved track area is an essential safety consideration. There are several ways of doing this. One is a static speed limit. This can be a signboard warning a driver or an electronic signal sent to an on-board computer. A speed restriction may be also enforced by signalling: a signal does not switch into a permissive aspect until a train is detected to occupy some preceding detection circuit for a duration time. A combination of such time duration and track length gives an upper train speed limit.

Physical layout properties. A range of properties pertinent to safety requires analysis of land topography over which track is build. As one example, it must be ensured that physical signals have certain minimum sighting distance giving a sufficient time for a driver to react. Sometimes tracks are so close together that carriages of a train going through a curve may come into a contact with carriages of a train located on a parallel track. A signalling engineer must identify and protect such areas (known as *fouling points*) via signalling rules. Further examples include gradients at stopping points (e.g., signals) that may be unsafe for heavy trains, parts of track susceptible to landslips, debris on the track due to nearby trees and overpasses, and so on. An important consideration is the spacing between signals and speed restriction signs: it must be possible for all trains to brake within the given limits to meet signal or speed limit restriction. Signal positioning and speed restriction would be wastefully conservative if one does not consider specific properties of traffic, in particular train acceleration and braking performance.

Quality of service. It is never sufficient to consider the safety aspect of a railway in isolation from its performance. Indeed, setting all signals permanently to red (stop) state trivially satisfies all the safety concerns discussed above. As a less extreme example, there could be a signalling mistake preventing or hindering train progress but not violating safety properties. Typically, when signalling a station or a junction, an engineer would have access to a provisional timetable. A timetable defines traffic class and station calling and dwelling times. It must be ensured that signalling is able to accommodate such traffic with some extra margin for unaccounted or delayed traffic. Simulation of train runs is the common way to check quality of service requirements.

3 Safety verification

Figure 2 depicts the interaction between a signaling engineer and a verification tool. There is a conceptual barrier between a railway model that an engineer interacts with and the model handled by formal verification tools.

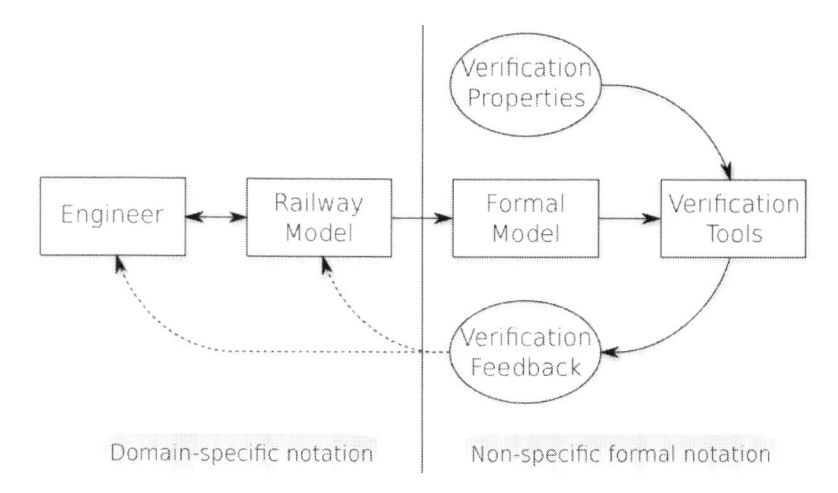

Fig. 2. Formal verification in railway domain: principal actors and flow

Let us now look into the main approaches to signalling verification approaches used in the industry and proposed by academia. In practice, several techniques are often combined to complement each other's strengths.

Manual review. Just as compilation of control tables is often a manual process, verification may also be accomplished via a carefully set up but otherwise manual review procedure. In most cases, to facilitate legibility, control tables are written in a highly structured tabular form following a common standard, i.e., UK Railway Group Standard GK/RT 0202 (RGSOnline 2014) although historic and regional peculiarities are not uncommon. One possible arrangement is having one company to design signalling and a competing company to verify it. The reasoning is that this way both parties are incentivized to do their best.

Manual review is a slow process with very high requirements to reviewers' expertise. It does not deliver any objective proof of safety. At the same time, it does not suffer from any limitations of a formal verification process.

Simulation. Railway industry widely employs railway simulation tools. These range from coarse-grained simulation of a national railway network to a detailed simulation of various aspects of mechanical performance of specific engines and carriages in a combination with specific rail and ballast types. Verification concerns span from analysis of digital communication protocols connecting trains and regional control to stressing of tunnels and bridges by passing trains.

Simulation is widely applied for timetable optimisation and interactive 3D simulation is sometimes used for driver training.

RailSys (RailSys 2014) and OpenTrack (OpenTrack 2014) are two of the well-known simulation suites applied in timetable optimisation and general analysis of signalling performance.

The main attraction of simulation is that it does not require deep understanding of railway functioning. Simulation tools present many aspects of railway performance in an intuitive, visual manner helping to quickly obtain the big picture of overall layout and signalling performance. There is, however, no guarantee of safety as simulation can only ever consider a tiny proportion of all scenarios.

Model checking. The safety challenge of railways and the fact that collision and derailment properties may be dealt with within the setting of discrete, inertia-less train movement makes railway safety verification especially appealing for formal method practitioners. The principal idea of railway model checking is quite simple: a model of train movement laws is combined with the definitions of track topology and signalling rules. A model-checking tool attempts to go through all or many execution scenarios to confirm that unsafe scenarios are ruled out. The list of modelling notations used in this setting is practically endless. Notable examples include Coloured Petri nets (Janczura 1998), process algebra CSP (Winter 2002), a continuation work based on the model-based notation ASM (Winter and Robinson 2003), an algebraic language Maude (Hagalisletto et al. 2007) and the B Method together with ProB model checking tool (Leuschel and Butler 2003).

Almost all model-checking approaches allow automatic instantiation of template models making application of model checking relatively straightforward for engineers. Many tools are able to report a sequence of steps leading to a safety violation. While model checkers are able to analyse many more scenarios than a simulator this comes at a price of reduced expressiveness (i.e., inability to reason about track geometry) and proof certificate is generally not ultimate: there could be a false negative (i.e., the absence of an error report in case an error is present but not discovered) when a model is too large to analyse exhaustively.

Theorem proving. Model checking imposes limitations on the model size and performs best with a relatively limited logical language. Theorem proving overcomes these limitations and offers potentially unlimited opportunities for verifying safety with the utmost level of rigour. Theorem proving is not necessarily an all-manual process: there is a large and successful community developing automated theorem provers (TPTP 2014). At the moment, automated prover support is best in the domain of first order logic and set theory; an attempt at reasoning about continuous train dynamics is likely to require an intervention by a highly skilled verification expert - the kind of people mostly found in academia. From our experience, even reasoning about track geometry is surprisingly difficult as this is a problem outside of the typical application domain of verification tools. One success story with theorem proving is the on-going application of B method in the railway domain (Essame and Dolle 2007). J.-R. Abrial has published methodological guidelines on an economical use of basic logic and set theory to reason about railway safety in a discrete setting (Abrial 2006).

Theorem proving, even with excellent tool support, requires a high level of expertise in formal verification and mathematical modelling. The semantic gap between logic and railway concepts is formidable. This leads to generally low productivity (but we should notice efforts like the BART tool for automatic refinement of B models (Burdy 1999), difficulties in interpreting tool feedback, and posing verification statements in a manner convincing to a non-expert reviewer.

4 Safety verification in SafeCap

The purpose of the SafeCap Toolset is to enable railway engineers to analyse complex junctions by experimenting with signalling rules, signalling principles, track topology, safety limits (e.g., speed limits for points and crossings) while receiving an on-line feedback from automated verification and analysis tools.

We have built the Toolset around Eclipse - a mature and extensible IDE framework. We used Eclipse Modelling framework (EMF) to realise our Domain Specific Language (Iliasov and Romanovsky 2012). One important consideration was the ability to benefit from the extensive EMF ecosystem which offers a toolkit for model manipulation and the construction of graphical and textual editing tools. Apart from the editing tools, the main components of the Toolset are transformation patterns, model-based animation, simulation and verification (see Figure 3).

We have applied the Event-B modelling notation and its refinement methodology to develop a theory of safe railway. This theory explicitly describes train movements, signal operation and point's control. It does not, however, deal with any specific topology or control table. The proof of safety (we consider absence of collisions and derailments, and protection of flanks[1]) is done for some class of topologies and control tables. The proof of the Event-B model, although challenging, is done once and for all.

An important by-product is the set of axiomatic conditions characterising the class of *safe topologies* and *safe control tables*. To establish that a given track topology and control table are safe we only need to check that they do not contradict the mentioned axiomatic conditions. We do not need to redo the proofs of Event-B model. Safety verification is accomplished by putting together the definition of a concrete topology, control table and the axiomatic conditions derived by the Event-B model. If a *constraint solver* does not find a contradiction in logical statements encoded by this composition then the concrete topology and control table are deemed safe. Returning to the Event-B domain, the absence of contradic-

---

[1] Protection of the movement of a train across a junction that prevent any other unauthorised movement coming into contact with it.

tion established by a constraint solver means that our generic Event-B model of train behaviour is refined by a model instantiated with the given track topology and control table.

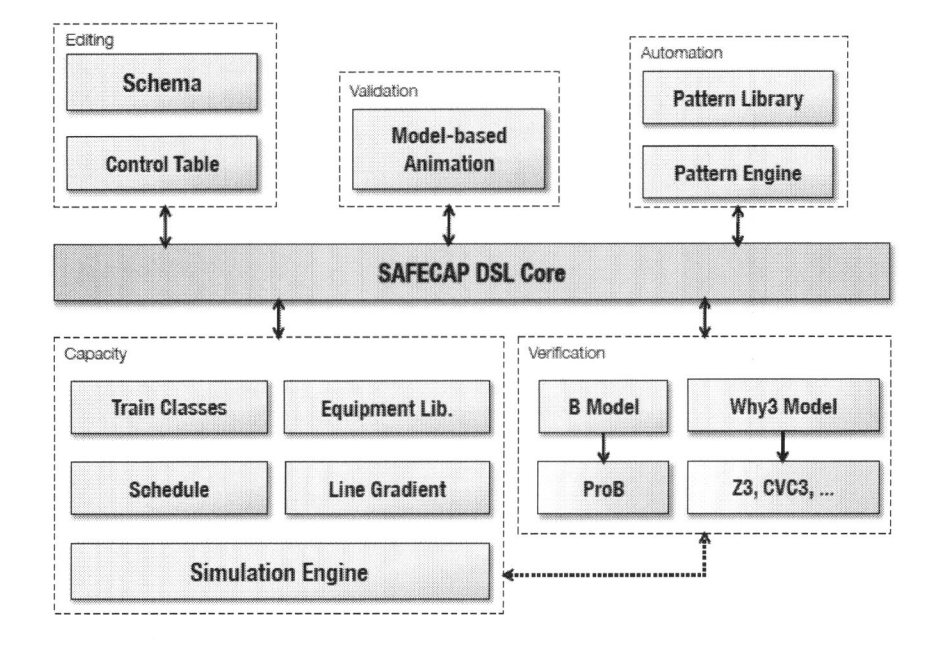

Fig.3. The architecture of the SafeCap Toolset

Schema topology and control table theories come in the form of a list of first order logic predicates; they do not define any state transitions or dynamic behaviour but rather well-formedness requirements to objects describing track topology and control table.

For constraint solving, we make use of two sets of formal notations and tools: B together with ProB (Leuschel and Butler 2003) and Why3 (Bobot et al. 2011). In the short term, we aim to benefit from their complementary strengths; in a longer term, the dual verification path provides a logical redundancy that makes a low-level encoding or tool bug unlikely to be left undiscovered.

4.1 Event-B

We apply the Event-B formal modelling notation (Abrial 2010) to specify and verify railway signalling. Event-B belongs to a family of state-based modelling languages that represent a design as a combination of state (a vector of variables) and state transformations (computations updating variables).

An Event-B development starts with the creation of a very abstract specification. A cornerstone of the Event-B method is the stepwise development that facilitates a gradual design of a system implementation through a number of correctness-preserving *refinement* steps. The general form of an Event-B model (or *machine*) is shown in Figure 4. Such a model encapsulates a local state (program variables) and provides operations on the state. The actions (called *events*) are characterized by a list of local variables (parameters) vl, a state predicate g called *event guard*, and a next-state relation S called *substitution* or event *action*.

```
machine M
  sees Context
  variables v
  invariant I(c, s, v)
  initialisation R(c, s, v')
  events
    E₁ = any vl where g(c, s, vl, v) then S(c, s, vl, v, v') end
    ...
end
```

Fig.4 Event-B machine structure

Event parameters and guards may be omitted leading to syntactic short-cuts starting with keywords when and begin.

Event g defines the condition when an event is *enabled*. Relation S is given as a generalised substitution statement (Abrial 1996) and is either deterministic (x := 2) or non-deterministic update of model variables. The latter kind comes in two notations: selection of a value from a set, written as x :∈ {2, 3}; and a relational constraint on the next state v', e.g., x | x' ∈ {2, 3}.

The invariant clause contains the properties of the system, expressed as state predicates, that must be preserved during system execution. These define the *safe states* of a system. In order for a model to be consistent, invariant preservation is formally demonstrated. Data types, constants and relevant axioms are defined in a separate component called *context*.

Model correctness is demonstrated by generating and discharging *proof obligations* - theorems in the first order logic. There are proof obligations for model consistency and for a refinement link - the forward simulation relation - between the pair of *abstract* and *concrete* models.

4.2 Discrete driving model

The discrete driving model is an Event-B model capturing train, signal and point behaviour. It proves that the described behaviour is contained within a certain safety envelope by formulating and proving, through a number of refinement steps, *safety invariants* corresponding to the first three verification objectives of Section 2. This model gives a formal definition of principal phenomena observed in railway operation: train movement, route reservation, point locking, route cancellation and so on.

To construct the proof we have used Event-B (Abrial and Mussat 1998) and the Rodin Toolset (Rodin 2014). Train driving rules are encoded by Event-B events - atomic state transitions - so that the overall model defines a state transition system. The safety properties are stated as a system invariant: a subset of possible states where the dangerous situations may not occur. The proof is done inductively by examining the effect of each event on a given safety property and discharging relevant proof obligation (first-order logic theorems).

The model in Figure 5 illustrates the notation and modelling style of Event-B. This particular model is the very first (abstract) model in the development chain.

The overall model is made of seven refinement steps with 470 verification conditions of which 301 were discharged automatically by Rodin theorem provers. Its development span over several months and several early versions were abandoned either due to misrepresentation of some railway concepts or unacceptable verification costs.

Apart from its role in the validation of first two layers, the discrete operational rules of the third layer are used to visually animate train movements over a given schema. There are two main applications for such an animation: replaying the results of model checking of discrete driving rules in order to pin-point the source of an error in a topology or a control table; and helping an engineer to understand how trains may travel through a schema with a given set of control rules.

4.3 Schema topology theory

The schema topology theory is responsible for verifying logical conditions expressed over track layout (i.e., track connections, point placement) and logical topology (i.e., routes and lines as paths through a schema). Few examples of verification conditions include the connectivity property (no isolated pieces of track), continuity of routes and lines, absence of cycles, correct traversal of points and valid placement of train detection circuit boundaries.

As a whole, these conditions express what we understand to be a valid track topology. We have tested them against a number of large-scale real-life layouts and we able to discover some problem in already informally validated track topologies. In addition, semi-automated alteration and generation of track layouts (e.g., via the improvement patterns we are developing in the tool) necessitates a careful and strict inspection of these basic properties. An automated verification process ensures high productivity and enables an engineer to explore a large range of designs within a short time.

```
machine route0
  sees ctx_line
  variables
    t_line   // Train/line association
    t_r_hd   // Train head position on a line
    t_r_tl   // Train tail position on a line
  invariant
    t_line ∈ TRAIN ↛ LINE
      // A train is mapped to the id of a route occupied by the head of a train
    t_r_hd ∈ TRAIN ↛ ℕ₁
      // correspondingly, t_r_tl(t) is the id of the route occupied by the tail of train t
    t_r_tl ∈ TRAIN ↛ ℕ₁
    dom(t_line) = dom(t_r_hd)
    dom(t_line) = dom(t_r_tl)
      // A train occupies a continuous route interval of route from tail till head
    ∀t·t ∈ dom(t_line) ⇒ t_r_tl(t) .. t_r_hd(t) ≠ ∅
    The routes a train occupies are the routes defined by the train line
    ∀t·t ∈ dom(t_line) ⇒ t_r_tl(t) .. t_r_hd(t) ⊆ dom(Line(t_line(t)))
    // Initially, there are no trains in the system
  initialisation
    t_line, t_r_hd, t_r_tl := ∅, ∅, ∅
  events
        // A train may appear in the system with this operation
    appear =
    any t, l where
      t ∈ TRAIN \ dom(t_line)   // a train must be not already in the system
      l ∈ LINE
    then
      t_line(t) := l   set the train line to l
      t_r_hd(t), t_r_tl(t) := 1, 1   // set head and tails routes
    end
        // Moves the head of a train from one route to another
    move_route_hd =
    any t where
      t ∈ dom(t_line)
      t_r_hd(t) < LineLen(t_line(t))   // train head must not be on the last line route
    then
      t_r_hd(t) := t_r_hd(t) + 1   // move the head one step forward
    end
        // Moves the tail of a train between routes
    move_route_tl =
    any t where
      t ∈ dom(t_line)
      t_r_tl(t) < t_r_hd(t)   // a tail must be strictly behind the head of the train
    then
      t_r_tl(t) := t_r_tl(t) + 1   // move the tail one step forward
    end
    ...
```

Fig.5. An Event-B model of abstract, route-level train movement (excerpt)

Figure 6 gives a sample of verification conditions written in the Classical B notation (Abrial 1996) and ready to be processed by model checking tool ProB (Leuschel and Butler 2003). Not shown is the encoding of domain specific elements (track graph, control tables) as sets, relations and functions of a B model. For a real-life example, such a model may be 6-14 thousand lines long. The same conditions and constructs are also generated in the Why3 theory notation. It is not a direct translation of the B model and we intentionally use a different representation of relations and functions to introduce a form of modelling diversity. At the moment, for the topology theory, ProB and Why3 verifications chains deliver broadly similar performance.

```
/* (1)   */  {} <<: NODE &
/* (2.a) */  {} <<: TRACK &
/* (2.b) */  TRACK <: NODE * NODE &
/* (2.c) */  elm(TRACK) = NODE & /* all nodes are connected by tracks */
...
/* (10) */ AMBIT : LA --> (POW(NODE) * POW(TRACK)) &
/* (11) */ ! (a, q, p) . (a : ran(AMBIT) & a = ( q |-> p ) => p <: q * q & {} <<: p) &
/* (12) */ ! (a, q, p) . (a : ran(AMBIT) & a = ( q |-> p) => p~ <: p) &
/* (13) */ ! (a, q, p) . (a : ran(AMBIT) & a = ( q |-> p) =>
                ! (n) . (n : q => closure(p)[{n}] = q) ) &
/* (14) */ union({p | # (a, q) . ( a : ran(AMBIT) & q <: NODE &
                p <: TRACK & a = ( q |-> p))}) = TRACK &
/* (15) */ ! (a, b, r, s, t, q) . (a : ran(AMBIT) & b : ran(AMBIT) & a /= b &
                a = (r |-> s) & b = (t |-> q) => s /\ q = {}) &
...
```

Fig.6. Schema well-formedness rules (an excerpt)

4.4 Control table theory

When the topology is verified we can define the conditions of operational safety. These are derived, via a formal proof, from a set of discrete (inertia-less) train movement rules and expressed as a set of constraints over signalling rules.

In SafeCap, we depart from the convention of associating control rules with trackside signals. Instead, we consider a more general situation where different signalling rules are applied depending upon the ultimate train destination or train type and attach control logic to a pair of line and route. This permits, for instance, to model, on the same track, an express train using two-aspect signalling and a freight train travelling over the same routes but in a three or four aspect mode. Such an arrangement may be used to achieve an optimal balance between headway and average speed in a heterogeneous traffic mix. Given the fact that in UK trackside signals are going to be made obsolete by 2030 (TSLG 2012) this represents a fairly modest scenario of using virtual signals to improve capacity.

The control table theory demonstrates such properties as the absence of potential collision (as may happen, for instance, when a proceed aspect is given while a protected part of track is still occupied) and derailment (due to incorrect point setting or point movement under a train). Other properties relate to the danger or circular dependencies between signals, dependencies between multi-aspect signals and operation of auto-signals, conformance with an Automatic Train Protection system, and verification of point and signal based flank protection. Certain properties, notably approach speed control via the timed occupation of a track section, are not verified at this stage, as the formalisation at this layer does not capture train inertia. Speed limit conformance and other time-related properties are formulated at the final, most detailed layer.

A list of sample control table theory conditions is given in Figure 7. For the shown rules, the outer quantification selects a pair of a line and a route that define a list of control rules (one per aspect). This model includes the topology model[1]. Constraint solving is the primary verification strategy: we try to detect a contradiction between concrete data structures defining topology and control tables and the verification conditions. Again, the model is given in both B and Why3 notations although this time the Why3 verification route is not successful for larger examples. This is due to the weakness in our axiomatization of the B mathematical notation in Why3. We are currently working on building a library of Why3 lemmas to support translation from B to Why3 and this, we believe, should deliver a significantly better result. In addition, for any mid to large-scale schema we currently have to exclude the verification of flank protection properties, as these require complicated computations over track topology. We are working on a program that would output a proof term for each instance of flank protection property so that a theorem prover or a constraint solver would only have to check the elementary steps of a prepared proof.

```
...
/* @label (CT.1): A permissive signal may be lit only when all route ambits are clear */
! (l, r). (l |-> r : CTO_DOM => ! (n). (n : 1 .. RASPECT(l, r)-1 =>
                    routeambits(r) <: CT_CLEAR(l, r, n) )) &
/* @label (CT.2): A route with an overlap may have permissive signal only
   when its overlap is reserved and confirmed as clear */
! (l, r). (l |-> r : ROVERLAP & r : dom(LINE(l)) => ! (n). (n : 1 .. RASPECT(l, r)-1 =>
                    TA[fst(ROUTE(LINE(l)(r)))] <: CT_CLEAR(l, r, n))) &
/* @label (CT.3.a): No point is set both normal and reverse */
! (l, r). (l |-> r : CTO_DOM => CT_NORMAL(l, r) /\ CT_REVERSE(l, r) = {} ) &
...
```

Fig. 7. Control table conditions (an excerpt)

5 Conclusions

The SafeCap offers an efficient tool for signalling engineers to design railway nodes (stations and junctions) while automatically checking the conformity of the signalling and topology against a range of validation criteria expressing operation-

---

[1] At the level, it is assumed that the topology theory has been verified and the topology constraints are turned into axioms

al safety and design integrity properties. Such level of automation enables rapid exploration of signalling designs in the pursuit of optimal capacity and performance stability.

The work on the SafeCap Toolset is now taken further in our new project SafeCap for FuTRO supported by Rail Safety and Standards Board. In this work we are developing a support for an integrated reasoning about capacity and energy of railway networks and nodes while ensuring whole systems safety.

Acknowledgments The work has been conducted as part of the UK EPSRC/Rail Safety and Standards Board SafeCap, EPSRC SafeCap-Impact and Rail Safety and Standards Board SafeCap for FuTRO projects. This work has been partially supported by the EPSRC/UK TrAmS-2 Platform Grant.

References

Abrial J-R (1996) The B-Book. Cambridge University Press
Abrial J-R (2006). Train systems. In Butler M J, Jones C B, Romanovsky A, Troubitsyna E (eds), Rigorous Development of Complex Fault-Tolerant Systems (FP6 IST-511599 RODIN project), LNCS 4157, Springer, 1-36
Abrial J-R (2010). Modelling in Event-B. Cambridge University Press
Abrial J-R, Mussat L (1998). Introducing Dynamic Constraints in B. In Proceedings of B'98: Recent Advances in the Development and Use of the B Method, LNCS 1393, Springer, 83-128
Bobot F, Filliatre J-C, Marche C, Paskevich A (2011). Why3: Shepherd your herd of provers. In Boogie 2011: First International Workshop on Intermediate Verification Languages, 53-64
Burdy L (1999). Automatic Refinement. In Proceedings of BUGM at FM'99
Essame D and Dolle D (2007). B in Large-Scale Projects: The Canarsie Line CBTC Experience. In Julliand J, Kouchnarenko O (eds.), B, LNCS 4355. Springer, 252-254
Fokkink W J, Hollingshead P R (1998). Verification of Interlockings: from Control Tables to Ladder Logic Diagrams. In Proceedings of the 3rd Workshop on Formal Methods for Industrial Critical Systems (FMICS'98)
Hagalisletto A M, Bjork J, Yu I C, Enger P (2007). Constructing and Refining Large-Scale Railway Models Represented by Petri Nets. IEEE Transactions on Systems, Man, and Cybernetics, Part C, 444-460
Iliasov A, Romanovsky A (2012). SafeCap domain language for reasoning about safety and capacity. Pacific-Rim Dependable Computing Conference (PRDC 2012). Niigata, Japan. IEEE CS
Janczura C W (1998). Modelling and Analysis of Railway Network Control Logic using Coloured Petri Nets. PhD thesis, School of Mathematics and Institute for Telecommunications Research, University of South Australia
Leuschel M, Butler M (2003). ProB: A Model Checker for B. In Keijiro A, Gnesi A, Dino M (eds.), Formal Methods Europe 2003, LNCS 2805. Springer, 855-874
OpenTrack simulator (2014). Website. Available at http://www.opentrack.ch/. Accessed 2014.
RailSys simulation platform (2014). Website. Available at http://http://www.rmcon.de. Accessed 2014
RGSOnline (2014). Railway Group Standards. Signalling Design: Control Tables. Rail Safety and Standards Board (RSSB). Available at http://www.rgsonline.co.uk/. Accessed 2014
Rodin (2014). Rigorous Open Development Environment for Complex Systems (RODIN). IST FP6 STREP project, online at http://rodin.cs.ncl.ac.uk/
Romanovsky A, Thomas M (2013). Industrial deployment of system engineering methods. Springer

TPTP (2014). Thousands of Problems for Theorem Provers. Available at www.tptp.org/. Accessed 2014
TSLG (2012). The Rail Technical Strategy (RTS). 2012. Available at http://www.futurerailway.org/RTS/Pages/Intro.aspx
Winter K (2002). Model Checking Railway Interlocking Systems. In Proceeding of the 25th Australian Computer Science Conference (ACSC 2002). Australian Computer Society, Inc. Darlinghurst, Australia, Australia
Winter K, Robinson N (2003). Modelling Large Railway Interlockings and Model Checking Small Ones. In Proceeding of the Australian Computer Science Conference (ACSC 2003). Australian Computer Society, Inc. Darlinghurst, Australia, Australia

A Comparison of SAE ARP 4761 and STPA Safety Assessment Processes

Nancy Leveson, Cody Fleming, John Thomas

MIT

Cambridge, MA, U.S.A.

Chris Wilkinson

Honeywell

Washington, DC, U.S.A.

Abstract *The increasing complexity of modern aircraft systems presents many challenges to the current process for aircraft safety assessment and certification. Automated features and equipment are becoming so complex that potential dysfunctional interactions and requirements flaws are much more difficult to recognize and prevent than in the past. We believe that the current process for assessing safety and certifying aircraft described in ARP 4761 is limited in its effectiveness for modern complex and software intensive systems and that a better approach is needed. This paper describes the ARP 4761 methodology and then a new accident causality model and associated hazard analysis technique is introduced and applied to the Wheel and Brake System example used in ARP 4761. We conclude with a comparison of the two approaches.*

1 Introduction

Various acceptable means for showing compliance with FAA safety requirements and the airworthiness regulations are contained in AC 25.1309-1A (FAA 1988). ARP 4761 (SAE 1996) is one industry standard, acceptable means of compliance for conducting a safety assessment though there is currently no advisory circular that specifically recognizes it. Nonetheless, ARP 4761 is a supporting part of the larger systems development process described by ARP 4754 (SAE 1996) (and its successor ARP 4754A (SAE 2010)). ARP 4761 is widely used and may be invoked by the regulators on a project by project basis through Issue Papers (FAA) or Certification Review Items (EASA).

© Nancy Leveson 2015. Published by the Safety-Critical Systems Club. All Rights Reserved

It can be argued that ARP 4754[1] has been quite effective on the loosely coupled aircraft and system designs that have prevailed in the industry. Aircraft and systems are increasingly closely coupled, software intensive, consequently complex and exhibit emergent behaviors. The traditional probabilistic risk analysis and hazard analysis methods recommended in ARP 4761 are not sufficiently effective on these kinds of systems since hazards and accidents may result from unsafe interactions among the components and not just because of component failure.

STPA (System-Theoretic Process Analysis) is a new hazard analysis method based on systems theory rather than reliability theory (Leveson 2012; Leveson 2013). STPA has its foundation in a new accident causality model called STAMP (System-Theoretic Accident Model and Process) (Leveson 2012) that extends the prevailing view of accidents as caused by component failures to include additional causes such as system design errors (including software and system requirements errors), human error considered as more than just a random "failure" and various types of systemic accident causes. As such, STPA is potentially more powerful than the traditional hazard analysis methods and approach used in ARP 4761.

The goal of a recent research project was to provide evidence to support this hypothesis by comparing the approach and types of results using the process described in ARP 4761 with those provided by STPA. Illustrations of the two approaches are provided by showing the results from the analysis of the Wheel Braking System (WBS) example provided in ARP 4761. Although neither example is complete, a direct comparison of the results is not appropriate. The example is useful in comparing the different philosophy behind the two approaches and the types of results that can be obtained. In this paper we provide a general description of the two approaches and a summary of the results from the comparison. More details can be found in the final research report (Leveson 2014).

2 The ARP 4761 Process

The ARP 4761 process has three parts—the Functional Hazard Analysis (FHA), the Preliminary System Safety Analysis (PSSA), and the System Safety Analysis (SSA) - which are performed at each relevant level of abstraction (or hierarchical level) for the aircraft and system under study.

Functional Hazard Analysis: The FHA is conducted at the beginning of the aircraft development cycle. There are two levels of FHA: the aircraft level FHA and the system level FHA. The aircraft-level FHA identifies and classifies the failure conditions associated with the aircraft level functions. The classification of these failure conditions establishes the safety requirements that an aircraft must meet.

---

[1] We will use ARP 4754 to include both ARP 4754 and ARP 4754A

The goal is to identify each failure condition along with the rationale for its severity classification. A standard risk assessment matrix is used, as shown in Table 1.

Both the failure of single and combinations of aircraft functions are considered. The failure condition severity determines the ARP 4754 development assurance level (DAL) (item development assurance level (IDAL) in ARP 4754A) allocated to the subsystem. Besides the DALs, there are some qualitative requirements generated in the aircraft level FHA, particularly those related to assuring independence of failures for aircraft level functions.

Later in the development process, the architectural design process allocates the aircraft-level functions to particular subsystems. A system-level[1] FHA considers the failures or combination of system or subsystem failures that affect the aircraft-level functions.

Table 1. Failure Condition Severity as Related to Probability Objectives and Assurance Levels (SAE ARP 4761 p. 14)

Probability (Quantitative)	Per flight hour 1.0 1.0E-3 1.0E-5 1.0E-7 1.0E-9				
Probability (Descriptive)	FAA	Probable	Improbable		Extremely Improbable
	JAA	Frequent	Reasonably Probable	Remote	Extremely Remote	Extremely Improbable
Failure Condition Severity Classification	FAA	Minor	Major	Severe Major	Catastrophic
	JAA	Minor	Major	Hazardous	Catastrophic
Failure Condition Effect	FAA & JAA	• slight reduction in safety margins • slight increase in crew workload • some inconvenience to occupants	• significant reduction in safety margins or functional capabilities • significant increase In crew workload or in conditions impairing crew efficiency • some discomfort to	• large reduction in safety margins or functional capabilities • higher workload or physical distress such that the crew could not be relied upon to perform tasks accurately or completely • adverse effects upon occupants	• all failure conditions which prevent continued safe flight and landing

---

[1] In ARP 4761, the "system-level" is the aircraft component or subcomponent level.

			occupants		
Development Assurance Level	ARP 4754	Level D	Level C	Level B	Level A

Note: A "No Safety Effect" Development Assurance Level E exists which may span any probability range.

The same procedure is used at both the aircraft-level and the system level (ARP 4761, p. 32-33):
1) Identification of all the functions associated with the level under study
2) Identification and description of failure conditions associated with these functions, considering single and multiple failures in normal and degraded environments.
3) Determination of the effects of the failure conditions.
4) Classification of failure condition effects on the aircraft (catastrophic, severe-major/hazardous, major, minor, and no safety effects.
5) Assignment of requirements to the failure conditions to be considered at the lower level classification.
6) Identification of the method used to verify compliance with the failure condition requirements.

Note the emphasis on failures and failure conditions. The classification of failure conditions establishes the safety requirements that an aircraft must meet (ARP 4761 p. 16). A fault tree analysis (FTA), Dependence Diagram (DD), Markov Analysis (MA), or other analysis methods can be used to derive lower level requirements from those identified in the FHA (ARP 4761 p. 17). The FHA also establishes derived safety requirements needed to limit the effects of function failure.

An example of part of a fault tree generated in the aircraft FHA for the WBS example is shown in Figure 1. The system-level FHA is similar and generates the following requirements for the WBS, which are provided to the PSSA:
1) Loss of all wheel braking during landing or rejected takeoff (RTO) shall be less than 5E-7 per flight
2) Asymmetrical loss of wheel braking coupled with loss of rudder or nose wheel steering during landing or RTO shall be less than 5E-7 per flight
3) Inadvertent wheel braking with all wheels locked during takeoff roll before V1 shall be less than 5E-7 per flight.
4) Inadvertent wheel braking of all wheels during takeoff roll after V1 shall be less than 5E-9 per flight.
5) Undetected inadvertent wheel braking on one wheel without locking during takeoff shall be less than 5E-9 per flight.

Preliminary System Safety Assessment (PSSA): The PSSA is used to complete the failure conditions list and the corresponding safety requirements using as inputs the aircraft and/or system FHA and the aircraft FTA. The PSSA involves a
> "systematic examination of a proposed system architecture to determine how failures can lead to the functional hazards identified by the FHA and how the FHA requirements can be met" (SAE ARP 4761 p. 40).

Probabilistic analysis is performed at the sub-system(s) level(s) to show that the failure probability meets the requirement passed down from the aircraft level FHA. Because a subsystem may consist of further subsystems, the process is continued to decompose the failure rate allocations to the component subsystems. Common Cause Analyses (CCAs) are also conducted in order to substantiate independence claims made in the FHA. Additional requirements are generated from the CCA for the WBS:

Fig.1. Aircraft FHA Preliminary Fault Tree (SAE ARP 4761 p. 182)

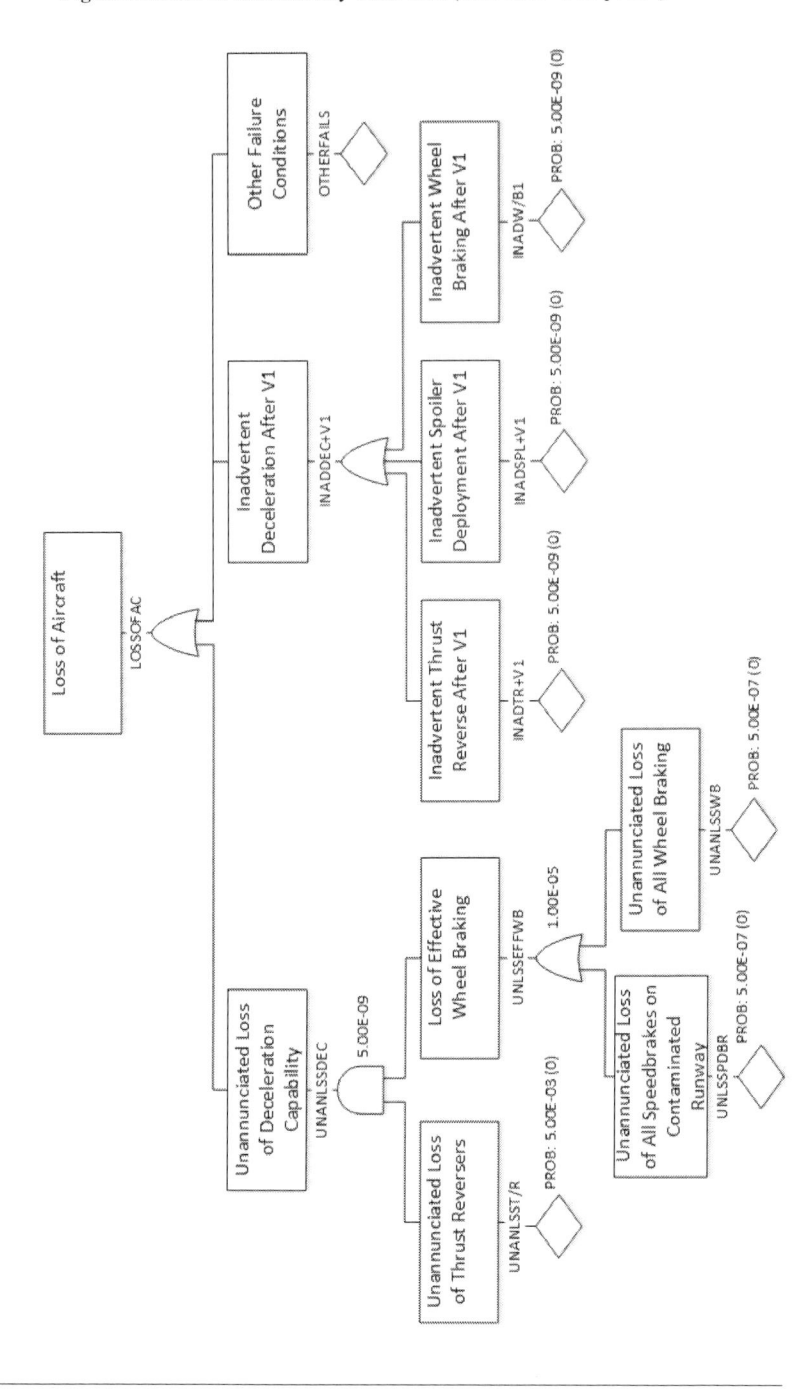

6) The wheel braking system and thrust reverser system shall be designed to preclude any common threats (tire burst, tire shred, flailing tread, structural deflection, etc.)
7) The wheel braking system and thrust reverser system shall be designed to preclude any common mode failures (hydraulic system, electrical system, maintenance, servicing, operations, design, manufacturing, etc.)

Design changes may be required to satisfy the top-level probabilistic failure requirement. For example, in the WBS example, the top level functional failure requirement was determined not to be met by a single and feasible Brake System Control Unit (BSCU). The modified fault tree results in derived lower-level requirements, e.g., the installation requirement that the primary and secondary hydraulic supply systems shall be segregated and that the BSCU shall comprise two independent, monitored channels. A CCA would be done to ensure that the failures are independent.

Item Level Requirements are generated from the fault trees and the design additions:

1) The probability of "BSCU Fault Causes Loss of Braking Commands" shall be less than 3.3E-5 per flight.
2) The probability of "Loss of a single BSCU shall be less than 5.75E per flight.
3) The probability of "Loss of Normal Brake System Hydraulic Components" shall be less than 3.3E-5 per flight.
4) The probability of "Inadvertent braking due to BSCU" shall be less than 2.5E-9 per flight.
5) No single failure of the BSCU shall lead to "inadvertent braking."
6) The BSCU shall be designed to Development Assurance Level A based on the catastrophic classification of "inadvertent braking due to BSCU."

The same process can be repeated at a lower level of detail, for example, the BSCU. The resulting requirements from such a BSCU analysis given in SAE ARP 4761 (p. 227) are:

Installation Requirements:

7) Each BSCU System requires a source of power independent from the source supplied to the other system.

Hardware and Software Requirements:

1) Each BSCU system will have a target failure rate of less than 1E-4 per hour.
2) The targeted probabilities for the fault tree primary failure events have to be met or approval must be given by the system engineering group before proceeding with the design.
3) There must be no detectable BSCU failures that can cause inadvertent braking.
4) There must be no common mode failures of the command and monitor channels of a BSCU system that could cause them to provide the same incorrect braking command simultaneously.

5) The monitor channel of a BSCU system shall be designed to Development Assurance Level A.
6) The command channel of a BSCU system may be designed to Development Assurance Level B.[1]
7) Safety Maintenance Requirements: The switch that selects between system 1 and system 2 must be checked on an interval not to exceed 14,750 hours.

System Safety assessment (SSA): Once the PSSA is completed at multiple levels of abstraction for the system, the final step of System Safety Assessment (SSA) begins. The SSA is a "[bottom-up] verification that the implemented design meets both the qualitative and quantitative safety requirements...defined in [both] the FHA and PSSA" (SAE ARP 4761, p. 21). For each PSSA carried out at any level, there should be a corresponding SSA. As with the PSSA, the SSA uses failure-based, probabilistic analysis methods.

Development Assurance Level (DAL): Because software and other components of an aircraft may not lend themselves to probabilistic assessment, a "development assurance level" or DAL is assigned to each component, depending on the safety-criticality of the component. RTCA documents that are often referenced in advisory circulars, policy memos, orders etc. as acceptable means of compliance include DO-178C (RTCA 2012) and DO-254 (RTCA 2000). These documents provide guidance to achieve design assurance of flight software and hardware respectively from Level A (flight critical) to Level E (no safety impact) inclusive. Note that both assume that there is an existing set of requirements and that these requirements are consistent and complete.

3 System-Theoretic Process Analysis (STPA)

STPA is a top-down, system engineering hazard analysis technique based on system theory (Leveson 2012; Leveson 2013). STPA implements a new accident causality model that assumes accidents are caused by inadequate enforcement of behavioral safety constraints on system component behavior and interactions. Rather than thinking of safety as a failure problem, it conceives of it as a control problem whereby safety is assured through controls built into the system design[2]. Note that failures are still considered, but they are considered to be something that needs to be controlled, as are design errors, requirements flaws, component interactions, etc.

The underlying model of causality, called STAMP (Systems-Theoretic Accident Model and Process), has been described elsewhere (Leveson 2012) and only a very brief description is provided here. STAMP is based on system theory, which was

---

[1] The allocations in 5 and 6 could have been switched, designing the command channel to level A and the monitor channel to level B.
[2] System is here considered to include the equipment hardware and software and any human controller

created to handle complex systems. In STAMP, safety is an emergent property that arises when the components of a complex system interact with each other within a larger environment. A set of constraints related to the behavior of the system components (physical, human, and social) enforces the safety property. Accidents occur when the interactions violate these constraints. The goal, then, is to control the behavior of the components and system as a whole to ensure the safety constraints are enforced in the system operating in its intended environment. Safety, then, is treated as a dynamic control problem rather than a component or functional failure problem. In contrast, the ARP 4761 process starts with functional decomposition and then considers the failure of the identified aircraft functions. The problem with an initial functional decomposition is that it leads to the neglect of emergent behavior characteristic of complex systems (Leveson 2012).

An important concept in STAMP (and also in system theory) is that of a process model. In systems theory, every controller contains a model of the controlled process (Figure 2). For human controllers, this model is usually called the *mental model*. This model includes assumptions about how the controlled process operates and the current state of the controlled process. It is used to determine what control actions are necessary to keep the system operating effectively and safely.

Accidents in complex systems often result from inconsistencies between the process (or mental) model used by the controller and the actual process state, which results in the controller providing unsafe control actions. For example, the autopilot software thinks the aircraft is climbing when it really is descending and applies the wrong control law, a military pilot thinks a friendly aircraft is hostile and shoots a missile at it; the software thinks the spacecraft has landed and turns off the descent engines prematurely; or the air traffic controller does not think two aircraft are on a collision course and therefore does not provide advisories to the aircraft to change course.

Analyzing what is needed in the software process model is an important part of the safe design for a system containing software. The same is true for accidents related to human errors. STAMP and STPA provide a way of identifying safety-critical information and potential operator errors and their causes, so they can be eliminated or mitigated.

Part of the challenge in designing an effective safety control structure is providing the feedback and inputs necessary to keep the controller's model consistent with the actual state of the controlled process. An important part of identifying potential paths to accidents and losses involves determining how and why the controls could be ineffective in enforcing the safety constraints on system behavior; often this is because the process model used by the controller is incorrect or inadequate in some way. The causes of such an inconsistency are identified in the new analysis techniques built on STAMP.

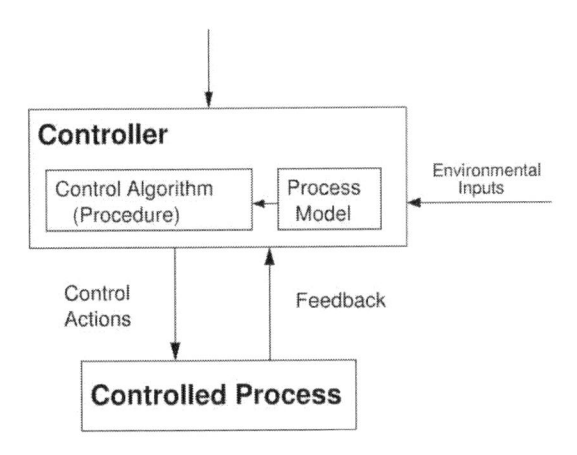

```
                    │
       ┌────────────▼──────────────────────────┐
       │ **Controller**                        │
       │  ┌──────────────┐    ┌─────────┐      │      Environmental
       │  │Control Algorithm│◄──│ Process │◄─────────── Inputs
       │  │ (Procedure)  │    │  Model  │      │
       │  └──────┬───────┘    └─────────┘      │
       └─────────┼─────────────────▲───────────┘
               Control           Feedback
               Actions
                 │                 │
       ┌─────────▼─────────────────┴───────────┐
       │        **Controlled Process**          │
       └────────────────────────────────────────┘
```

Fig. 2. A Simple Control Loop Showing a Process Model

The STPA process starts at the system level, as does FHA,[1] and iterates until the hazards have been adequately analyzed and handled in the design. The goal of STPA is similar to that of other hazard analysis methods; it tries to determine how the system hazards could occur so the cause(s) can be eliminated or mitigated by modifying the system design. The goal, however, is not to derive probabilistic requirements, as in ARP 4761, but to identify hazardous scenarios that need to be eliminated or mitigated in the design or in operations. Hazards are defined as they are in System Safety engineering, that is, as system states or sets of conditions that, when combined with some set of environmental worst-case conditions, will lead to an accident or loss event (Leveson 1995; Leveson 2012).

Humans are included as part of the system that is analyzed. STPA thus provides a structured method to identify human errors influenced by the system design, such as mode confusion and loss of situational awareness leading to hazards as well as the hazards that could arise due to loss of synchronization between actual automation state and the crew's mental model of that state.

STPA uses the beginning products of a top-down system engineering approach, including the potential losses (accidents) and hazards leading to these losses. The hazard of particular relevance to the WBS can occur when the aircraft operates on or near the ground and may involve the aircraft departing the runway or impacting object(s) on or near the runway. Such accidents may include hitting barriers, other aircraft, or other objects that lie on or beyond the end of the runway at a speed that causes unacceptable damage, injury or loss of life. The full report (Leveson 2014) identifies hazard H4 which we take as an example. H4 is decomposed into the following deceleration-related hazards:

---

[1] ARP 4761 labels the "aircraft level" what we call the "system" level, where the system is the largest unit being considered. What the ARP labels the "system level" is, in more standard system engineering terminology, the components or subsystems. The difference is not important except for potential confusion in communication.

H4-1: Inadequate aircraft deceleration upon landing, rejected takeoff, or taxiing

H4-2: Deceleration after the V1 point during takeoff

H4-3: Aircraft motion when the aircraft is parked

H4-4: Unintentional aircraft directional control (differential braking)

H4-5: Aircraft maneuvers out of safe regions (taxiways, runways, terminal gates, ramps, etc.)

H4-6: Main gear wheel rotation is not stopped when (continues after) the gear is retracted

The high-level system safety constraints (SCn) associated with these hazards are a simple restatement of the hazards in terms of requirements or constraints on the design.

SC1: Forward motion must be retarded within TBD seconds of a braking command upon landing, rejected takeoff, or taxiing.

SC2: The aircraft must not decelerate after V1.

SC3: Uncommanded movement must not occur when the aircraft is parked.

SC4: Differential braking must not lead to loss of or unintended aircraft directional control

SC5: Aircraft must not unintentionally maneuver out of safe regions (taxiways, runways, terminal gates and ramps, etc.)

SC6: Main gear rotation must stop when the gear is retracted

The next step in STPA is to create a model of the aircraft functional control structure. The STPA analysis is performed on this functional control structure model. While a general control structure that includes the entire socio-technical system, including both development and operations, can be used, in this example we consider only the aircraft itself. Figure 3 shows a very high-level model of the aircraft, with just three components: the pilot, the automated control system (which will probably consist of multiple computers), and the physical aircraft components. For complex systems, such as aircraft, levels of abstraction can be used to zoom in on the pieces of the control structure currently being considered. This type of top-down refinement is also helpful in understanding the overall operation of the aircraft and to identify interactions among the components.

The role of the pilot, as shown in the Figure 3 control structure, is to manage the automation and, depending on the design of the aircraft, directly or indirectly control takeoff, flight, landing, and maneuvering the aircraft on the ground. The pilot and the automated controllers contain a process model of the system they are controlling. The automation is controlling the aircraft so it must contain a model of the current aircraft state. The pilots also need a model of the aircraft state, but in addition they need a model of the state of the automation and a model of the airport environment in which they are operating. Many pilot errors can be traced to flaws in their understanding of how the automation works or of the current state of the automation.

Pilots provide flight commands to the automation and receive feedback about the state of the automation and the aircraft. In some designs, the pilot can provide direct control actions to the aircraft hardware (i.e., not going through the automated system) and receive direct feedback. The dotted lines represent this direct feedback. As the design is refined and more detailed design decisions are made, these dotted line links may be eliminated or instantiated with specific content. The pilot always has some direct sensory feedback about the state of the aircraft and the environment.

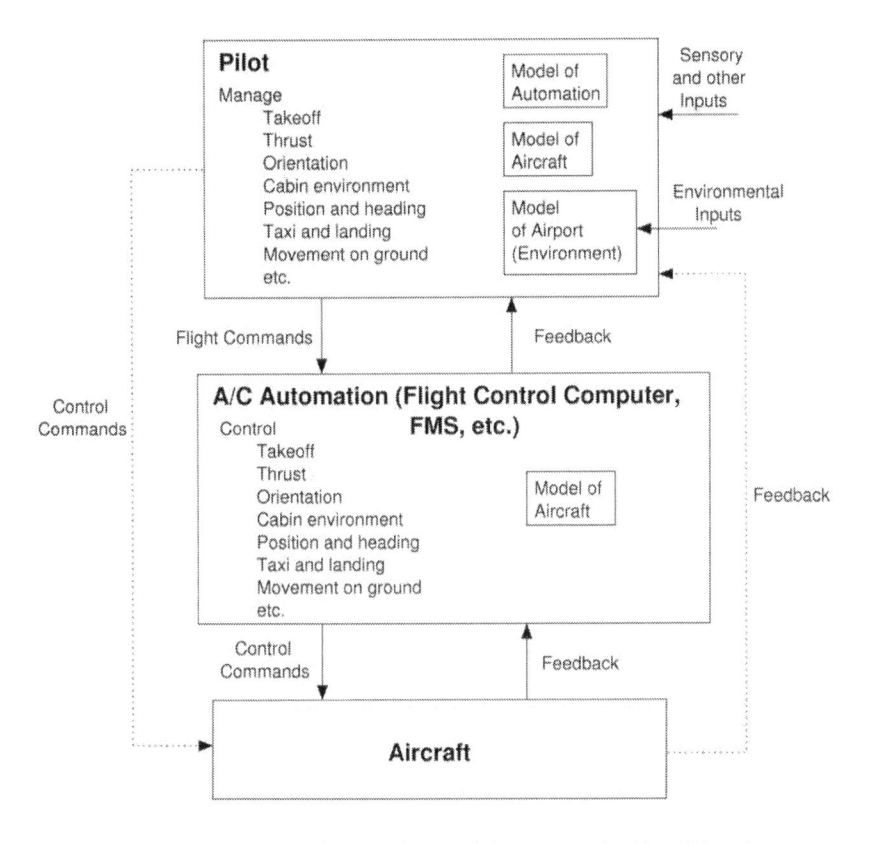

Fig. 3. A High-Level Control Structure at the Aircraft Level

Figure 4 zooms in on the control model for the ground control function, which is the focus of the example in ARP 4761. There are three basic physical components being controlled, the reverse thrusters, the spoilers, and the wheel brakes. By including the larger functional control structure than simply the WBS, STPA can consider interactions (both intended and unintended) among the braking components related to the hazard being analyzed.

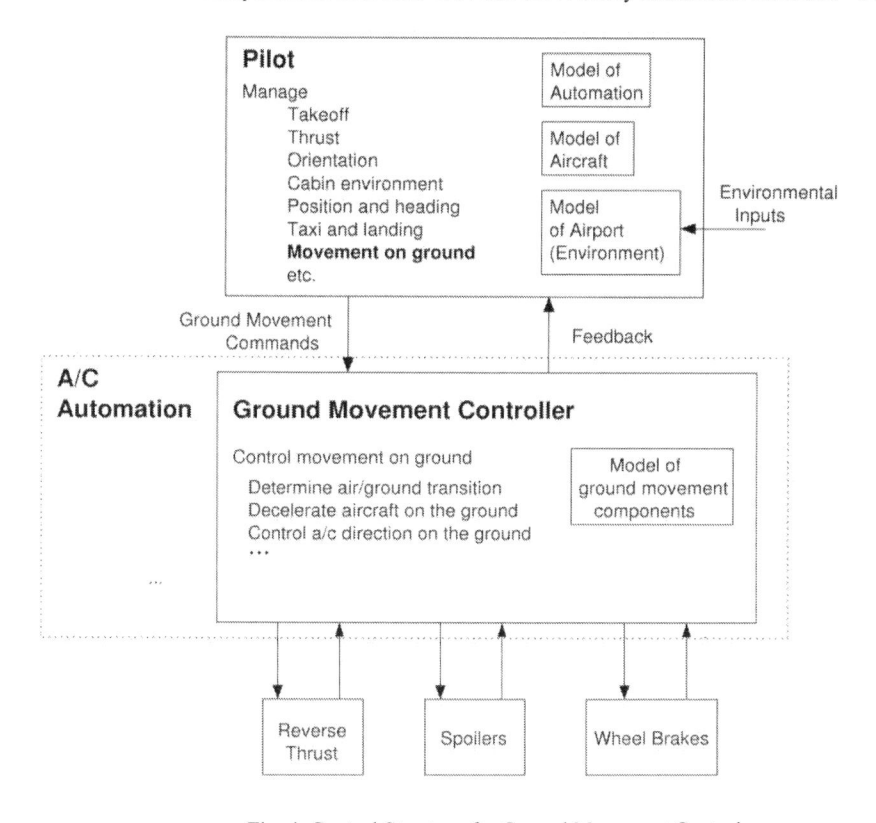

```
┌─────────────────────────────────────────────────────────┐
│  Pilot                          ┌─────────────┐         │
│                                 │ Model of    │         │
│  Manage                         │ Automation  │         │
│       Takeoff                   └─────────────┘         │
│       Thrust                    ┌─────────────┐         │
│       Orientation               │ Model of    │         │
│       Cabin environment         │ Aircraft    │         │
│       Position and heading      └─────────────┘   Environmental
│       Taxi and landing          ┌─────────────┐    Inputs
│       Movement on ground        │ Model       │◄───     │
│       etc.                      │ of Airport  │         │
│                                 │ (Environment)│        │
│                                 └─────────────┘         │
└─────────────────────────────────────────────────────────┘
           │                              ▲
       Ground Movement                 Feedback
       Commands                           │
           ▼                              │
 A/C     ┌───────────────────────────────────────────┐
 Automation │  Ground Movement Controller              │
         │                                            │
         │  Control movement on ground    ┌─────────┐ │
         │  Determine air/ground transition│Model of │ │
         │  Decelerate aircraft on the ground│ground movement│
         │  Control a/c direction on the ground│components│
         │  ...                            └─────────┘ │
         └───────────────────────────────────────────┘
              │            │            │
              ▼            ▼            ▼
         ┌────────┐   ┌────────┐   ┌────────────┐
         │ Reverse│   │Spoilers│   │Wheel Brakes│
         │ Thrust │   │        │   │            │
         └────────┘   └────────┘   └────────────┘
```

Fig. 4. Control Structure for Ground Movement Control

Finally, figure 5 shows a more detailed model of the WBS functional control structure. This model of the functional structure differs from the model of the physical structure of the WBS found in ARP 4761 and in many similar hazard analysis processes. STPA starts without a specific design solution to potential problems. Instead, it starts from the basic required functional behavior and identifies the ways that that behavior can be hazardous. Designers can later decide on particular design solutions, such as redundancy, if that turns out to be necessary to satisfy the safety requirements derived through this analysis.

The goal of the STPA analysis is to identify hazardous behaviors so they can be eliminated or controlled in the system design, which results in identifying behavioral (functional but not necessarily probabilistic) safety requirements for the various system components, including the software and human operators.

For example, one problem we identified is that the BSCU receives brake pedal commands from both pilots, but the pilots never receive any feedback about what the other pilot is doing. This feedback is important not only for manual braking (pilots may both assume the other is controlling the pedal), but also because if either pilot touches the pedal when Autobrake is active, it will automatically disarm the Autobrake system. A failure or fault oriented process that does not include analyzing pilot contributions to accidents, such as ARP 4761, might not derive a requirement that the crew should be alerted that Autobrake has been deactivated.

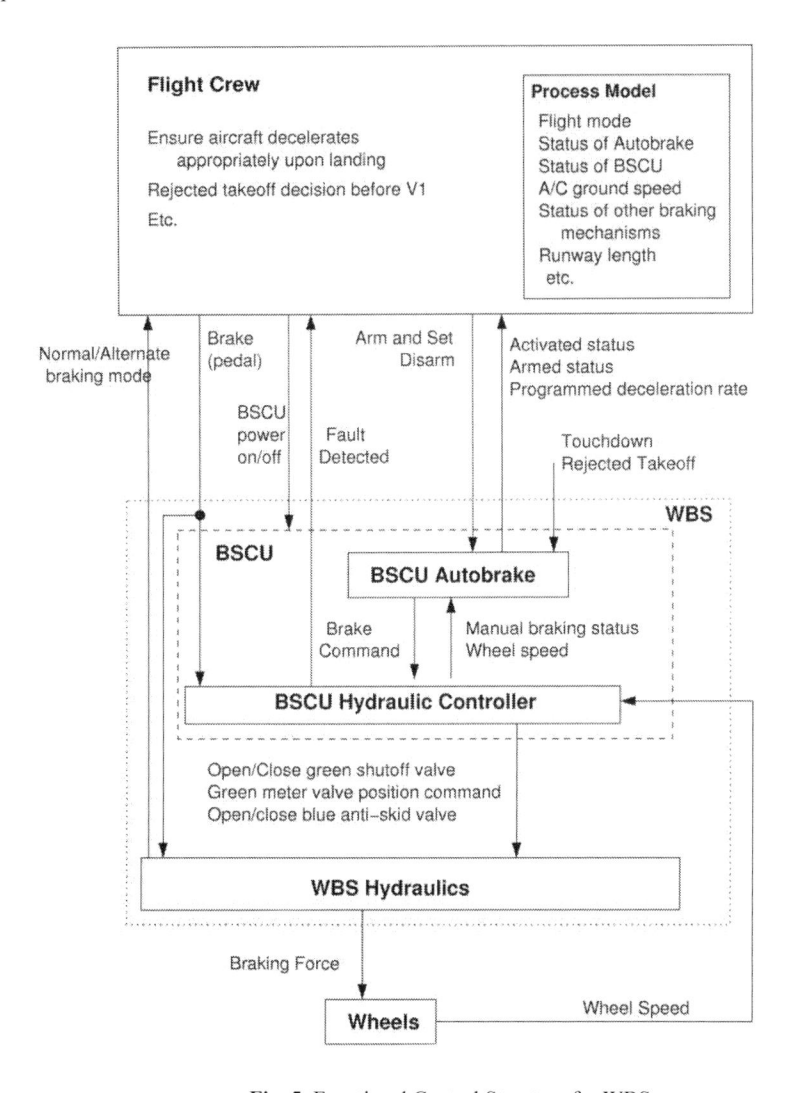

Fig. 5. Functional Control Structure for WBS

The rest of the STPA process can be decomposed into two main steps: (1) identifying unsafe control actions that can lead to system hazards and (2) identifying causal scenarios for the unsafe control actions. The scenarios include component failures but also additional factors such as direct and indirect interactions among system components (which may not have "failed"). The identified causal scenarios serve as the basis for developing system and component safety requirements and constraints.

The first step in STPA identifies potential hazardous control actions. At this stage in the analysis, it is immaterial whether control actions are provided manually or automatically. Our purpose is to define the hazardous control actions from any source.

We have developed automated tools based on a mathematical formalization of STPA to assist in the Step 1 analysis but details of those are beyond the scope of this paper (Thomas 2013).

The results of Step 1 are used to guide the generation of scenarios in Step 2 and can also be used to create requirements and safety constraints on the system design and implementation. For example, a safety constraint on the pilot might be that manual braking commands must be provided to override Autobrake in the event of insufficient Autobraking. Such constraints on humans clearly are not enforceable in the same way as constraints on physical components, but they can be reflected in the design of required pilot operational procedures, in training, and in performance audits. Some requirements that are considered to be error-prone or unachievable by human factors experts might result in changes in the braking system design.

We have found it convenient to document the unsafe control actions in a tabular form. The entries in the tables include both the control action (found in the control structures) and the conditions under which it will be hazardous. The first column lists control actions that can be given by the controller and the four following columns list how those control actions could be hazardous in four general categories. These hazardous control actions are referred to as unsafe control actions (UCA). The four UCA categories are:

- **Not providing causes hazard**: Not providing the control action under specific conditions will lead to a hazard.
- **Providing causes hazard**: Providing the control action under specific conditions will lead to a hazard.
- **Too soon, too late, out of sequence causes hazard**: The timing of the control action is critical relative to another control action.
- **Stopped too soon, applied too long causes hazard**: Applicable only to continuous control actions.

Unsafe control may depend on the operational phase, so the applicable phase is noted in the table. For example, not providing braking input in cruise is not hazardous whereas it is in the landing phase. In the full report we labeled the UCAs with a reference code (e.g. CREW.1a1), some of which we will use as examples in the causal analysis step (Step 2). In Tables 2 to 4, we show examples of unsafe control actions generated for the flight crew, the BSCU Autobrake Controller, and the BSCU Hydraulic Controller.

Control Action By Flight Crew:	Not providing causes hazard	Providing causes hazard	Too soon, too late, out of sequence	Stopped too soon, applied too long
CREW.1 Manual braking via brake pedals	CREW.1a1 Crew does not provide manual braking during landing, RTO, or taxiing when Autobrake is not providing braking (or insufficient braking), leading to overshoot [H4-1, H4-5]	CREW.1b1 Manual braking provided with insufficient pedal pressure, resulting inadequate deceleration during landing [H4-1, H4-5	CREW.1c1 Manual braking applied before touchdown causes wheel lockup, loss of control, tire burst [H4-1, H4-5]	CREW.1d1 Manual braking command is stopped before safe taxi speed (TBD) is reached, resulting in overspeed or overshoot [H4-1, H4-5]
		CREW.1b2 Manual braking provided with excessive pedal pressure, resulting in loss of control, passenger/crew injury, brake overheating, brake fade or tire burst during landing [H4-1, H4-5	CREW.1.c2 Delayed manual braking applied too late (TBD[1]) to avoid collision or conflict with another object and overloads braking capability given aircraft weight, speed, distance to object (conflict), and tarmac conditions [H4-1, H4-5]	CREW.1d2 Manual braking applied too long, resulting in stopped aircraft on runway or active taxiway [H4-1]

Table 2. Unsafe Control Actions for Flight Crew

---

[1] Where we did not know enough about braking system design to write specific requirements, we used "TBD" to indicate the need for more information by aircraft designers.

A Comparison of SAE ARP 4761 and STPA Safety Assessment Processes

Control Action BSCU:	Not providing causes hazard	Providing causes hazard	Too soon, too late, out of sequence	Stopped too soon, applied too long
BSCU.1 Brake command	BSCU.1a1 Brake command not provided during RTO (to V1), resulting in inability to stop within available runway length [H4-1, H4-5]	BSCU.1b1 Braking commanded excessively during landing roll, resulting in rapid deceleration, loss of control, occupant injury [H4-1, H4-5]	BSCU.1c1 Braking commanded before touchdown, resulting in tire burst, loss of control, injury, other damage [H4-1, H4-5]	BSCU.1d1 Brake command stops during landing roll before TBD taxi speed attained, causing reduced deceleration [H4-1, H4-5]
	BSCU.1a2 Brake command not provided during landing roll, resulting in insufficient deceleration and potential overshoot [H4-1, H4-5]	BSCU.1b2 Braking command provided inappropriately during takeoff, resulting in inadequate acceleration [H4-1, H4-2, H4-5]	BSCU.1c2 Brake command applied more than TBD seconds after touchdown, resulting in insufficient deceleration and potential loss of control, overshoot [H4-1, H4-5]	BSCU.1d2 Brake command applied too long (more than TBD seconds) during landing roll, causing stop on runway [H4-1]

Table 3. Unsafe Control Actions (BSCU Autobrake Controller)

Control Action Hydraulic Controller:	Not providing causes hazard	Providing causes hazard	Too soon, too late, out of sequence	Stopped too soon, applied too long
HC.1 Open green shut-off valve (i.e. allow normal braking mode)	HC.1a1 HC does not open the valve to enable normal braking mode when there is no fault requiring alternate braking and Autobrake is used [H4-1, H4-5]	HC.1b1 HC opens the valve to disable alternate braking mode when there is a fault requiring alternate braking [H4-1, H4-2, H4-5] HC.1b2 HC opens the valve to disable alternate braking when crew has disabled the BSCU [H4-1, H4-2, H4-5]	HC.1c1 HC opens the valve too late (TBD) after normal braking is possible and needed (e.g. for Autobrake functionality) [H4-1, H4-2, H4-5] HC.1c2 HC opens the valve too late (TBD) after the crew has enabled the BSCU [H4-1, H4-2, H4-5]	HC.1d1 HC holds the valve open too long (TBD time) preventing alternate braking when normal braking is not operating properly [H4-1, H4-2, H4-5] HC.1d2 HC stops holding the valve open too soon (TBD) preventing normal braking when it is possible and needed (e.g. for Autobrake functionality) [H4-1, H4-2, H4-5]

Table 4. Unsafe Control Actions (BSCU Hydraulic Controller)

The results of Step 1 can be used to produce general safety requirements for subsystems, training, etc. They will be refined into more detailed requirements in Step 2 when the causes of the unsafe control actions are identified. Some example requirements for the flight crew derived from the unsafe control actions are:

FC-R1: Crew must not provide manual braking before touchdown [CREW.1c1]
Rationale: Could cause wheel lockup, loss of control, or tire burst.

FC-R2: Crew must not stop manual braking more than TBD seconds before safe taxi speed reached [CREW.1d1]
Rationale: Could result in overspeed or runway overshoot.

FC-R3: The crew must not power off the BSCU during autobraking [CREW.4b1]
Rationale: Autobraking will be disarmed.

etc.

Example requirements that can be generated for the BSCU:

BSCU-R1: A brake command must always be provided during RTO [BSCU.1a1]
Rationale: Could result in not stopping within the available runway length

BSCU-R2: Braking must never be commanded before touchdown [BSCU.1c1]
Rationale: Could result in tire burst, loss of control, injury, or other damage

BSCU-R3: Wheels must be locked after takeoff and before landing gear retraction [BSCU.1a4]
Rationale: Could result in reduced handling margins from wheel rotation in flight.

Finally, some examples of requirements for the BSCU hydraulic controller commands to the three individual valves:

HC-R1: The HC must not open the green hydraulics shutoff valve when there is a fault requiring alternate braking [HC.1b1]
Rationale: Both normal and alternate braking would be disabled.

HC-R2: The HC must pulse the anti-skid valve in the event of a skid [HC.2a1]
Rationale: Anti-skid capability is needed to avoid skidding and to achieve full stop in wet or icy conditions.

HC-R3: The HC must not provide a position command that opens the green meter valve when no brake command has been received [HC.3b1]
Rationale: Crew would be unaware that uncommanded braking was being applied

Step 2 involves identifying causes for the instances of unsafe (hazardous) control identified in Step 1. It also identifies the causes for a hazard where safe control was provided but that control was improperly executed or not executed by the controlled process. Figure 6 shows some of the factors that should be considered in this process. Notice that the unsafe control actions (upper left hand arrow from the controller to the actuator) have already been identified in Step 1.

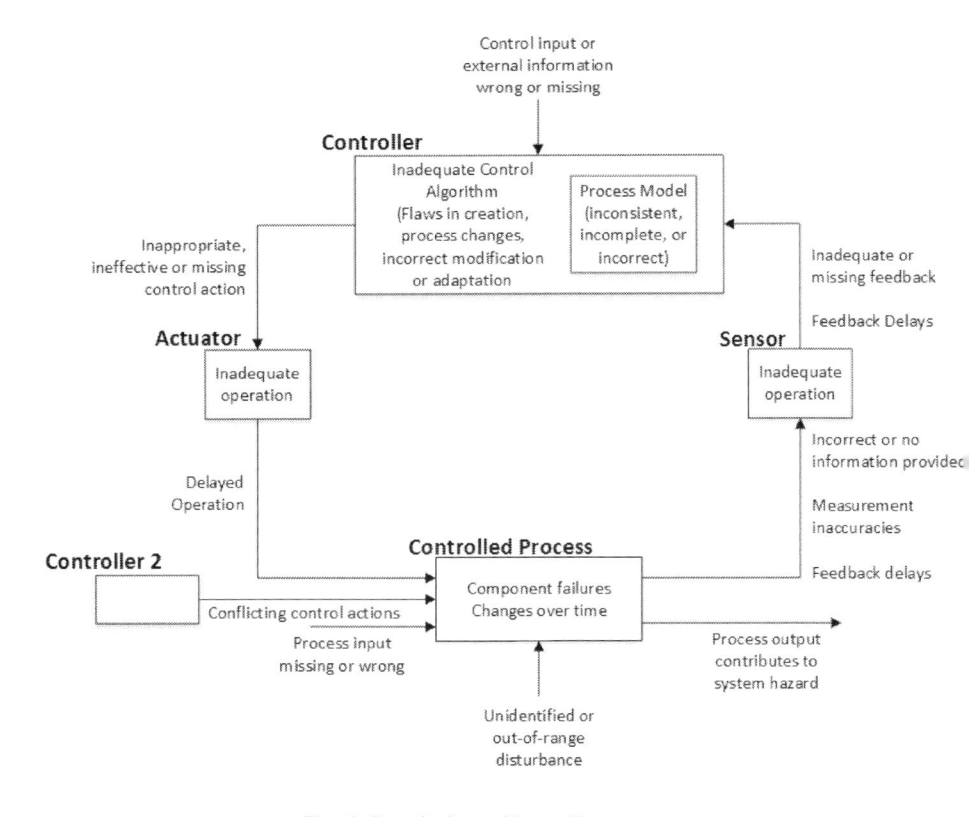

Fig. 6. Generic Control Loop Flaws

This process differs from a FMEA in that not all failures are considered, but only causes of the identified unsafe control actions. It is similar to the scenarios leading to a hazard that are identified in fault tree analysis, but more than just component failure is identified and indirect relationships are considered. The use of a model (the functional control structure) on which the analysis is performed and a defined process that the analyst follows are less likely to lead to missing scenarios and allows the analysis to be revised quickly following design modifications arising from the hazard analysis.

Only one example causal scenario generated by Step 1 is shown for an unsafe flight crew control action and one for the BSCU.

A Comparison of SAE ARP 4761 and STPA Safety Assessment Processes 75

UNSAFE CONTROL ACTION – CREW.1a1: Crew does not provide manual braking when there is no Autobraking and braking is necessary to prevent H4-1 and H4-5.

Scenario 1: Crew incorrectly believes that the Autobrake is armed and expect the Autobrake to engage (process model flaw). Reasons that their process model could be flawed include:

a) The crew previously armed Autobrake and does not know it subsequently became unavailable, AND/OR

b) The feedback received is adequate when the BSCU Hydraulic Controller detects a fault. The crew would be notified of a generic BSCU fault but they are not notified that Autobrake is still armed (even though Autobraking is no longer available), AND/OR

c) The crew is notified that the Autobrake controller is still armed and ready, because the Autobrake controller does not detect when the BSCU has detected a fault. When the BSCU detects a fault it closes the green shut-off valve (making Autobrake commands ineffective), but the Autobrake system itself does not notify the crew.

d) The crew cannot process feedback due to multiple messages, conflicting messages, alarm fatigue, etc.

Possible new requirements for S1: The BSCU hydraulic controller must provide feedback to the Autobrake when it is faulted and the Autobrake must disengage (and provide feedback to crew). Other requirements may be generated from a human factors analysis of the ability of the crew to process the feedback under various worst-case conditions.

UNSAFE CONTROL ACTION – BSCU.1a2: Brake command not provided during landing roll, resulting in insufficient deceleration and potential overshoot

Scenario 1: Autobrake believes the desired deceleration rate has already been achieved or exceeded (incorrect process model). The reasons Autobrake may have this process model flaw include:

a) If wheel speed feedback influences the deceleration rate determined by the Autobrake controller, inadequate wheel speed feedback may cause this scenario. Rapid pulses in the feedback (e.g. wet runway, brakes pulsed by anti-skid) could make the actual aircraft speed difficult to detect and an incorrect aircraft speed might be assumed.

b) Inadequate external speed/deceleration feedback could explain the incorrect Autobrake process model (e.g. inertial reference drift, calibration issues, sensor failure, etc.).

Possible Requirement for S1: Provide additional feedback to Autobrake to detect aircraft deceleration rate in the event of wheel slipping (e.g. fusion of multiple sensors)

4. Comparing STPA and ARP 4761

Because the goals are so different, it is difficult to compare the ARP 4761 process and STPA, particularly in terms of comparing the detailed results. We have, however, created a detailed comparison of the different underlying assumptions about accident causation, philosophical differences, and differences in the results of the two types of analyses. The complete comparison is too extensive to fit the limits of this conference paper. Table 5 summarizes the most important philosophical differences in the two processes. For a detailed discussion, the reader is referred to the full report (Leveson 2014).

Table 5. Differences between the Two Processes

	ARP 4761 Safety Assessment Process	STPA Hazard Analysis Process
Underlying Accident Causality Model	Assumes accidents are caused by chains of component failures and malfunctions. Based on analytic reduction.	Assumes accidents are caused by inadequate enforcement of constraints on the behavior and interactions of system components. Based on systems theory.
	Focuses the most attention on component failures, common cause/mode failures.	Focuses on control and interactions among components, including interactions among components that have not failed as well as individual component failures.
	Consideration of limited (mostly direct) functional interactions among components at the aircraft level.	Identifies indirect as well as direct unsafe functional relationships among components at any level
	Safety is to a large degree treated as a function/component reliability problem	Safety is treated as a different (and sometimes conflicting) system property than reliability.
Goals	Safety assessment.	Hazard analysis.
	Primarily quantitative, i.e., to show compliance with FAR/JAR 25.1309. Qualitative analyses (e.g., CCA and DAL) are used where probabilities cannot be derived or are not appropriate.	Qualitative. Goal is to identify potential causes of hazards (perform a hazard analysis) rather than a safety assessment. Generates functional (behavioral) safety requirements and identifies system[1] and component design flaws leading to hazards.

---

[1] Here we are using "system" in the general sense to denote the entire system being considered, such as the aircraft or even a larger transportation or airlines operations system in which the aircraft (and its subsystems) is only one component.

A Comparison of SAE ARP 4761 and STPA Safety Assessment Processes 77

	ARP 4761 Safety Assessment Process	STPA Hazard Analysis Process
Results	Generates probabilistic failure (reliability) requirements for the system and components. Also generates some qualitative aircraft level requirements and derived requirements resulting from design or implementation decisions during the development process. Derived requirements are not directly traceable to higher level requirements although they can influence higher level requirements.	Generates functional safety requirements. Identifies design flaws leading to hazards.
	Likelihood (and severity) analysis	Worst case analysis.
Role of humans (operators) in the analysis	Crew and other operators are not included in analysis except as mitigators for the physical system component failures. Crew errors are treated separately from and not addressed by ARP 4761.	Crew and operators are included as integral parts of the system and the analysis.
Role of software in the analysis	Does not assign probabilities to software. Instead identifies a design assurance level (DAL) and assumes rigor of assurance equals achieving that level.	Does not assign probabilities to software. Instead treats software in same way as any controller, hardware or human. Impact of behavior on hazards analyzed directly and not indirectly through design assurance.
	Software requirements assumed to be complete and consistent.	Generates functional software safety requirements to eliminate or control system hazards related to software behavior.
	Safety assessment considers only requirements implementation errors through IDALs	All software behavior is considered to determine how it could potentially contribute to system hazards
Process	Iterative, system engineering process that can start in concept formation stage	Iterative, system engineering process that can start in concept formation stage
Cyber Security and other system properties	Not addressed by ARP 4761.	STPA-Sec integrates safety and security analysis (not shown in this paper). Other emergent properties can also be handled.

5. Conclusions

In the reality of increasing aircraft complexity and software control, we believe the traditional safety assessment process used in ARP 4761 omits important causes of aircraft accidents. We need to create and employ more powerful and inclusive approaches to evaluating safety that include more types of causal factors and integrates software and human factors directly into the evaluation. STPA is one possibility, but the potential for additional approaches should be explored as well as improvements or extensions to STPA. There is no going back to the simpler, less automated designs of the past, and engineering will need to adopt new approaches to handle the changes that are occurring.

Acknowledgments This research was partially supported by NASA Aviation Safety Program under contract NNL10AA13C

References

Leveson N, *Engineering a Safer World*, MIT Press, 2012

Leveson N, STPA Primer, 2013, http://sunnyday.mit.edu/STPA-Primer-v0.pdf. Accessed 1 December 2014

Leveson, N, Wilkinson C, Fleming C, Thomas J, Tracy I, A Comparison of STPA and the ARP 4761 Safety Assessment Process, MIT PSAS Technical Report, Rev. 1,Oct. 2014, http://sunnyday.mit.edu/papers/ARP4761-Comparison-Report-final-1.pdf. Accessed 2 December 2014

RTCA, DO-254/ED-80: Design Assurance for Airborne Electronic Hardware, 2000

Thomas J, Extending and Automating a Systems-Theoretic Hazard Analysis for Requirements Generation and Analysis, Ph.D. Dissertation, Engineering Systems Division, MIT, June 2013

RTCA, DO-178C/ED-12C: Software Considerations in Airborne Systems and Equipment Certification, 2012

SAE, ARP 4761: Guidelines and Methods for Conducting the Safety Assessment Process on Civil Airborne Systems and Equipment, ARP 4761, Dec. 1996

SAE, ARP 4754A: Guidelines for Development of Civil Aircraft and Systems, 2010

SAE, ARP 4754, Certification Considerations for Highly-Integrated or Complex Aircraft Systems, 1996

FAA, Advisory Circular: AC 25.1309-1A, System Design and Analysis, AC 25.1309-1A, 1988

Can Evidence-Based Software Engineering Contribute To Safer Software?

K.R. Wallace

BAE Systems

Portsmouth, UK

Abstract *Over the past decade evidence-based software engineering (EBSE) has emerged to offer an adaptation of the methods of evidence based medicine for the needs of software engineering. Whilst potentially complementary to established practices, evidence to date of successful translation of the approach from the academic environment to industrial practice is scarce. With an emphasis on structured arguments, rigorous analysis and bodies of evidence software safety is one engineering sub-discipline apparently well placed to benefit from adopting the approach. This paper assesses whether EBSE has or can, in practice, make a contribution to engineering safer software.*

1 Introduction

Introduced by Kitchenham and colleagues in 2004 evidence-based software engineering (EBSE) has, over the past decade, evolved into a recognised approach to software engineering research. Proponents of EBSE advocate a number of benefits in transforming the practice of software engineering to be evidence-driven, appealing in doing so to the proven benefits of the approach when applied to disciplines such as medicine, education and psychology. Critics might counter that prior to embracing EBSE achieving the status of a recognised engineering discipline would be a more realistic and beneficial goal, notwithstanding all the progress that recent years have undoubtedly seen in this regard.

Whether software engineering (SE) is, or can ever become, capable of standing shoulder-to-shoulder with other established engineering disciplines is a long running debate. Certainly many would recognise, if not empathise, with Sutton's (2008) characterisation that:

> Software has long been the odd man out in business: It operates in ways that are different than, and often incompatible with, the disciplines of other industries, even when it is teamed with those disciplines under the same enterprise. It generally underperforms other industries on productivity improvement, integration success, quality and customer satisfaction.

Notwithstanding these differences what is undeniable is the extent to which software now underpins or enables the contemporary world such that even basic aspects of societal function are now dependent upon the correct operation of software. One corollary of such ubiquity is the steady encroachment of software into sectors, such as aerospace, defence, energy, medicine and transportation, where safety is by necessity a very significant and substantial consideration. Regardless of domain, assuring the software contribution to safety-critical or safety-related systems is recognised as challenging both technically and in terms of the time and effort involved. This is particularly the case when certification or the presentation of a safety argument is required to satisfy either a regulatory body or other competent authority. Factor in increasing system complexity, fast-paced software technology evolution driven principally by innovation, time-to-market and the demands of consumer-led mass markets, and the scale of the challenge becomes apparent.

In appealing to rigorous and repeatable methods for the provision of unbiased bodies of evidence upon which to make informed decisions EBSE would, in principle, appear well suited for adoption when considering the contribution of software to system safety. The aim of this paper is to assess whether the proposed benefits of EBSE can, in practice, be realised. In particular this paper considers whether a decade of EBSE has succeeded in fulfilling the intent:

> EBSE aims to improve decision making related to software development and maintenance by integrating current best evidence with practical experience and human values.

articulated by (Dyba et al. 2005).

The assessment presented herein has been achieved through application of EBSE techniques, principally the Systematic Review (SR) method. Originating within Evidence-Based Medicine (EBM) arguably the most notable result of EBSE, to date, has been the adoption and adaption of the SR method to the needs of software engineering.

2 Evidence-Based Software Engineering

This section provides a brief summary of the salient aspects of EBSE of relevance to this paper, primarily the use of SRs; currently the principal method employed by EBSE. Interested readers may wish to refer to recent work reported by (Kitchenham and Brereton 2013) and (Zhang and Babar 2013), and references therein for further details of the current state of practice of SRs and the relationship with EBSE. It should be noted that in much of the literature the terms Systematic Review (SR) and Systematic Literature Review (SLR) are used rather interchangeably, however, in this paper the term SR will be used exclusively.

SRs are an established and accepted means of aggregating evidence from primary (empirical) studies to achieve robust conclusions in respect of particular questions; thereby facilitating, in theory at least, informed and unbiased decision-making processes. In order to achieve such robustness, aggregation of evidence has to be both comprehensive and repeatable. Accordingly, much of the focus for the EBSE research community has, to date, been upon ensuring that SR methods are capable of achieving this outcome when applied to software engineering.

This paper employs the demonstrated ability of SR methods to achieve robust and repeatable results (MacDonell et al. 2010) as a starting point. The product of a SR, referred to as a *secondary* study, is itself capable of being reviewed typically as part of the study of SR techniques, the results of such investigations giving rise to *tertiary* studies (Kitchenham et al 2010).

The SR reported herein assesses the contribution of EBSE to software safety by reviewing, principally, secondary studies (i.e. SRs) for evidence of significant or substantial contributions to the practice of software engineering in respect of software safety. The present study can, therefore, be classed as being tertiary in nature. At present almost all existing tertiary studies are introspective: their focus being upon evaluation of SR methods and associated techniques. In contrast the present work has sought to extract information and results relating to one particular topic, namely safety aspects of software, principally by means of consideration of existing secondary studies. The dominance of secondary studies in the present work reflects a lack of other relevant products arising from EBSE that were eligible for inclusion in the present SR.

One recent and highly relevant SR in the field of safety is the work of (Nair et al. 2014) who have applied the method to derive a data set which includes a taxonomy for evidence employed to achieve safety certification. As will be discussed the present work has made explicit use of this taxonomy in the conduct of the assessment reported herein. Accordingly this study has been excluded from the group of secondary studies analysed for the present assessment.

3 Methods

The SR process reported herein can be summarized thus:

Phase 1: Manually search electronic databases for relevant literature and retrieve the metadata for each publication, as captured in the associated citation

Phase 2: Manually apply a multi-stage filtering process whereby a set of exclusion criteria are applied in succession to the retrieved metadata

Phase 3: Analyse and assess the studies remaining following filtering to establish whether there is significant or substantial evidence of results originating from EBSE which can be applied to improve safety aspects of software.

Usually a critical step in the SR process is to define the set of Research Questions (RQ) to which answers are then sought from the existing literature. This facilitates an increasing refinement of the studies reviewed to ensure relevance to the research questions. In contrast, in seeking to identify a contribution of EBSE to safety aspects of software the present study poses a more general question. Accordingly the phases of the SR reported herein have been constructed to ensure that the filtering process reflects the generality of the contribution to this topic that potentially exists in the literature.

Table 1. Results for Phase 1: Database search

Database	Title Keyword	AND Index Keyword	AND Author Keyword	Count
ACM DL	Evidence	-	-	314
	Review	-	-	1654
	Evidence	-	Software	15
	Review	-	Software	62
IEEE Xplore	Evidence	-	-	2780
	Review	-	-	6763
	Evidence	Software	-	175
	Review	Software	-	624
	Evidence	-	Software	44
	Review	-	Software	140
	Review	Software	Systematic	74
	Review	Software	Mapping	1
	Review	Software	Literature	53
Science Direct	Evidence	-	-	38117
	Review	-	-	53644
	Evidence	-	Software	27
	Review	-	Software	99

3.1 Phase 1: Extraction

Data extraction took the form of a manual keyword search in the ACM Digital Library (ACM DL), IEEE Xplore and Science Direct databases. Recognising the introduction of EBSE in 2004 (Kitchenham et al. 2004) these searches were limited from that year until early September 2014. To reduce duplication resulting from the indexing of other databases by the ACM DL use of this database was confined to studies published by either the ACM or affiliated organisations. No equivalent restrictions were necessary for the two other databases. The results of the keyword search are shown in Table 1, the totals in bold identifying the search results employed in subsequent phases. The availability of the index keyword search facility in the IEEE Xplore database was employed to refine the search results produced by this database. No equivalent facilities were available in either the ACM DL or ScienceDirect databases.

3.2 Phase 2: Filtering

Stage 1 filtering utilised the facilities of the respective databases to remove intra-database duplicates. Removal of inter-database duplicates (i.e. the same publication cited in more than one database) was not performed until Stage 5 of the filtering process. On completion of this stage citations and abstracts for each publication identified were retrieved from the respective databases.

Stage 2 employed a manual review of each publication abstract and keywords to identify evidence of the use of, affinity for, or explicit recognition of EBSE and associated methods such as SR. Hence an article stating it was a systematic review was included at this stage whereas an article stating it was a 'thorough review' or a 'literature review' but which made no acknowledgement of EBSE or associated techniques was excluded. Publications which employed a language other than English for either the title or abstract were also excluded at this stage.

At Stage 3 publications were again manually reviewed and included if any of the following criteria applied:

1. There was explicit identification of safety in the title, abstract or keywords
The abstract indicated industrial or other real-world application
The abstract indicated the provision of recommendations, advice or guidelines potentially capable of informing practice
The abstract indicated results either quantitative or qualitative in nature which might be used as a basis for making decisions in regard to safety aspects of software, or which could contribute to such decisions.

Conversely for Stage 3 publications were excluded if:

1. The scope of study was explicitly stated as being exclusively research focused
The abstract recorded only research conclusions/recommendations
The title or abstract identified the study as being of a pedagogic nature

The abstract identified the study as being introspective in nature; the subject of investigation being either EBSE as a topic itself or related methods or techniques
Results reported in the abstract were ambiguous, equivocal, or tentative or, alternatively, were identified as being preliminary in nature
The study was of a hybrid nature wherein the purpose of any EBSE aspect of the study, as identified in the abstract, was to act as the basis for the introduction of a novel technique, method or other equivalent original contribution
The publication was an advocacy or opinion-based article.

For this stage the inclusion criteria took precedence over those for exclusion.

Stage 4 repeated the manual review process of previous stages. At this stage inclusion was on the basis of affinity of content, as identified in the abstract and keywords, to any of the evidence taxonomy terms reported by Nair et al. For this stage an exact match of term or phrase was not necessary, partial matching of terms or equivalent terms being sufficient to merit continued inclusion of a source. Accordingly this stage was intentionally biased towards continued inclusion.

Stage 5 involved the collation of all remaining results, removal of inter-source duplicates and separation of the results into two categories of publication venue: *Journal* and *Other*.

In practical terms data extraction and filtering involved an initial set-up phase permitting familiarisation with the facilities offered by the respective databases. This enabled a viable process to be established which included efficient techniques to address identified limitations in the functionality provided by each database. Thereafter a first pass through the process was conducted as a 'pipe-cleaning' exercise. This pass confirmed that execution of the process for all publications identified in Phase 1 could be achieved in a period equivalent to eight hours. A second pass through the process was then performed to obtain the data used for the assessment. The two passes through the complete process were conducted on consecutive days.

On completion of the filtering process the results obtained were compared with the studies reported in the tertiary studies of (Kitchenham et al. 2010) and (da Silva et al. 2011). This was intended to permit identification of possible limitations of the process and facilitate investigation of the root cause of omission of any publications which should have satisfied all the filtering criteria. The results of this comparison confirmed there had been no inappropriate exclusion of studies.

As a second validation of the process the results were compared with a set of sources collected previously in connection with other investigations utilising SRs (Wallace 2014), these sources approximating to the quasi-gold standard (QGS) suggested by (Zhang and Babar 2011). This comparison identified two additional sources which, while failing to meet the filtration criteria, were considered of sufficient relevance to be included in the subsequent analysis phase. The predominant cause of publication omission was the lack of inclusion of 'software' as an author or index keyword in the publication metadata.

The results of Stages 1 – 5 filtering are given in Table 2. For Stage 5 the counts for both the Journal and Other categories of publication venue are provided, the latter figure being in parentheses.

Table 2. Stage 1 - 5 results.

Source	Stage 1	Stage 2	Stage 3	Stage 4	Stage 5
ACM DL	74	38	9	6	0 (5)
IEEE Xplore	218	97	25	18	3(15)
ScienceDirect	124	60	21	13	13(0)
Totals	416	195	55	37	16(20)

3.3 Phase 3: Analysis

Prior to detailed analysis the results obtained at Stage 5 were refined by category using, in the first instance, the evidence taxonomy of Nair et al. previously employed at Stage 4. While it proved practical to assign a portion of the studies to categories in this taxonomy, for the majority there was no obvious appropriate category. Accordingly further categories were derived from the topics reported in the identified studies. These additional categories were then added to the taxonomy of Nair et al. to create an 'Augmented Taxonomy'.

Each of the studies to be analysed were then categorised against either the basic or augmented taxonomy. The results of this process are given in Table 3. This table also includes the separation of studies into the two publication venue categories: in this instance the latter category consisted exclusively of studies published in conference proceedings. As the requirements for publication in a peer-reviewed journal are recognized to be more stringent than other venues only this category of publication was subjected to further analysis.

Table 3. Stage 5 results by category.

Category	ID	Nair et al	Augmented Taxonomy	Journal	Other	Totals
Automated Static Analysis	ASA	●		1	0	1
Defects/Faults	DEF		●	2	1	3
Requirements Engineering	SRE	●		2	1	3

Category	ID	Nair et al	Augmented Taxonomy	Journal	Other	Totals
Risk	RISK	•		0	1	1
Software Architecture	SWA	•		1	2	3
Software Development Lifecycle	SDLC		•	1	3	4
Software Process Improvement	SPI		•	1	2	3
Software Process Simulation	SPS		•	1	0	1
Software Quality	SWQ		•	1	4	5
Software Robustness	SWR		•	1	0	1
Software Variability	SWV		•	1	0	1
System Security	SYS		•	0	1	1
User Experience/Interface	UX		•	0	1	1
V&V	VVR	•		5	3	8
Basic Totals		5	-	9	7	16
Augmented Totals		-	9	17	19	36

The remaining publications, as listed in Table 4, were reviewed for evidence of references to safety, specifically software safety as an identified topic of interest in the study. Of these: 6 equating to 35% of the total number reviewed contained relevant references, these studies being identified in Table 5.

The extent to which each publication identified in Table 5 considered primary studies which explicitly addressed safety aspects of software was quantified by:

Counting the number of primary studies which were identified in the publication as being directly related to safety
Counting any other primary studies cited in the publication which addressed safety aspects of software.

Counting of both categories was restricted exclusively to the inclusion of safety and software as keywords in either the study title or journal title as cited in the study. No review of the primary study itself was undertaken.

Table 4. Journal studies reviewed.

Title	Authors	Classification
A systematic review of search-based testing for non-functional system properties	Afzal W et al. (2009)	VVR
A systematic literature review on the industrial use of software process simulation	Ali N B et al. (2014)	SPS
A systematic review of the application and empirical investigation of search-based test case generation	Ali S et al. (2010)	VVR
A systematic review on the relationship between user involvement and system success	Bano M and Zowghi D (2014)	SWQ
A systematic review of software architecture evolution research	Breivold H P et al. (2012)	SWA
Empirical studies of agile software development: A systematic review	Dybå T and Dingsøyr T (2008)	SDLC
Variability in software systems – A systematic literature review	Galster M et al. (2014)	SWV
A systematic literature review on fault prediction performance in software engineering	Hall et al. (2012)	DEF
A systematic literature review of actionable alert identification techniques for automated static code analysis	Heckman S and Williams L (2011)	ASA
Testing scientific software: A systematic literature review	Kanewala U and Bieman J M (2014)	VVR
Analyzing an automotive testing process with evidence-based software engineering	Kasoju A et al. (2013)	VVR
On strategies for testing software product lines: A systematic literature review	Machado I d C et al. (2014)	VVR
A systematic literature review of stakeholder identification methods in requirements elicitation	Pacheco C and Garcia I (2012)	SRE
Software fault prediction metrics: A systematic literature review	Radjenović D et al. (2013)	DEF
A systematic review of software robustness	Shahrokni A and Feldt R (2013)	SWR
Evaluation and measurement of software process improvement – A systematic literature review	Unterkalmsteiner M et al. (2012)	SPI
A systematic literature review to identify and classify software requirement errors	Walia G S and Carver J C (2009)	SRE

Table 5. Analysis of journal studies which identify safety

Source Identifying Safety	Classification	Primary Studies Referenced	Which Consider Safety	AND Which Are Software Focused
Afzal W et al. (2009)	VVR	35	4	4
Ali S et al. (2010)	VVR	64	0	0
Galster M et al. (2014)	SWV	196	1	Unknown
Kasoju A et al. (2014)	VVR	48	2	2
Shahrokni A and Feldt R (2013)	SWR	144	6	4
Walia G S and Carver J C (2009)	SRE	149	2	2
		Unique studies	14	12

The results of this final analysis are also given in Table 5. Due to the variable nature of identification of primary studies reported in these sources the figures provided can only be indicative.

3.4 Phase 3: Assessment

Consistent with the low number of studies which explicitly considered safety, in quantitative terms the extent to which safety constituted a contributory topic area in each study was also low. The review by Afzal et al. of non-functional system testing contained the largest number of citations to primary studies relating to safety, however even this number represented only 11% of the total number of primary studies examined, performance (execution time) being the highest proportion at approximately 43%, security and usability each accounting for 20% of the primary studies. This review explicitly identified the various non-functional areas by topic and provided accompanying data on the extent of primary literature contributing to each of the identified topics.

The studies by Kasoju et al. and Shahrokni and Feldt. contained the next highest proportions of primary studies addressing safety of software, however, at approximately 4% for each neither contribution can be regarded as being significant. For reference the comparable figure for the study of Nair et al. is approximate 16%. The publication venues for primary studies are given in Table 6.

Table 6. Venue for primary studies

Publication Venue	Count
Journal	4
Conference Proceedings	5
Book Chapter	1
URL	1
Other	1

Complementary to the quantitative measures qualitative review of each publication was undertaken to establish any results of significance in respect of software safety.

Afzal et al. provides a context for testing of software safety relative to other non–functional properties such as security, human factors, performance and usability. As such this study is structured in a manner that clearly identifies the extent, in quantitative terms, of safety consideration, albeit the actual extent of the identified coverage is, as noted, limited. As the contributory literature extended only until 2007, however, the study cannot account for evidence from either research or practice, which has emerged subsequently.

The study by Ali et al., on search-based software-testing (SBST) is potentially complementary to that of Afzal el al. In this instance, however, the focus was towards an understanding of the state of research in respect of SBST up until 2007, rather than translation into industrial practice. Nevertheless the conclusion of these authors in 2010 that:

> The number of papers which contain well-designed and reported empirical studies in the domain of test- case generation using SBST is very small. As a result, there is a limited body of credible evidence that demonstrates the usefulness of SBST techniques for test case generation.

suggests that this topic is unlikely to produce results of practical use in the near future, irrespective of whether safety is a consideration or not.

In this respect, however, the potential for significant or substantial developments occurring subsequently to impact on the findings reported in the respective studies, given the passage of time since publication of each of them, has to be acknowledged.

The papers by Galster et al. and Shahrokni and Feldt are both recent and address emerging topics (software variability and robustness respectively). As such neither is in a position to provide definitive statements on their respective topics of study that could impact significantly on safety aspects of software, and unsurprisingly neither do. A principal theme in both is the lack of evidence pertaining to practical applications in industrial settings: a common observation in the wider

EBSE literature reflecting the known gap between software research and practice more generally (Beecham et al. 2014). In this respect the study by Shahrokni and Feldt provides data to illustrate the extent of this issue for the topic of software robustness.

The publication of Kasoju et al. is the sole study to report on the industrial application of EBSE, indeed as the authors note:

> … to our knowledge this is the first time they [EBSE techniques] have been used in combination for solving a problem in a concrete case study.

Thus this study provides an initial template for the application of EBSE in an industrial setting (automotive) wherein, although it is not the topic of principal interest to the study, safety represents a significant consideration.

The final study by Walia and Carver, which addressed requirements errors, resulted in a proposed taxonomy of errors with the potential to be applied in practice. This review arguably provides the most promising contribution to practice in the form of the proposed taxonomy and classes of contributory error therein.

A general observation in respect of all the foregoing studies is that half are concerned with V&V aspects, whereas only one is concerned with requirements. If this ratio accurately reflects the balance of effort between these two topics, at least within SE research, it raises questions regarding the contributory factors to such a ratio and whether, from the perspective of practice, this represents or approaches an optimal ratio. Such questions are, however, beyond the scope of the current work and have, therefore, been left for future consideration.

4 Threats To Validity

Consistent with recommended practice in the conduct of SRs, threats to the validity of the results presented can be identified, namely:

use of a restricted number (3) of source databases
a bounded execution time for the data extraction and filtering process (equivalent to an 8 hour working day)
exclusion in the final assessment of any studies not published in peer-reviewed journals
use of a single reviewer (the author)
the use of a recently introduced taxonomy derived using the SR method
the use of a quantitative assessment reliant on the presence of keywords in a small population of primary studies.

which are addressed individually.

The present study deliberately set the threshold of publication quality, including venue, to be high. This approach is likely the absolute minimum that, for example, any safety argument appealing to results of EBSE would have to demonstrate. Similarly it is highly likely that in an industrial context informed decision making involving consequential economic outcomes would seek equivalent assurance/confidence, prior to embarking on any programme of change or implementation attributable exclusively or substantially to results obtained from EBSE. Additional databases might have increased the number of contributory publications, though whether any such increase could also have satisfied the noted quality threshold is open to question. Similar arguments apply in respect of the exclusion of studies not published in peer-reviewed journals. It is assumed that only in very exceptional cases will high quality SRs published in conference proceedings not thereafter be further refined and developed sufficiently to enable publication in a peer-reviewed journal.

The amount of time taken to conduct SRs has repeatedly been identified as a significant difficulty in the conduct of SRs (Kitchenham and Brereton 2013). Tertiary studies, by virtue of being de-coupled from empirical studies and employing standard methods as they do, gain a degree of immunity from this issue. Moreover as the quality threshold was deliberately set high then the extent of such de-coupling will be even more pronounced in the present study. Accordingly the risk that the strictly bounded execution time of the current SR led to the exclusion from the assessment of a study offering a significant contribution is considered to be both low probability and impact.

At least some of the difficulties in the conduct of SRs are attributable to the lack of experience of those conducting the reviews. The benefits of having an experienced practitioner with a research background are considerable, as wider experience in applying EBSE within Naval Ships has demonstrated. Conversely the time bounding of the process reduced the 'dwell' time on particular abstracts thereby reducing the potential for seduction and hence introduction of consequential bias towards studies which included well written and attractive abstracts. Such bias acting to disadvantage studies which, though of more relevance, were less effectively presented.

In the absence of the taxonomy of Nair et al., the present study would either have proceeded in an un-structured manner or would alternatively have had to devise another equivalent taxonomy. Regardless of the validity of content of the taxonomy employed neither of the other two alternatives can credibly be regarded as more appropriate options than use of an existing approach, particularly one derived by means of the SR process.

The work of Nair et al. aside, the absence in the studies analysed of significant or substantial results of a compelling nature, could easily lead to a simple binary outcome of the assessment. In this regard assessing the extent of an EBSE contribution to any given topic in SE represents a more general problem, to which there appears to be at, this time, no tried and tested solution. If EBSE is to be given the opportunity to contribute to SE then binary outcomes, particularly if applied across many different topics in SE, are unhelpful and have the potential to mask early indicators of more significant and substantial results in the future.

Given the apparent lack of a proven measure of EBSE contribution, the quantitative assessment approach employed herein was devised solely as an initial method for achieving a more sensitive measure of the possible value of EBSE. Whether this measure is practical, effective and accurate can only be established though wider application and review of the results obtained in doing so.

5 Conclusions

This paper posed the question of whether EBSE can contribute to safer software. Before considering a response in the light of the evidence presented herein it is instructive to consider how the EBSE research community might answer the question by reference to the findings of tertiary studies which have reviewed the products of EBSE research. In this regard (Kitchenham et al. 2010) note that:

> 'Currently, the topic areas covered by SLRs are limited'.

with (da Fabio et al. 2011) concluding the following year that:

> ... the majority of the SLRs ... fail to provide guidelines for practitioners, thus decreasing their potential impact on software engineering practice

In reviewing the available SR literature it is evident that in the intervening period since these assessments were made there has been an increase in the number of topic areas addressed which, in this instance, might offer a contribution in respect of safety aspects of software. That this is the case can be confirmed by reference to Table 7, which identifies the year of publication of all the studies in the Journal category considered in the present work.

Table 7. Year of publication of journal studies analysed, Nair et al. (2014) included

Year	2008	2009	2010	2011	2012	2013	2014
Studies	1	2	1	1	4	3	6
%	6%	11%	6%	6%	22%	17%	33%

Despite this welcome trend, the low number of studies indicates the absence of a substantial contribution to software safety by EBSE, particularly taking into account the fact that less than half of these studies incorporate explicit consideration of safety.

Undoubtedly the data set and taxonomy of Nair et al. represents a significant contribution to the topic, even though the scope of that work extends beyond SE. Similarly the industrial application of EBSE reported by Kasoju provides an early indication that EBSE may now be capable of moving beyond the confines of research and finding real-world application in a sector where safety is a significant consideration. Moreover the substantive body of introspective work addressing the methods of EBSE themselves provides significant confidence in the validity of EBSE methods when they do eventually find practical application.

In summary, a decade of EBSE has resulted in credible methods and a growing body of knowledge regarding an increasing number of areas of SE. In respect of software safety, however, substantial evidence of a contribution remains absent. Nevertheless there has been some recent and significant progress with the potential for practical application beginning to emerge. It is likely that the extent to which EBSE can, in the future, make a significant and substantial contribution to software safety will increase, though the timescales over which this outcome can be achieved remain uncertain.

With regard to recommendations: it is suggested that practitioners would benefit from gaining and thereafter maintaining familiarity with the practice and benefits of EBSE. Having achieved such awareness it is further recommended that they thereafter sustain an ongoing awareness of the results of EBSE, giving consideration to how these results might be applied in their particular circumstances when seeking to achieve safer software and to provide credible evidence to this effect.

Acknowledgments The author is grateful to Steve Anderson, Paul Sagar and Trevor Seager for lively discussions on EBSE. Special thanks are due to Jane Mynott at the BAE Systems ATC Library for her valued assistance to this work.

References

Afzal W, Torkar R, Feldt R (2009) A systematic review of search-based testing for non-functional system properties. Information and Software Technology 51:6:957 – 976

Ali N B, Petersen K, Wohlin C (2014) A systematic literature review on the industrial use of software process simulation. Journal of Systems and Software (in press)

Ali S, Briand C, Hemmati H, Panesar-Walawege R K (2010) A systematic review of the application and empirical investigation of search-based test case generation. IEEE Transactions on Software Engineering 36:6:742 – 762

Bano M, Zowghi D (2014) A systematic review on the relationship between user involvement and system success. Information and Software Technology (in press)

Beecham S, O'Leary P, Baker S, Richardson I, Noll J (2014) Making Software Engineering Research Relevant. IEEE Computer, 47:4: 80 – 83

Breivold H P, Crnkovic I, Larsson M (2012) A systematic review of software architecture evolution research. Information and Software Technology 54:1:16 – 40

Dybå T, Dingsøyr T (2008) Empirical studies of agile software development: A systematic r view. Information and Software Technology 50:9-10:833 – 859

Dybå T, Kitchenham B A, Jorgensen M, (2005) Evidence-based software engineering for practitioners. IEEE Software 22:1:58 – 65

da Silva F Q B, Santos A L M, Soares S, França A C A, Monteiro C V F, Maciel F F (2011) Six years of systematic literature reviews in software engineering: An updated tertiary study. Information and Software Technology 53:9:899 – 913

Galster M, Weyns D, Tofan D, Michalik B, Avgeriou P (2014) Variability in software systems – A systematic literature review. IEEE Transactions on Software Engineering 40:3:282 – 306

Hall T, Beecham S, Bowes D, Gray D, Counsell S (2012) A systematic literature review on fault prediction performance in software engineering. IEEE Transactions on Software Engineering 38:6:1276 – 1304

Heckman S, Williams L (2011) A systematic literature review of actionable alert identification techniques for automated static code analysis. Information and Software Technology 53:4:363 – 387

Kanewala U, Bieman J M (2014) Testing scientific software: A systematic literature review. Information and Software Technology 56:10:1219 – 1232

Kasoju A, Petersen K, Mäntylä M V (2013) Analyzing an automotive testing process with evidence-based software engineering. Information and Software Technology 55:7:1237 – 1259

Kitchenham B, Brereton P (2013) A systematic review of systematic review process research in software engineering. Information and Software Technology 55:12: 2049 – 2075

Kitchenham B, Dybå T, Jorgensen M (2004) Evidence-based software engineering. 26th International Conference on Software Engineering (ICSE 2004), Edinburgh, UK

Kitchenham B, Pretorius R, Budgen D, Brereton O P, Turner M, Niazi M, Linkman S (2010) Systematic literature reviews in software engineering - A tertiary study. Information and Software Technology 52:8:792 – 805

MacDonell S, Sheppard M, Kitchenham B, Mendes E (2010) How reliable are systematic reviews in empirical software engineering ? IEEE Transactions on Software Engineering 36:5:676 – 686

Machado I d C, McGregor J D, Cavalcanti Y C, de Almeida E S (2014) On strategies for testing software product lines: A systematic literature review. Information and Software Technology 56:10:1183 – 1199

Nair S, de la Vara J L, Sabetzadeh M, Briand L (2014) An extended systematic literature review on provision of evidence for safety certification. Information and Software Technology 56:7:689 – 717

Pacheco C, Garcia I (2012) A systematic literature review of stakeholder identification methods in requirements elicitation. Journal of Systems and Software 85:9:2171 – 2181

Radjenović D, Heričko M, Torkar R, Živkovič A (2013) Software fault prediction metrics: A systematic literature review. Information and Software Technology 55:8:1397 - 1418

Shahrokni A, Feldt R (2013) A systematic review of software robustness. Information and Software Technology 55:1:1 – 17

Sutton J M (2008) Welcoming software into the industrial fold. CROSSTALK: The Journal of Defense Software Engineering, May 2008

Unterkalmsteiner M, Gorschek T, Islam AK M M, Cheng C K, Permadi R B, Feldt R (2012) Evaluation and measurement of software process improvement – A systematic literature review. IEEE Transactions on Software Engineering 38:2:398 – 424

Wallace K R (2014) Safe and Secure: Re-engineering a software process set for the challenges of the 21st challenge.9th IET Conference on System Safety and Cyber Security, Manchester, UK, 2014

Walia G S, Carver J C (2009) A systematic literature review to identify and classify software requirement errors. Information and Software Technology 51:7:1087 – 1109

Zhang H, Babar M A (2013) Systematic reviews in software engineering: An empirical investigation. Information and Software Technology 55:12: 1341 – 1354

Zhang H, Babar M A, Tell P (2011) Identifying relevant studies in software engineering. Information and Software Technology 53:6:625 – 637

Applicability of Formal Methods for Safety-Critical Systems in the Context of ISO 26262

S. Kandl[(1)], M. Elshuber[(1)], S. Gulan[(2)], T. Nguyen[(3)], S. Rieger[(2)], P. Schrammel[(4)], R. Sisto[(5)]

(1) Vienna University of Technology Vienna, Austria

(2) TWT GmbH Science&Innovation Stuttgart, Germany

(3) Infineon Technologies Austria AG Villach, Austria

(4) University of Oxford Oxford, UK

(5) Politecnico di Torino Torino, Italy

Abstract *Formal methods are a means for verification and validation with the main advantage that a system property can be verified for the overall system (including all possible system states). The drawbacks of formal methods are the additional effort for the formalisation of the requirements and for building a model of the system, and, the limitations due to computational restrictions (handling the state-space explosion). ISO 26262 "Road Vehicles - Functional Safety" is a standard for the assessment of the development process for safety-relevant components in the automotive domain. The standard addresses formal methods for the specification of safety requirements and for the product development at software level. Formal methods for the hardware development or at system level are (by now) not explicitly foreseen by the standard. In this work we will give an overview on the basic principles and the state-of-the-art of formal methods (in detail, model checking). Then we will present different approaches for the application of formal methods at system level including some preliminary evaluation results for an industrial use case. Based on these experiences we will discuss the applicability of formal methods in the context of ISO 26262 (i.e., for automotive components) in view of the limitations of formal techniques for applications in the automotive domain.*

1 Introduction

In computer science, formal methods are a particular kind of mathematical techniques for the specification, development and verification of software and hardware systems. The use of formal methods for software and hardware design is motivated by the expectation that, as in other engineering disciplines, performing appropriate mathematical analysis can contribute to the reliability and robustness of a design. *Formal* in this context means some kind of mathematical description

© S. Kandl et.al. 2015. Published by the Safety-Critical Systems Club. All Rights Reserved.

and reasoning (the basis for this is automata theory and predicate logic). In the beginning of model checking it was possible to process models with between 10^4 and 10^5 states, refinements of the OBDD-based techniques[1] have pushed the state count up to more than 10^{120} states (Clarke et al. 2000). An abstraction technique called counterexample-guided abstraction refinement (CEGAR) (Clarke et al. 2003) enhanced the performance of model checking techniques at another order of magnitude. Modern complex embedded systems, considering all the details, have far more states than the given benchmark. Thus the application of formal methods requires incorporating only the relevant parts of the system in the model for verification. So, the applicability of formal methods at system level depends on the way how the crucial aspects of the system (necessary to verify the considered properties) are presented in the model without risking a state-space explosion. ISO 26262 (ISO 2010) addresses mainly the verification of safety requirements (a subset of all functional requirements), therefore we will focus on the verification tasks for safety-related requirements.

In this work we present different approaches for the application of formal methods for a case study at system level (involving hardware and software). First, we will elaborate formal methods mentioned by the standard ISO 26262. Then we will give a short overview on the basic principle of model checking. Subsequently, we will describe the different approaches and we will give some preliminary results regarding the applicability of the proposed techniques for an industrial use case (Airbag, see Section 3.1). Based on the experiences so far, we will conclude with recommendations for the use of formal methods in the context of ISO 26262.

2 Formal Methods

In this section we give a brief overview on formal methods defined by the standard ISO 26262 and explain the basic principles of model checking.

2.1 Formal Methods in ISO 26262

The applicability of formal methods in general (static analysis, model checking) is defined by several terms in the standard ISO 26262. Following definitions refer to the standard (ISO 2010):

Formal notation [Part 1 – 1.47]: Description technique that has both its syntax and semantics completely defined.
EXAMPLE: Formal notations include Z (Zed), NuSMV, Prototype Verification System (PVS), and Vienna Development Method (VDM).

---

[1] OBDD means Ordered Binary Decision Diagram, a special data structure for a compressed representation of sets or relations, used as an efficient way to store the automaton model.

Applicability of Formal Methods for Safety-Critical Systems in the Context of ISO 26262   97

Formal verification [Part 1 – 1.48]: Method used to prove the correctness of a *system* against the specification in *formal notation* of its required behaviour.

Semi-formal notation [Part 1 – 1.117]: Description technique whose syntax is completely defined but whose semantics definition can be incomplete.
EXAMPLE: System Analysis and Design Techniques (SADT); Unified Modeling Language (UML).

Semi-formal verification [Part 1 – 1.118]: *Verification* that is based on a description given in *semi-formal notation (verification* in this context is defined as: determination of completeness and correct specification or implementation of requirements from a phase, or sub-phase).
EXAMPLE: Use of test vectors generated from a semi-formal model to test that the *system* behaviour matches the model.

For the software-level, formal methods are prescribed by the standard, for instance: On the software level the table of "Methods for the verification of software unit design and implementation" lists semi-formal verification as a highly recommended method (++), and formal verification as a recommended method (+) for ASIL D-systems; O means no explicit recommendation, see Table 1 ((ISO 2010), *Part 6: Product development: software level*, Table 9).

Part 4: Product development at the system level does not explicitly mention, neither, formal methods, nor, formal notations. Thus it seems that the standard does not prescribe this kind of verification techniques for systems.

Table 1: Methods for the Verification of SW Unit Design and Implementation (ISO 2010)

	Methods	ASIL			
		A	B	C	D
1a	Walk-through	++	+	O	O
1b	Inspection	+	++	++	++
1c	Semi-formal verification	+	+	++	++
1d	Formal verification	O	O	+	+
1e	Control flow analysis	+	+	++	++
1f	Data flow analysis	+	+	++	++
1g	Static code analysis	+	++	++	++
etc.	etc.				

However, *Part 8: Supporting processes*, which is a cross-cutting part that applies to the whole development cycle, mentions formal methods when it addresses the specification and verification of safety requirements:
Semi-formal notations are highly recommended, and formal notations are recommended for the specification of safety requirements for ASIL C-systems and

ASIL D-systems, see Table 2 ((ISO 2010), *Part 8: Supporting processes*, Table 1).

Table 2: Specifying Safety Requirements (ISO 2010)

	Methods	ASIL			
		A	B	C	D
1a	Informal notations for requirements specification	++	++	+	+
1b	Semi-formal notations for requirements specification	+	+	++	++
1c	Formal notations for requirements specification	+	+	+	+

Moreover, semi-formal verification is highly recommended, and formal verification is recommended for the verification of safety requirements for ASIL C-systems and ASIL D-systems, see Table 3 ((ISO 2010), *Part 8: Supporting Processes*, Table 2).

Table 3: Methods for the Verification of Safety Properties (ISO 2010)

	Methods	ASIL			
		A	B	C	D
1a	Verification by walk-through	++	+	+	+
1b	Verification by inspection	+	++	++	++
1c	Semi-formal verification	+	+	++	++
1d	Formal verification	O	+	+	+

Although the use of formal methods at system level is (yet) only partially addressed by the standard, the assessment process by the standard ISO 26262 should not be restricted to recommended methods defined in the standard, but should also consider state-of-the-art methods that are applicable system development. Therefore one aim of ongoing research is to evaluate methods that improve the overall system quality and to give recommendations for an improvement of the standard. In general, it should be the overall goal in developing, testing, and verification of a safety-relevant system to ensure its proper functioning and its reliability with all available methods. The big advantage of formal verification is that we can prove a property once for all possible system states, whereas in verification by testing we have to provide a huge test set to check the system behaviour in different system states.

In the following, we give a short introduction into the basic principle of model checking and present different approaches of formal methods at system level, addressing the ways and possibilities to apply them to a real industrial use case.

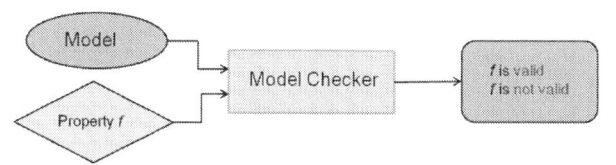

2.2 Basic Principle of Model Checking

The term *formal methods* comprises many techniques, methods, and tools, like static analysis, theorem proving, model checking, etc. (D'Silva et al. 2008). For this paper we will focus on model checking. For the application of model checking we need a *model* of the system under test (SUT) and *properties* of the SUT (see Figure 1).

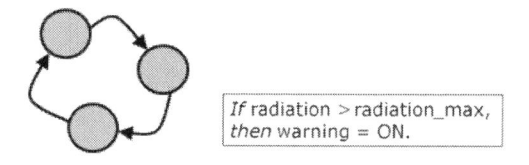

Fig.1: Model and Property of the SUT

In formal verification we prove whether a property of the SUT is *valid* in the given model or not. Figure 2 shows *model checking,* which is one very common technique for formal verification.

Fig.2: Principle of Model Checking

Assuming that the model maps the behaviour of the SUT, all the requirements of the SUT should be valid in the model. In the case of a violation, the reason for the violation has to be identified. First, the correctness of the model with respect to the SUT has to be checked. Secondly, also the formulation of the requirements should be revised. If both, the model, and the requirements are confirmed to be correct, a violation of a property indicates an implementation error in the SUT. In this case the model checker produces a counterexample (that can be used as a test case).

2.2.1 Formalisation of Requirements

The goal of formal verification is to prove the compliance of a system to its specification. The specification consists of a set of properties, the system requirements. For the use of formal methods the system requirements have to be *formalised*.

In general, we distinguish two types of properties:

1. **Safety properties** state that nothing bad happens. Many important properties in practice fall into this category: state properties (*invariants*) such as the absence of runtime errors (e.g., overflows and division-by-zero) and mutual exclusion, and trace properties like deadlock freedom or guarantees with respect to response times and deadlines. The violation of safety properties can be shown by a finite execution trace (counterexample).
2. **Liveness properties** state that something good eventually happens, i.e., a desirable action is eventually executed. Examples for such properties are termination or fair choice. Such properties cannot be violated by finite execution traces.

Properties are usually specified using *temporal logics*. We shortly discuss two such logics here, LTL and CTL, following the NuSMV Tutorial (NuSMV 2014).

Linear Temporal Logics (LTL) characterizes an execution path of a system (formalised as a transition system). Typical LTL operators are: **F** p ("in the future p"), **G** p ("globally p"), p **U** q ("p until q"), **X** p ("next p").

An LTL formula is considered true in a given state if it is true for all the paths starting in that state. It is true for the model if it is true for all initial states of the model.

Computational Tree Logics (CTL) enables the specification of properties that take into account the non-deterministic, branching evolution of a system (e.g., due to external inputs (user, sensor)), which forms an (infinite) tree. The paths in the tree that start in a given state are the possible alternative evolutions of the system from that state. In addition to the operators in LTL, CTL operators have path quantifiers which express that a property should hold for all paths (**A**) or some paths (**E**).

LTL and CTL have in general different expressive power, but also share a significant intersection that includes most of the common properties used in practice.

Most safety requirements can be expressed by invariants, i.e., properties that hold in any state of the system, and hence, they can be expressed using formulas of the form **G** p in LTL or **AG** p in CTL.

3 Description of the Approaches

In this section we describe the different approaches for the application of formal methods to an industrial use case.

3.1 Use Case – Airbag

The increasing number of airbags in a vehicle and the requirement to comply with stricter safety requirements while costs must be reduced has forced automotive airbag system design to a new System-on-a-Chip (SoC) design approach. Our use case is a modern airbag system, i.e., a heterogeneous sophisticated system comprised of sensors, an airbag Electronic Control Unit (ECU) – which consists of an airbag SoC chipset and an airbag embedded main micro-controller (μC) – and actuators. The airbag ECU is the hardware within a multiple airbag system that controls the deployment of airbags within a car.

3.2 Verification of SystemC with SPIN

The model checker SPIN is designed for checking concurrent systems and the input language Promela (**Pro**cess **Me**ta **La**nguage) is used for specifying such systems. SPIN is a widely used and matured model checker. According to our experience SPIN supports many useful features for the purpose of formal verification at system level (for a list of features see (SPIN 2014)). SPIN can be used to verify properties given in Linear Temporal Logic (LTL) and expects the input model to be written in the input language Promela. This can be done manually from most models but this would require a qualified engineer to do it. Doing so requires a strict documentation to establish a correct and traceable transformation, this not only expensive but also error prone.

Our approach concentrates on the automated translation from a SystemC-model to Promela. SystemC is a library for C++ that allows describing hardware in C++. It contains a series of macros and base classes allowing the system designer to use constructs similar to Hardware Description Languages (HDLs). Moreover it does not constrain the engineer to be restricted to HDL coding rules. Instead it allows the engineer to use C++ code to implement some logic. Furthermore SystemC provides a scheduler allowing compiling the model into an executable program for simulation. These features qualify SystemC for describing system level models. It allows the description of common hardware aspects, like signals and FIFOs but also to use plain C/C++ code to model software aspects of the model. However, the engineer has to keep in mind that only an abstraction is created. The translation from SystemC to Promela consists mainly of two major parts, see Figure 3

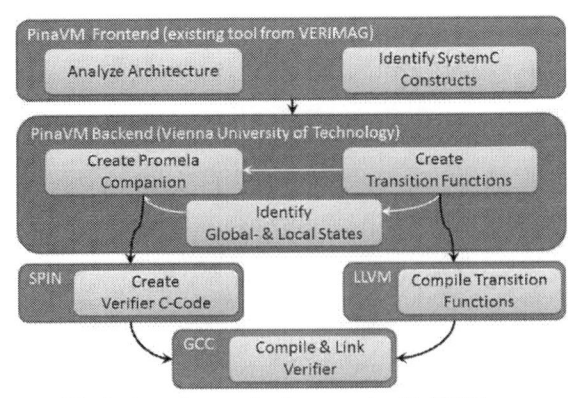

Fig.3: Overview on the Framework with SPIN[1]

1. **Analysis of the model structure given in SystemC:**
   For analyzing the structure of SystemC models an existing tool called PinaVM is used. PinaVM is in an academic development state, thus showing its functionality as a proof-of-concept. The tool needs further improvements to support additional features required to fit for industrial use cases. Therefore it will be extended with a special focus to the Airbag use case.

2. **Actual translation of the models functional blocks to Promela:**
   Once the model analysis is completed PinaVM hands over information on the kind and number of existing SystemC modules, threads, events and communication channels. This information is used to generate a Promela model to accurately represent the original SystemC model. However, in contrast to translate each and every line of code to Promela, we group functional blocks end export them to separate functions. These functions are compiled and called from the Promela model in order (1) to speed up the verification process by executing native machine code and not simulating it and (2) to reduce the memory requirements by keeping as many variables as possible local to those functions, and thus hidden from the verifier.

For technical details about this approach, please, refer to (Elshuber et al. 2013).

3.2.1 Experiences

We applied our approach to simple demonstration examples with success (Elshuber et al. 2013). We converted an example and checked for termination, and found performance gains in comparison to a classical line-by-line translation. In addition to termination checks, we were also able to check LTL properties like (**GF** p) on

---

[1] PinaVM is a tool used to analyse the SystemC-model, SPIN is a model checker, LLVM is a compiler infrastructure, and GCC is a compiler collection for several programming languages.

the same demonstration example. The limiting factor is not the complexity of the LTL property; currently it is the kind of SystemC- and C-constructs used in the model. Some constructs are translated by the compiler in a way making it hard to analyze it by our tool, thus the creation of the transition functions fail. To show the usability of our approach we are currently translating the airbag use case manually.

The main target of our work is to find a way to apply formal verification in the design phases of the architecture of safety-related computer systems. This is essential, especially for safety-related applications, in which we want to find methodologies to increase the confidence in the product as well as to reduce the development risks emerging from late detection of serious design errors.

Although airbag application systems are not new, the high integration of a large amount of functionalities into one single device with stricter safety requirements while still being cost-effective has brought the airbag system design to a new height of innovation. In addition, new sensor communication protocol developments have driven completely new platform development (in both HW/SW, e.g.: HW – airbag SoC chipset and SW -- embedded SW of the main µC) to design teams. Especially, the verification of such a safety-related system becomes a real challenge mainly because of:

- Verification for the airbag SoC HW has to cover real-time embedded mixed signal domains.
- Most of the SoC functionalities can only be verified at the top level of the chip including many of the requested interactions requiring the embedded SW. The classical mixed-signal simulation approach becomes a bottle neck due to simulation performance issues.
- For new sensor protocol development, many verification scenarios of the sensor interfaces, such as, long-term verification run with checking of millions of sensor data frames are not suitable using computer-based simulation.
- As well, reducing time-to-market and right-the-first-time design in automotive electronics industry -- one of the key requirements in projects to win customer and market share -- have posed a great challenge to the design and verification team.

3.3 Bounded Model Checking of SystemC using CBMC

This section presents a bounded model checking approach for the verification of a SystemC-model. The approach connects the Scoot tool (Blanc et al. 2008), (Blanc and Kroening 2008), (Scoot 2014) and CBMC tools (Clarke et al. 2004),

(CBMC 2014) to enable SystemC verification (the name of the tool CBMC means C Bounded Model Checking). The verification proceeds as follows (see Figure 4):

1. The requirements are formalised and added as observer processes to the SystemC model.
2. The resulting SystemC model is passed to our tool:
   a. The SystemC sources are analysed and transformed into a static, optimized simulator in C++.
   b. This simulator is then model-checked with bit-level accuracy.
   c. If the verification fails, i.e., a property is violated, a counterexample is returned.

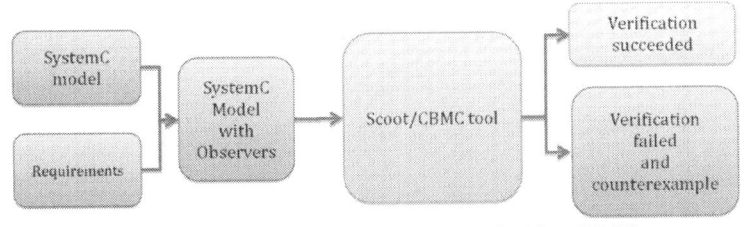

Fig. 4: Bounded Model Checking Approach Using CBMC

Formalisation of the Requirements as Property Observers

The requirements have to be formalised as temporal logic properties. These properties are then rewritten as observer processes in SystemC code and added as observer components in parallel to the SUT. An observer is a component that takes the system under test's inputs and outputs (or state) as its inputs and an output that says whether the system's behaviour is correct (or erroneous) as depicted in Figure 5. The observer implements a property as follows: A property typically consists of two parts:
1. An assumption about the inputs, i.e. which input values are realistic.
2. An assertion about the system under test's outputs (or state if accessible), i.e. what is the expected behaviour.

The semantics is implication: Assumption => Assertion. I.e., if the input values are invalid then any behaviour of the system is allowed and the output of the observer will be correct.

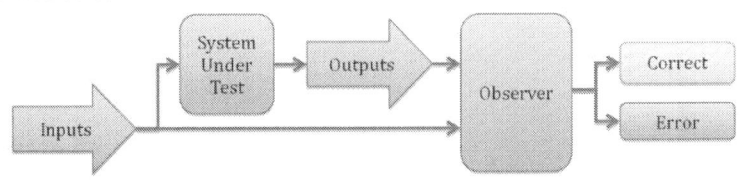

Fig.5: Property Formalised as an Observer

SystemC Simulation with Scoot and Model Checking Engine CBMC

SystemC is a C++ library for discrete event simulation of concurrent processes consisting of predefined types and components. Scoot statically analyses the module hierarchy, the port bindings and processes, and uses information about the dependency relations between processes to reduce the number of inter-leavings using partial order reduction. Based on this information, Scoot can generate an efficient static simulator (C++ source file). It is also possible to explore the (reduced number of) inter-leavings. The resulting simulator is then compiled using a C++ compiler and run. We use CBMC as a bounded model checking engine for C and C++. CBMC verifies execution paths up to a bounded length and returns counterexamples if a bug is found. It is fully automatic, has no false alarms (all counterexamples are true bugs), and uses exact bit-level semantics (including floating-point arithmetic). It is incomplete in the sense that it can analyse programs only up to a bounded depth (limited loop unwinding).

3.3.1 Experiences

We performed a preliminary evaluation of the approach on the Airbag model. For this purpose we manually translated the SystemC model to C with a non-deterministic scheduler. We used two properties from the requirements:
- Property 1 "If the squib[1] has been activated an impact has been detected."
- Property 2 "The time between the impact and the activation of the squib must not exceed 2ms.", see Figure 6.

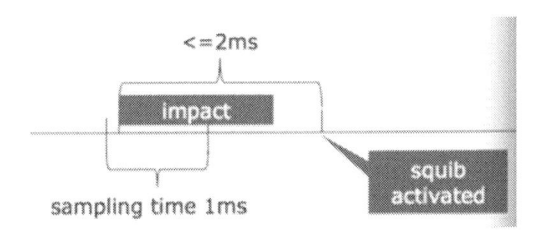

Fig.6: Property 2 (for the Bounded Model Checking Approach)

For property 1, we introduced a bug into the model that violates the property and showed that CBMC can find the bug in 1.2 seconds. The bug is very shallow and can be detected after two loop unwindings[2], each of them corresponding to a time step of 20ms of system execution.

---

[1] A squib is a small explosive device used for the inflation of the Airbag.
[2] Loop unwinding is a technique to transform a loop to optimise the execution speed of a program.

We have checked property 2 under two assumptions:
1. All sensors start sampling at time 0. Then we prove that squib activated in less than 680μs. For proving this property it is sufficient to check for a number of unwindings corresponding to more than 680μs. We checked it for 40 unwindings (800μs) for which CBMC takes 37 minutes.
2. With assumption that each sensor can start sampling with an initial delay <1ms, we have to prove that the squib is activated in less than 2ms (100 unwindings). This is very demanding for the bounded model checker: We obtained the proof for 105 unwindings after 28 hours.

These experiments were performed on an Intel Xeon 2.40 GHz with 90GB RAM.

CBMC is a powerful tool for checking functional safety and timing properties as we have shown above. Its advantage is its precision and trustworthiness with respect to the system that it is analyzing (e.g. floating point numbers). However, this also results in disadvantages regarding the performance - it is not clear whether such high precision is actually required for system-level verification. Verifying properties that ask to check long delays, for instance, require many unwindings of the model, which is very expensive.

A Scoot-based frontend for translating from SystemC to (simpler) C++ or C is under development. SystemC is a particularly inappropriate language for system-level formal verification due to the complexity of C++ it is based on, which makes it extremely difficult to verify without abstractions, and hence, verification will lose guarantees regarding the artefact being verified.

3.4 Modelling and Verification of Real-Time Aspects

Informally, a *timed automaton* (TA) (Alur and Dill 1994) is a finite automaton that is augmented by a set of integer-valued *clocks*. These run *uniformly* (i.e., at the same rate) and can be individually *reset* to zero – intuitively, clocks are considered to be independent stopwatches. While a TA is "running", its clocks' values must satisfy certain conditions that allow the automaton to proceed with its execution. A TA may thus be seen as a type of state-chart with particular restrictions and tailored semantics.

The running behaviour of a TA is restricted by side conditions in terms of the clock values. A side condition may be associated with a state, in which case it is called invariant, or with a transition, in which case it is called a guard. The TA is allowed to reside in a particular state or take a transition if-and-only-if its clocks satisfy the associated invariant or guard. Moreover, each transition may reset a number of clocks to zero – provided that, before taking the transition, the clocks satisfy its guard in the first place.

In Figure 7 a simple example is shown for modelling a watchdog-timer. Consider a system processing a stream of data items which shall be at most 2ms apart.

If an item arrives with more than 2ms delay, an alarm shall be triggered. The system/automaton consists of two states, PROCESS and ALARM, it accepts the events data and reset – which indicate the arrival of a data item and turning off the alarm– and has the clock w, which counts the time in milliseconds. The arrival of each data item resets w to zero. However, once w reaches 2ms, the system enters the alarm state – this is enforced by prohibiting remaining in the processing state or taking the transition that resets w. However, the alarm state can be left by the reset event, which also resets the clock.

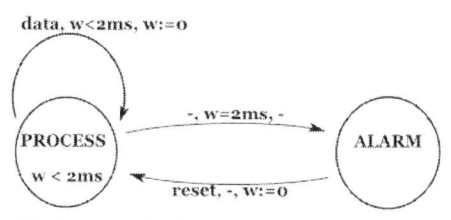

Fig.7: Example Timed Automaton - Watchdog

A useful extension of the TA - formalism is to allow for TAs that run in parallel and communicate – or *synchronize* – on certain events.

A state-of-the-art implementation of timed automata with capabilities for real-time model checking is the UPPAAL-toolkit (Uppaal 2014). This toolkit is under constant development and comes with an academic as well as a commercial licence. Moreover, there is a comparably large body of literature featuring UPPAAL, providing introductory and industrial examples (Behrmann et al. 2014).

The UPPAAL-tool employs a variant of TA that has the same expressive power as the canonical model from. It supports networks of TAs and provides some features that ease intuitive modelling; in particular communication by shared variables and the specification that time must not pass in a particular state. The logic used by UPPAAL for expressing real-time properties is a restriction of TCTL.

3.4.1 Experiences

The tool UPPAAL was used to create a generic model of the squib-actuator (a safety-related component of the Airbag-use cases) and its diagnostic system and to validate timing requirements for the Airbag.

The model contains information on the duration of each step and the task is to verify whether the timing requirements are met. For example, measuring squib resistance takes 70ms, i.e., the system remains a specific state for 70ms. This information is integrated into the TAs by means of a clock and clock-constraints. After several iterations from a monolithic to a modular model, i.e., from a single TA to a network of TAs, we arrived at the following UPPAAL-model consisting of four components, see Figure 8.

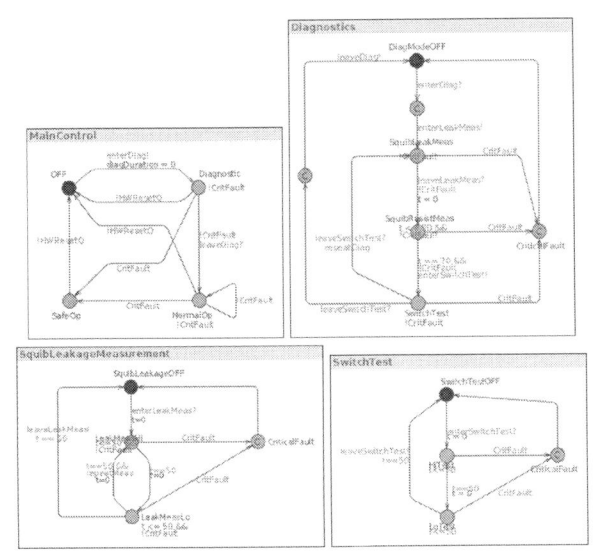

Fig.8: Uppaal-Model for the Squib-Actuator Diagnostics

For the use case-example, the overall execution time of the diagnostic subsystem is of interest. To get information about the diagnostic duration, we query the verifier for properties involving this clock. For example, to check whether the diagnostic process takes no more than 500 time units, given that no critical fault occurs, the following query may be provided to the verifier:

$$A \ [\ ]\ ((\mathit{MainC.diagDuration} > 500\ \&\&\ !\mathit{critFault})\ \textbf{imply}\ \mathit{MainC.NormalOp})$$

In the given model, this query fails and the verifier provides a counterexample that can be viewed in the simulator. Assuming that the translation from the statechart to the TA-model is correct, this proves that the initial model does not satisfy the timing-requirements on the diagnostic subsystem. Therefore, this fault is already caught on the model level.

In another example we show that the TA-model respects the property that upon registering a critical fault, the state *SafeMode* is entered immediately. This property is verified by UPPAAL.

UPPAAL provides a convenient means to model discrete dynamical systems with timing properties. It serves to formally verify conformance of these models with timing requirements and to prove violations of these properties by providing counterexamples. The graphical editor and simulator provide easy access to the tool's functionality and serve to gather an intuitive understanding of the model's dynamics. In some cases, workarounds are needed in order to model certain behaviour. The concept of a network of communicating TAs is well suited for a modular approach to modelling, which further improves readability and understanding by the safety engineer. With UPPAAL it was possible to check timing properties for the Airbag-use case on a high-level model of the Squib-actuator.

3.5 Abstraction and Refinement Checking

Abstraction and refinement are two strictly related concepts that can be exploited in several ways in formal verification. This section presents the application of abstraction to model checking and introduces refinement checking.

Refinement checking is a verification technique similar to model checking, where two models are compared by exhaustive exploration of their behaviours, in order to check whether one is a correct refinement of the other one. Here correctness means that a particular refinement relationship binds the two models. Like model checking, refinement checking can be automated, but unlike model checking, refinement checking does not require that properties are specified in (temporal) logics. This difference is interesting because temporal logics are often regarded as too difficult to use and out of the reach of many system engineers, while formally specifying system models, e.g. by state charts, is generally more affordable for them. Refinement checking can be useful in the context of ISO 26262, as a formal verification technique to be applied in the left hand side of the V development cycle.

Refinement checking is a verification technique that takes two models as input, an abstract model M^A and a refined model M. Refinement checking checks that they are in refinement/abstraction relationship, i.e. that M is a refinement of M^A or, equivalently, that M^A is an abstraction of M. The output is a yes/no response. In case the answer is no, counterexamples are generated showing why the two models are not in the supposed refinement/abstraction relationship. Different refinement/abstraction relationships can be used for this kind of verification.

PAT, Topcased, SysML

PAT (Process Analysis Toolkit) (PAT 2014) is a formal verification toolset for concurrent and real-time systems. It includes a simulator, a model checker, and a refinement checker for the CSP# language (Sun et al. 2009), which is an extension of the well-known formal specification language CSP (Concurrent Sequential Processes). The PAT model checker can check LTL properties on CSP# models while the refinement checker supports the verification of refinement relationships (in particular trace refinement) on CSP# models.

PAT is known to be an efficient, state-of-the-art tool for refinement checking. Its efficiency derives mainly from its ability to perform refinement checking on-the-fly, i.e. without storing the whole set of states and transitions in memory. In addition, all PAT verification tools use all the most important known optimization techniques: partial order reduction, symmetry reduction, process counter abstraction, parallel model checking. For this reason we deem PAT is a very good candidate for experimenting with refinement checking and for evaluating the extent to which this technique can really be applied.

Topcased (Topcased 2014) is an eclipse-based modeling environment specifically targeted to the design and development of critical embedded systems including hardware and software. Topcased promotes model-driven engineering and formal methods as key technologies. For this reason it is particularly interesting for our purposes. Topcased supports SysML (OMG 2012) modeling and it is open to support other UML-based modelling languages (e.g. EAST-ADL). As SysML and EAST-ADL are both used in the design of automotive systems, Topcased is potentially interesting for the automotive world.

SysML is a UML-based modeling language that inherits from UML several features (but not all, because the more software-oriented features of UML were kept out of SysML) and extends UML with other features that are specific for system modeling. One of the key extensions of SysML with respect to UML is the possibility to model requirements.

For an overview on the workflow of this approach refer to Figure 9:
When requirements are specified in SysML in the Topcased-environment, behavioural models are associated with them. Then, the plug-in automatically extracts these models using XSLT transformations, and creates corresponding CSP formal models on which refinement is then verified.

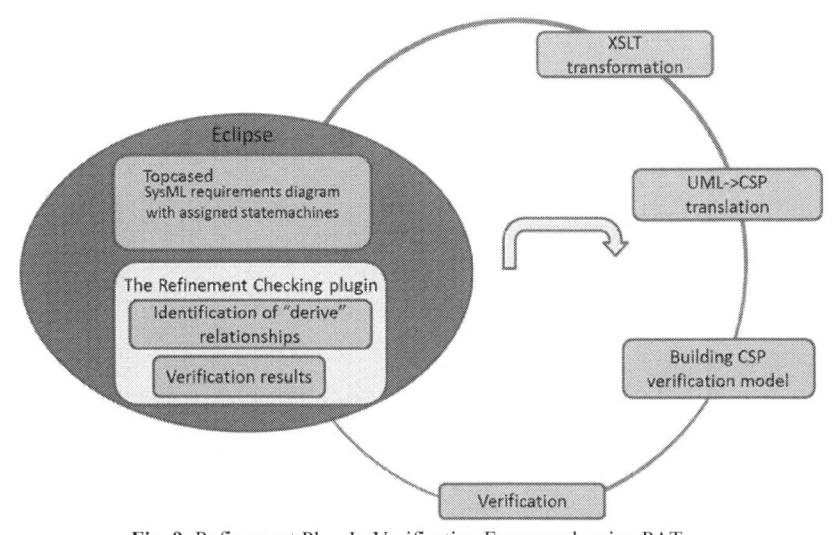

Fig. 9: Refinement Plug-In Verification Framework using PAT

3.5.1 Experiences

Due to some confidentiality issues, until now only partial evaluation has been possible on the use case. This has been done starting from the an initial set of requirements given by the industrial partner about their use case, and trying to complete these requirements with other requirements not coming from the industrial partner. The additional requirements have been guessed, trying to be coherent with the description of the use case given by the industrial partner and with the System-C model. Then, each requirement has been formalised by means of a state machine.

This formalisation work identified some critical issues. Probably, in the proposed approach, this is the most difficult step to be performed by a requirements engineer, because it obliges the engineer to identify what is the admitted behaviour implied by the requirement, focusing only on the requirement that is being considered (i.e. without introducing extraneous elements in the behavioural model). Another difficulty is the fact that in our approach behavioural models have to be expressed in an event-based fashion, while in the normal practice state-based representations are more common. Translating from a state-based representation to an event-based one may sometimes prove difficult. Yet another possible difficulty that was identified stands in the SysML behavioural model formalisms (e.g. UML state machines), for which some restrictions apply in the Topcased plug-in. These restrictions have been mostly inherited from the tools used for the conversion from SysML to CSP. For example, for state machines, the conversion is performed by PAT, which does not admit some of the state machine constructs. These restrictions may reduce the usability of the approach. However, in the future, they could be removed by extending the modules that generate CSP. The set of requirements so formalised has been completed with derive relationships and the coherence of the derived requirements has been checked using the refinement checking plug-in. This check detected some minor errors made during the initial formalisation of some requirements, thus proving the usefulness of the check itself.

In conclusion, so far the applicability of the proposed approach has been only partially evaluated on a real use case. This evaluation, although limited, has made it possible to check that the proposed approach works in practice and to identify some criticalities of the approach itself, some of which have already been removed. In future work, we would like to proceed with the experimentation work. Furthermore, the tool prototype will be improved regarding other identified issues that can be fixed with reasonable effort.

3.6 System Property Monitoring with STL

This section describes results from the application of Signal Temporal Logic (STL) for system property monitoring, in detail checking the correctness of an automotive sensor interface (part of the Airbag-use case). It was decided to investigate this formal method, as the STL implementation as an assertion, promises to strengthen the communication between different disciplinary teams, ensuring a clear and common understanding between teams on the system properties and requirements.

Assertion based monitoring and STL

Assertion-based monitoring is a promising technology for verification of Analogue and Mixed-Signal (AMS) designs, i.e. designs that consist of interacting digital and analog components. It successfully exports some well-established ingredients from digital verification to the AMS domain, while retaining the relative simplicity and scalability of the simulation-based verification. In essence, assertion-based monitoring frameworks consist of an assertion language used to formalise the requirements that describe the correct interaction between analogue and digital components, including timing constraints due to the communication delays. The formal assertions are then translated into monitors, a special checking tool that read simulation traces of the design-under-test and check for the assertion satisfaction/violation.

STL (Maler and Nickovic 2004) is an assertion language extending Linear Temporal Logic (LTL) (Maler and Nickovic 2013). LTL enables declarative, formal and compact specification of reactive system requirements. Its original use was for evaluating sequences of states and events in digital systems. STL extends LTL to specification of properties involving both digital and real-valued variables defined over dense time. Monitoring of STL was implemented in the tool Analog Monitoring Tool (AMT) (Pnueli 1997). The offline monitoring flow based on using STL for formalizing assertions and monitoring them with AMT is depicted in Figure 10. This specification language has been successfully used in the past for monitoring in various application domains, such as analog circuits, biochemical reaction, synthetic biological circuits and music. STL has also been extended in several other directions.

3.6.1 Experiences

The assertion-based monitoring methodology from Figure 10 and the AMT tool were applied on the Airbag system application with the focus on the new airbag sensor interface using the new DSI3 standard, promoted by the DSI consortium. For details refer to (Nguyen and Nickovic 2014).

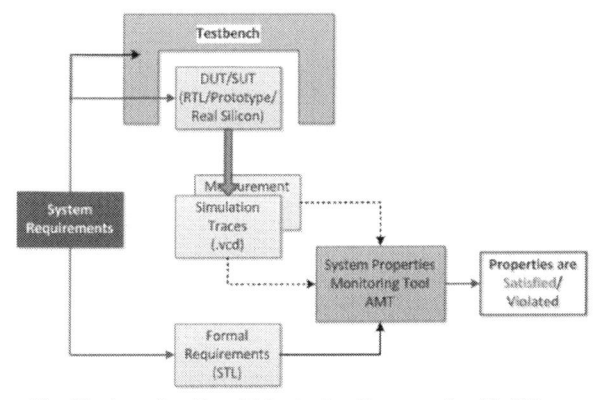

Fig. 10: Assertion-Based Monitoring Framework with STL

It was found that the monitoring itself represents a negligible overhead to the design simulation, while automatically providing useful debugging information to the designer as well as reducing time and error prone due to manual inspection of the simulation results. While STL is a rigorous, unambiguous and powerful specification language, it is often not very intuitive to the engineers, and especially to analog designers. Therefore, the future plans of the development group for following topics are highly appreciated and will make the proposed concept more practical and useful for future automotive applications:

- Graphical language for specifying common STL patterns
- STL assertion libraries:
- More flexible syntax for STL that allows declaration of variables and constants outside of the assertions
- Diagnostics for assertion violations:
- Assertion language extensions
- Online hardware FPGA implementation of STL monitors.

4 Summary and Conclusion

In (Hall 1990) seven myths about formal methods are listed. Although this article is more than 20 years old, the content seems to be still up-to-date.

Formal methods can be a very powerful means to assist the verification and validation of a system. Due to the many available methods, techniques and tools, it is a non-trivial task to select an appropriate method for a specific verification issue. Whereas formal methods are already more established in the high safety-critical domain (like avionics, railway, or defence), until now they are not really commonly used in automotive. Reasons for this are the higher complexity of applications in that domain and that the additional effort for applying formal methods may be not reasonable for this kind of applications.

In this paper we presented different approaches for formal methods for an industrial use case from automotive (Airbag). The approaches based on SystemC are still under development caused by the integration of the intricate semantics of C++ (the basis for SystemC). The described prototypes are applicable to simple examples as a proof-of-concept, but up to now, not fully working for the use case. Other approaches (like Uppaal for verifying timing properties, or SysML-modelling for the verification of high-level requirements, or using STL for hardware monitoring) have been successfully applied to components of the use case.

Based on the evaluation results, we can conclude with following recommendation. Regardless the scalability of formal methods for an industrial use case from automotive, we experienced that only the application of a specific approach improves the overall system development process as the test engineers are forced to analyse and revise the system in a formal (i.e., unambiguous and rigorous) way. The additional effort is hardly justifiable for a single use case, but accumulates for several developments. The approaches presented in this work intend to provide a basis for future research for formal methods at system level in the context of ISO 26262 and the preliminary results should encourage test engineers from automotive to consider the integration of formal methods into their development and testing environment.

Acknowledgments This work has been partially funded by the ARTEMIS Joint Undertaking and the National Funding Agencies of the participating countries for the project VeTeSS under the funding ID ARTEMIS-2011-1-295311.

References

Edmund M. Clarke, Orna Grumberg, and Doron A. Peled (2000): Model Checking. The MIT Press, 2000.
Clarke, E.M., Grumberg, O., Jha, S., Lu, Y., Veith, H. (2003): Counterexample-guided abstraction refinement for symbolic model checking. J. ACM 50(5), 752–794, 2003.
ISO (International Organization for Standardization) (2010): ISO 26262: Functional safety road vehicles. Version 2010-12-04.
Vijay D'Silva, Daniel Kroening, Georg Weissenbacher (2008): A Survey of Automated Techniques for Formal Software Verification. IEEE Trans. on CAD of Integrated Circuits and Systems 27(7): 1165-1178, 2008.
NuSMV Tutorial. (last visited: 09-22-2014): Online: http://nusmv.fbk.eu/NuSMV/tutorial/v25/tutorial.pdf.
SPIN. (last visited: 09-22-2014): Online: http://spinroot.com/spin/what.html.
Martin Elshuber, Susanne Kandl and Peter Puschner (2013): Improving System-Level Verification of SystemC Models with SPIN. FSFMA2013: 74-79.
Nicolas Blanc, Daniel Kroening, Natasha Sharygina (2008): Scoot: A Tool for the Analysis of SystemC Models. TACAS 2008: 467-470.
Nicolas Blanc, Daniel Kroening (2008): Race analysis for SystemC using model checking. ICCAD 2008: 356-363.
Scoot Website. (last visited: 09-22-2014): Online: http://www.cprover.org/scoot/
Edmund M. Clarke, Daniel Kroening, Flavio Lerda (2004): A Tool for Checking ANSI-C Programs. TACAS 2004: 168-176.

CBMC Website. (last visited: 09-22-2014): Online: http://www.cprover.org/cbmc/.
R. Alur, D. Dill (1994): A theory of timed automata. Theoretical Computer Science 126: 183-235, 1994.
Uppaal. (last visited: 09-22-2014): Online: http://www.uppaal.org/.
G. Behrmann, A. David, K.G. Larsen (last visited: 09-22-2014): A Tutorial on UPPAAL 4.0. Department of Computer Science, Aalborg University, Denmark. Online: http://www.it.uu.se/research/group/darts/papers/texts/new-tutorial.pdf.
PAT. (last visited: 09-22-2014): Online: http://www.comp.nus.edu.sg/~pat/.
Jun Sun, Yang Liu, Jin Song Dong and Chun Qing Chen (2009): Integrating Specification and Programs for System Modeling and Verification. The 3rd IEEE International Symposium on Theoretical Aspects of Software Engineering (TASE 2009), pages 127 - 135, Tian Jing, China, July, 2009.
Topcased. (last visited: 09-22-2014): Online: http://www.topcased.org/.
OMG (2012): OMG Systems Modeling Language (OMG SysML™, Version 1.3), 2012.
Oded Maler and Dejan Nickovic (2004): Monitoring temporal properties of continuous signals. In FORMATS/FTRTFT, pages 152 – 166, 2004.
Oded Maler and Dejan Nickovic (2013): Monitoring properties of analog and mixed-signal circuits. STTT, 15(3):247 – 268, 2013.
Amir Pnueli (1977): The temporal logic of programs. In FOCS, pages 46 – 57, 1977.
T. Nguyen and D. Ničković (2014): Assertion-Based Monitoring in Practice – Checking Correctness of an Automotive Sensor Interface, Formal Methods for Industrial Critical Systems, pp. 16-32, 2014.
Hall, A. (1990): Seven myths of formal methods. Software, IEEE 7(5) (Sept 1990) 11–19.

Functional Safety by Design – Magic or Logic?

Derek Fowler

JDF Consultancy LLP

Reading, UK

Abstract *The paper considers how we should set about designing safety-related systems (as defined in standards such as IEC 61508) to be safe. Using two transportation examples, it considers the degree and extent to which adherence to industry-specific process standards (the 'magic' approach of the title) would lead us to a complete, safe solution; deducing that this approach would lead to an incomplete solution, the paper shows how we need to rationalize what we mean by safety in the particular context, before determining a more holistic and 'logical' approach to developing a functionally safe design.*

1 Introduction

In his thought-provoking paper (Amey 2001), Peter Amey warns of the dangers of being seduced by highly complex technology and placing blind faith in 'magic' tools and development processes for assurance of safety, instead of using logical reasoning as to why a system can be considered to be safe.

Almost 14 years later, it is appropriate to consider how much these warnings have been heeded. Certainly, (Bieder and Bourrier 2013) poses the rhetorical question: "how desirable or avoidable is proceduralization?" [of safety management], for a very wide range of safety-related sectors. It concludes that, in some cases, 'a kind of point of no return has been reached… in the march to more rules'.

This paper, therefore, picks up the thread and starts by considering two items – an electric toaster and a car airbag – in everyday use[1] and examines *logically* what makes the items safe or not.

This leads on to the consideration of safety standards that are representative of practices in two industry sectors: rail and aviation / air traffic management (ATM), and to an examination of whether they offer magical (highly proceduralized) or logical (rationally-based) approaches to safety assessment.

---

[1] In the latter case 'use' might be considered to be passive!

© Derek Fowler 2015. Published by the Safety-Critical Systems Club. All Rights Reserved

Finally, practical experience from recent rail and ATM projects is used to provide a framework for a more logical, and thereby more complete, approach to the design of safety-related systems. An example of its application in the rail sector is then described, followed by an overview of how it is being applied to a current airport-development project.

2 What is safe – logically?

Consider first, the humble pop-up electrical toaster. Functionally, its purpose is clear – simply, to make toast. From a safety perspective, whether it makes toast or not, and how well the toaster performs (i.e. whether the toast that is does make is good or bad), are of no concern whatsoever – providing the device doesn't kill or injure anyone in the process of making toast.

The toaster could, however, still be considered to be safety-related because it has hazardous failure modes that could lead to, inter alia, burns, electrocution or even widespread death / serious injury through house fire. What characterises *these* hazardous states is that they emanate *entirely* from the toaster itself - i.e. no toaster, no hazards.

Now consider a car airbag. It too has hazardous failure modes that could, for example, impart high levels of kinetic energy into an occupant of the car, obscure the driver's vision and/or interfere physically with the safe handling of the car. Generally, these hazards would arise from misuse (e.g. not wearing a seatbelt[1]) or as a result of the airbag deploying at an inappropriate moment. Again, these hazardous states emanate entirely from the device itself.

What distinguishes an airbag from a toaster (from a narrow, purely safety viewpoint) is that its *purpose* is clearly safety-related – i.e. to reduce the likelihood of death, or serious injury, in the event of a head-on collision. The hazards in this case do *not* emanate from the device; rather, they are already present in the *environment* in which the airbag operates (i.e. a car journey) and, for this reason, are sometimes known as *pre-existing* hazards.

Not unreasonably, there is an expectation that car airbags make car journeys safer (albeit as a last resort[2]); therefore, to consider whether an airbag is safe (or not) *solely* on the basis of the likelihood that it would actually kill or maim, without considering its ability to save life or prevent serious injury in the event of a crash, would be completely irrational (Fowler and Pierce 2012).

Logically, the rationale for fitting airbags to cars must be that they are *designed* to prevent far more deaths or serious injuries than they cause - a safety case would, of course, have to prove (not merely assert) this to be true. The positive contribution of the airbag comes from attributes that define what the airbag does

---

[1] Some airbags are actually inhibited if the corresponding seatbelt is not engaged
[2] As is usually the case for a *protection* system – see section 3.1 herein

(i.e. its *functional* properties) and how well it does it (its *performance*) whereas its negative contribution to safety is determined by its *failure* properties — examples of these positive and negative properties, for an airbag, are given in (Fowler and Pierce 2012).

3 Standards

This section looks at safety standards from two industry sectors: rail and aviation / air traffic management and considers whether they fit the 'magical' or 'logical' description. But first, IEC 61508 (IEC 2010) is used as a benchmark for this discussion.

3.1 IEC 61508

IEC 61508 (IEC 2010) is based on the simple (but effective) concept of an *equipment under control* (EUC) - i.e. an inherently hazard-creating system for which *control* and/or *protection* systems are designed in order to mitigate these hazards such that the resulting EUC risk is reduced to an acceptable level. Control & protection systems provide what are known collectively as *safety functions* (SF), and the amount by which they are required to reduce the EUC risk is called *necessary risk reduction* (NRR). Being a generic standard, IEC 61508 does not specify what an *acceptable* level of risk is.

The risk relationships involved are illustrated in Figure 1.

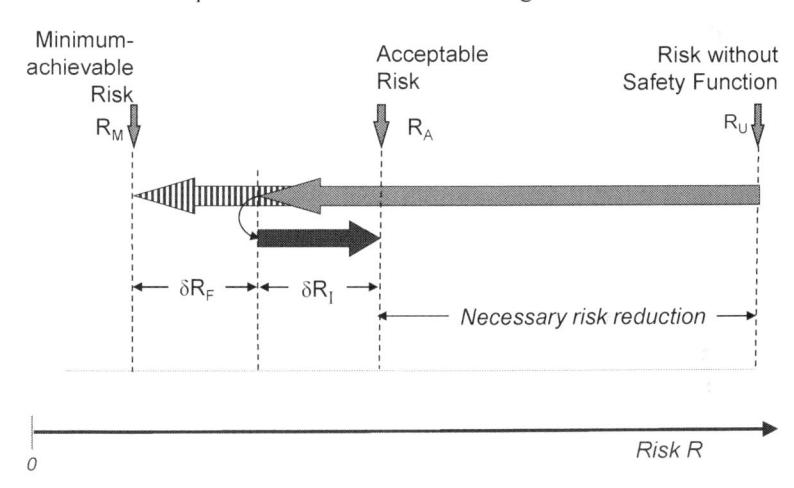

Fig. 1. Risk Reduction I IEC 61508

In this single-axis risk graph:

R_U is the unmitigated risk associated with the EUC

R_A is the acceptable level of risk

R_M is the minimum level of risk that could be achieved if (hypothetically) the SFs never failed in any way; it is determined by the functionality and performance of the SFs, and is not zero because functionality and performance always finite and there are invariably some EUC risks that SFs cannot eliminate completely.

δR_F is the risk increase created when the SFs fail to operate - i.e. fails to mitigate the risk associated with the EUC; δR_F is a function of R_U-R_M and of the frequency and duration of failure.

δR_I is the *new* risk introduced by incorrect operation of the SFs – e.g. doing the wrong thing and/or at the wrong time; it is independent of R_U and R_M.

Since the marginal risks δR_F and δR_I are limited by R_A (fixed by regulation, typically) and R_M (dependent on the functionality and performance of the SFs), it would seem logical for a safety assessment to consider how, and by how much, a SFs could *reduce* risk in the absence of failure of itself before worrying about what happens when the SF fails.

An alternative way (Fowler and Pierce 2012) of considering the above is through the logic of a *fault tree*, a very simplified example of which is shown in Figure 2.

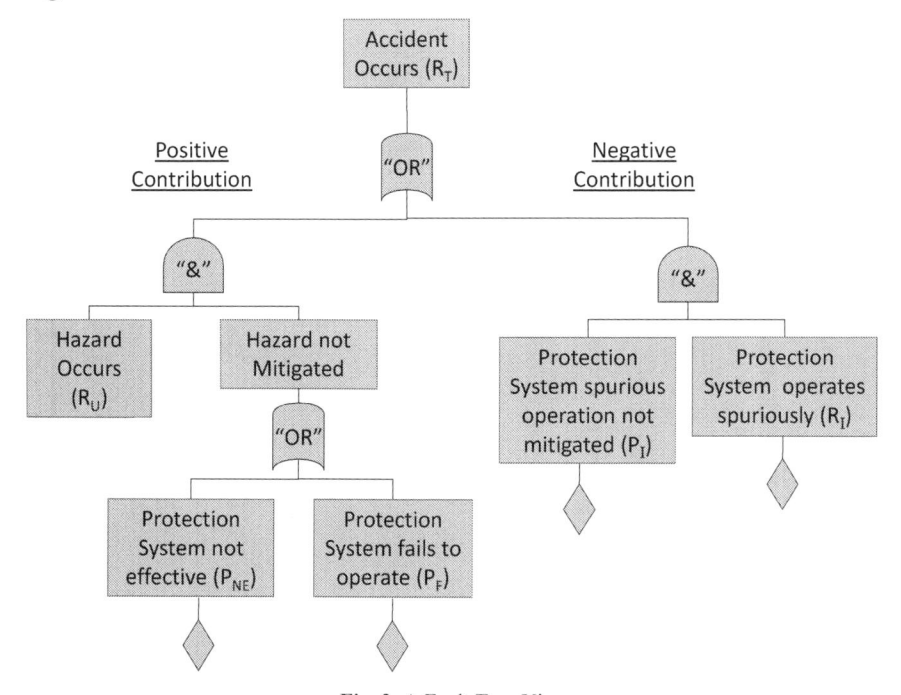

Fig. 2. A Fault Tree View

In this example, of a protection system, an accident occurs when either the *pre-existing* hazard[1] occurs *and* the consequences of that hazard are not mitigated by the protection system, *or* when the protection system operates incorrectly *and* the effects of this are not mitigated by something else.

If the pre-existing hazard were not mitigated at all, then the accident rate (R_T) would be the same as the hazard-occurrence rate – i.e. the hazard-occurrence rate is the unmitigated EUC risk (R_U) defined in the previous slide. Non-mitigation of the pre-existing hazard would occur if either:

the protection system were not effective against that particular hazard – as indicated by $R_M \neq 0$ in Figure 1; or

the protection system failed to operate at all – as indicated by δR_F in Figure 1.

The risk of a protection-system-generated accident (δR_I in Figure 1) is $\sim P_I \times R_I$.

The problem with Fault Trees is that they are not magic. Their 'number-crunching' ability makes them attractive (or, to some, a menace[2]) but the logic needs to be determined by the user; it is unusual to find a Fault Tree that models anything other than failure, the example in Figure 2 being an exception, not the norm. Yet, as explained in detail in (Fowler and Pierce 2012), the *failure* analysis of a system can lead only to *failure*-related properties of that system – i.e. its reliability and integrity. In in order to specify *success*-related properties of a system (i.e. its functionality and performance) we must look *outside* the system, into the environment in which it is intended to operate - in IEC 61508 terms, we must look at the EUC (the source of R_U).

To summarise thus far, it is fundamental to the IEC 61508 approach that the purpose of SFs is to *reduce*, to an acceptable level, the risks associated with the (*pre-existing*) hazards created by an EUC. Of course, we must also take into account any risks inadvertently *created* by the SFs themselves but, logically, there would be no point in addressing these risks until we *first* demonstrated the potential of the SFs to provide what IEC 61508 calls "necessary risk *reduction*", in the absence of failure of the SF itself.

If the above rationale for the ordering of safety activities is not sufficiently compelling, then consider the following definition of *failure*, from (IEEE 1992):

> '...the inability of a system or component to perform its required function within the specified performance requirement ...' (IEEE 1992).

This suggests that before we can analyse the failure of a system we first need to specify the functionality and performance required of it – because failure is deviation from success, *not* the other way around. If we don't do this, our failure analysis is likely to be incomplete and/or incorrect.

---

[1] See section 2

[2] It is worth remembering, when decomposing a Fault Tree through many layers of 'or' gates, that the last Ice Age finished only 1e8 hours ago, and a failure rate of 1.2e-14 per hour equates to a mean time between failures as great as the age of the Universe!

Given the apparently unassailable logic set out in this section, it would be reasonable to expect industry-specific standards to set out clear guidance on how to design control & protection systems that not only have an extremely low probability of killing or seriously injuring anyone but *also* operate in a way that would actually reduce substantially risk of death or serious injury that was inherent in the host system EUC). The next two subsections consider how well these expectations are met in the standards for the rail and aviation sectors, respectively.

3.2 Railway Standards

The railway industry has a long history of the deployment of control & protection systems to achieve high levels of safety for rail transport.

Traditionally, the safety assessment of any new, or substantially-modified, European railway system would follow CENELEC standards (CENELEC 1999), which lay down the processes to be followed for analysing the causes and consequence of hazards associated with the *system under consideration* – i.e. the control & protection systems that we trying to specify. Such a failure-based approach might be fine (and necessary) as far as it goes, but the problem (as explained in section 2 above) is that it leads only to the specification of the reliability and integrity of the system concerned, and tells us nothing about its required functionality and performance – i.e. the properties that actually make the railway *safer*, as opposed to less safe.

A search of the CENELEC suite of standards reveals a well-hidden "gem" in the form of clause 5.4 of (CENELEC 2012), which states that Section 2 of the Technical Report:

> '…shall contain all the evidence necessary to demonstrate correct operation of the system/sub-system/equipment under fault-free conditions (that is, with no faults in existence), in accordance with the specified operational and safety requirements.'

Invaluable as this *normative* statement is, there is no guidance given anywhere in the CENELEC standards on *how* to do it. So it is not unusual, and is certainly the author's own (albeit somewhat limited) experience of the rail industry, for safety engineers to generate countless safety requirements concerned with how often (per ice age) various failure modes of the system(s) under consideration can be allowed to occur, and few (if any) safety requirements about what the system must *do*, and how well it must do it, when it *isn't* failing – ie for the state in which the system spends the vast majority of its operational life!

It is tempting to ask "Where is design – where is the *logic* - in this?" but first it is important to look at this issue from a historical perspective. The railway has developed (very successfully) product standards for the functionality and performance of electro-mechanical / pneumatic signalling systems, interlocking systems and automatic train-protection systems, underpinned by what are often called signalling principles covering, inter alia, their installation and operation. These standards, almost exclusively the province of signalling engineers, served the industry very well for many decades, and it was in the past necessary for safety engineers to ensure *only* the reliability and integrity of the control & protection systems.

However, the problem for a 21st-centrury rail operator wanting, say, to increase capacity by replacing its traditional fixed-block signalling system with a moving-block, communications-based train-control (CBTC) system, is that there are few standards for such products – partly because the widespread use of software has enable a wide range of technical solutions to CBTC signalling requirements.

Ignorance of this problem, and blind faith in the "magic" of CENELEC *process* standards. could lead (indeed, in the author's experience *has led*) to irrational conclusions concerning the safety of such systems.

3.3 Aviation and Air Traffic Management Standards

Reference (Fowler 2013) provides a detailed critique of the safety assessment methodology, and corresponding safety regulations, published by the European Organisation for the Safety of Air Navigation (Eurocontrol). That critique is highly relevant to the subject of this paper, and is summarised as follows.

Air Traffic Management (ATM), like aviation in general, is a highly proceduralized activity. This ensures the necessary consistency in the provision of ATM services, and predictability in the response of aircraft, in order to ensure a safe environment, internationally.

Until the mid-1990s, safety was an implicit but key feature of ATM operations. Technical systems were functionally quite basic, and the primary role of ATM – that is, reducing the risks that are inherent in aviation, especially the risk of mid-air collisions, or high-energy collisions, between aircraft on the ground – was vested in highly skilled, well-motivated human operators (controllers) working to well-defined procedures.

With the increased use of technology, and the introduction of formal safety management system, in ATM during the second half of the 1990s, Eurocontrol started on the development of its very detailed Safety Assessment Methodology – SAM (currently Eurocontrol 2007). This was followed, in the early 2000s, by a series of six Eurocontrol Safety Regulatory Requirements (ESARRs).

The decision was made to base the SAM on the well-established commercial-aircraft-certification procedures ARP 4754 (SAE 2010[1]) and ARP 4761 (SAE 1996) – more particularly the latter. This proved later to be a very significant decision and therefore it is interesting to consider the ARP 4754 / 4761 approach, as outlined in Figure 3.

---

[1] The version of ARP 4754 used in the SAM development was the one prior to that listed in the reference section herein. All other references to the standard are to the 2010 version.

The stated purpose of ARP 4754 is to 'provide designers, manufacturers, installers, and certification authorities with a common international basis for demonstrating compliance with airworthiness requirements applicable to highly integrated or complex systems'. The main aircraft systems-development process, set out in ARP 4754, and outlined on the right-hand side of Figure 3, follows classic systems-engineering principles and provides links to the supporting safety-assessment process shown on the left, and expanded in ARP 4761.

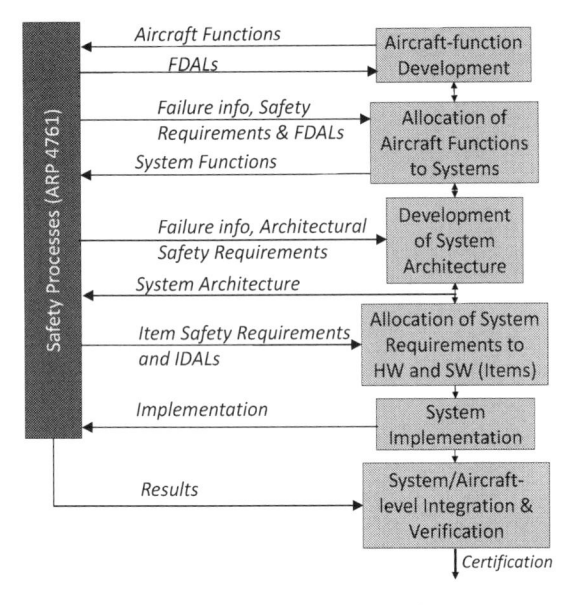

Fig. 3. ARP 4754 / 4761 overview

The relationship between the two standards is as follows:

the design, development and implementation of the aircraft systems is detailed in ARP 4754; it is concerned with the *functionality and performance* of the aircraft functions, and their underlying systems, in relation to the airworthiness of the whole aircraft

at each level of representation of the aircraft functions / systems, information from the ARP 4754 process is fed across to the ARP 4761-defined safety process and analysed from the viewpoint of the possible consequences and causes of failure of the aircraft functions / systems

the results of the safety analyses are then back into the function / system development process in two forms:

- safety requirements for reliability & integrity, and for *additional* functionality & performance, to provide respectively for the control and/or mitigation of the aircraft functions / systems, and
- *development assurance levels* (DALs) to ensure that appropriate rigour is applied to the design, development, implementation and validation & veri-

fication of the aircraft functions, systems and 'items' (hardware and software).

The results of applying ARP 4754 and 4761 are included in the eventual aircraft-certification application. Together, using good *systems-engineering* practice, they provide a thorough, well-integrated process for ensuring that the *aircraft functions*[1] carry out what required of them, to the required level of performance and with the required reliability and integrity.

In the absence of a mature IEC 61508 at the time, adapting this well-proven commercial aircraft certification process for use in ATM would have been a sensible strategy were it not for the fact that, by basing the SAM almost exclusively on ARP 4761, what was put in place for ATM was a safety assessment process that:

addressed only the accident-*inducing* properties (i.e. reliability and integrity) of ATM systems

largely ignored ATM's accident-*prevention* properties (i.e. functionality and performance, which would have been covered had an adaptation of ARP 4754 also been put in place (which it was not)

largely missed the key element of the socio-technical ATM system and by far the greatest cause of ATM system failure – i.e. the human operator (EUROCONTROL 2002).

Such apparent oversights are understandable in a historical context since at that time (late 1990s) the main 'functionality' of ATM was vested in the Controller, had evolved in that way over many years, and no one really knew how to carry out a safety assessment of a human operator. The problem, as we saw for the rail industry in section 3.2, above, technological advances have changed the situation markedly, leaving the failure-based, equipment-focused SAM incapable of meeting the current and future needs of ATM system safety assessment.

The limitations of the SAM have been recognised, for some years, by the European Commission's Single European Sky ATM Research (SESAR) programme[2] (Fowler et al 2011) but not universally among the wider European ATM community. The fact that the SAM continues to provide a "magic" solution to only a small part of the overall safety-assessment problem has not been helped by European safety regulations whose origins lie in the early development of the SAM itself (Fowler 2013).

In relation to 'What is safe - logically?' the whole problem with the SAM approach is indicated in a single statement that defines a hazard as 'what could go wrong with the system and what could happen if it did' (Eurocontrol 2010, FHA - Safety Objectives Specification). As we saw in section 2 above, that might be fine for toasters but not for systems whose purpose is to make the world (of aviation) safer!

---

[1] In most cases IEC 61508 would rightly consider the aircraft functions to be *safety functions*

[2] A collaborative project to completely overhaul European airspace and its air traffic management (ATM) systems

To conclude this discussion on safety standards, the following is one of the most important statements made in ARP 4754:

> It should be understood that the level of rigor in the Development Assurance of an aircraft/system function or item is established by assignment of a Development Assurance Level, be it a FDAL to a Function or IDAL to an Item.

In other words, the ARP 4761 safety-assessment process determines the rigour that has to be applied during the ARP 4754 development of the system and its elements.

This would also seem to be the intention behind IEC 61508 and yet, in the author's experience, very few (if any) systems / safety engineering standards make this vital point with such clarity or, in some cases, address it at all at the *system* level. There are some *software* standards – e.g. DO-178C / ED-12C (RTCA / EUROCAE 2012), and Part B, Section 3 of UK CAA safety regulation CAP 670 (CAA 2014) – which are very much based on this principle; however, if they are not linked closely to an adequate *system*-level process, there is a danger of developing high-integrity software against the wrong *system*-level requirements.

4 The way forward – an outline solution

The section outlines a solution to the above shortcomings, of rail and ATM safety assessment, which is based on work carried out on major projects in each sector. Since rail and ATM are both concerned with public transportation, it is not surprising to find a lot of commonality in the approach – hence the section starts with a generalised framework. Also, because both sectors are well served by processes for analysing the failure (i.e. accident-inducing) properties of control & protection systems, the description focuses almost entirely on how to specify their *failure-free* (i.e. accident-prevention) properties and design the underlying system(s).

A worked example is then given of the application of the generic framework, to the upgrade of a complex urban railway signalling system, and is loosely based on a real-life re-signalling project.

The section finishes with a shorter note on the potential application of the framework to ATM, based on a current capacity-enhancement project at a major international airport.

4.1 A generalised framework - overview

The framework, outlined in Figure 4, is based on three, related viewpoints:

three classical phases of system development: definition, high-level design and implementation (including detailed design)
three / four levels of representation of the system

three Safety Arguments: one for each development phase.

Since the primary objective is to develop control & protection systems that prevent (or at least reduce the consequence of) accidents, it follows logically that the *starting* point for the process has to be those hazards, *outside* of the control & protection systems, that could lead to an accident – what IEC 61508 might call EUC hazards, but are known herein as *pre-existing* hazards.

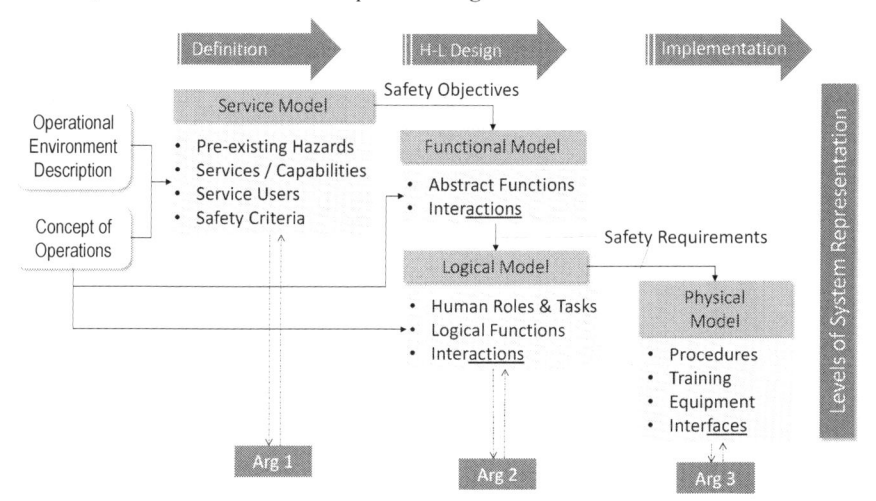

Fig. 4. Generic Model

Firstly, as the pre-existing hazards are specific to an operational environment (OE), a description of the OE is a vital input to the process. It is also important to keep clearly in mind what the equivalent of the EUC is for the particular OE.

Services (or Capabilities) are *what* the control & protection systems provide in the OE in order to mitigate the pre-existing hazards – in effect, these Services / Capabilities are very high-level SFs.

The Service-level Model is a description of *when* and *how* the Services / Capabilities operate to mitigate the pre-existing hazards. This description will often be derived from a Concept of Operations and it needs to be in a form that lends itself to analysis in order to demonstrate the completeness, correctness and coherency of the set of Services / Capabilities and of the main output – the Safety Objectives.

The term Safety Objectives is used at this level (rather than Safety Requirements, as used at the lower levels of the framework) because they describe *what* has to be achieved at the Service level in order to reduce the risk from the pre-existing hazards to an acceptable level. Unlike safety requirements, they do not state *by whom* or *by what* it has to be achieved. It might not be obvious here but experience has shown that the distinction is very helpful in ensuring completeness, correctness and coherency of the Service-level safety specification.

The Functional Model (FM) is an abstract representation of the design of the system that is entirely independent of the logical design (see below) and of the eventual physical implementation of the system. It describes, for each of the Services / Capabilities, what safety-related functions are performed and the data that is used by, and produced by, those safety functions. It is not always essential to produce a FM but, in the case of functionally-rich control & protection systems, experience has shown that to get sufficient assurance of the completeness of the logical design of the ATM system, with respect to the Service-level specification, it is necessary to bridge the two with a detailed functional representation of the system.

A Logical Model (LM) is a high-level, architectural representation of the system design that it is entirely independent of the eventual physical Implementation of that design. The LM describes the main human roles tasks and machine-based functions and explains what each of those "actors" provides in terms of functionality and performance. The LM normally does not show elements of the physical design, such as hardware, software, procedures, training etc - nor does it represent human-machine interfaces (HMIs) explicitly, these being *implicit* in every link between a human and machine actor in the model.

The Physical Model is the design of what actually is to be built and operated; it usually comprises the following elements: hardware and software, HMIs, communication media / protocols, operational and maintenance procedures and training material.

As mentioned in the context of the Service-level Model, each of the three lower-level representations of the system must be capable of, and subjected to, rigorous analysis in order to demonstrate (through *argument* and *evidence*) that an adequate set of safety properties (i.e. Safety Requirements) is specified at each level, and satisfied ultimately, in the implementation. Depending on the nature of the hazard mitigations provided by the system, such properties might cover function, performance, timing, capacity, accuracy, resolution, overload tolerance, and robustness to adverse external events.

Analysis techniques will, of course, depend on the properties to be analysed, and the degree of rigour required, but may range from basic mathematical modelling, through various UML techniques, formal methods, to fast-time and real-time (human-in-the-loop) simulations.

The eponymous "logic" in the whole approach is applied (and "magic" avoided) through the use of a rigorous, structured Safety Argument. The term "an argument-based approach" is coined in (Fowler et al 2011) and is based, at its highest level, on three claims:

Arg 1 - the system has been specified to be safe - for a given set of Safety Criteria, in the stated operational environment

Arg 2 - the system design satisfies the specification

Arg 3 - the implementation satisfies the design

The key feature of this approach is that it is the argument that drives the process, not the other way around.

A rigorous argument must be supported by equally rigorous evidence, based on the following (Fowler and Pierce 2012):

Direct evidence - which provides actual measures of the attribute of the *product* (i.e. any artefact that represents the system), and is the most direct and tangible way of showing that a particular assurance objective has been achieved
Backing evidence – which relates to the quality of the *process* by which those measures of the product attributes were obtained, and provides information about the quality of the direct evidence, particularly the amount of confidence that can be placed in it.

4.2 Example Application – an urban railway communications-based train control system

For the purposes of this example:

the track layout is given – i.e. is part of the OE
the equivalent of the "EUC" (i.e. that which has to be controlled and protected) is the movement of trains (including passengers) around the track
in order to improve the peak capacity, it has been decided to upgrade the railway from a conventionally-signalled, fixed-block system to a CBTC-based, moving-block system.

4.2.1 Service-level Specification

For our railway, the set of pre-existing hazards is as shown in Table 1. They are derived simply by considering what could go wrong with the movement of trains / passengers around the track, in the stated OE.

Table 1. Pre-existing hazards for a typical railway

ID	Pre-existing Hazard
Hp#1	Conflict between any pair of train missions
Hp#2	Conflict between train mission and track configuration
Hp#3	Train speed exceeding limitations of the track
Hp#4	High and/or uneven acceleration / deceleration of a train
Hp#5	Conflict between train profile and fixed structure
Hp#6	Presence of non-fixed obstacles or unauthorised persons on track
Hp#7	Conflict between train mission and workforce / vehicles on track
Hp#8	Passengers enter or leave a moving train
Hp#9	Passenger evacuation outside platform
Hp#10	Passenger embarkation / disembarkation at platform

ID	Pre-existing Hazard
Hp#11	Train encounters adverse rail-surface conditions
Hp#12	Conflict between train mission and trackside fire
Hp#13	Station fire / other emergency on a station
Hp#14	Conflict between train mission and flooding of track / tunnel
Hp#15	Conflict between train mission and structural failure
Hp#16	Exposure to potentially lethal voltage
Hp#17	Inadequate ventilation on train
Hp#18	Passengers too close to, or fall/jump off, platform edge
Hp#19	Fire, or other emergency, on board a train

Since the Capabilities are designed to mitigate the pre-existing hazards, it is a reasonably straightforward task to derive the former from the latter. The resulting Capabilities are listed in Table 2. For example, the pre-existing hazard (Hp#2) "Conflict between Train Mission and Track Configuration" would be mitigated by the capability (S#1) "Ensure Safe Route", provided, of course, the capability in question had sufficient functionality and performance to do what was required of it -- demonstrating this is a vital part of the verification assurance.

Table 2. List of Train Control Safety Capabilities

ID	Train Control Safety Capability
S#1	Ensure Safe Route
S#2	Ensure Safe Separation of Trains
S#3	Ensure Safe Speed
S#4	Ensure Safe Passenger Transfer at Stations
S#5	Supervise the Permanent Way [1]
S#6	Control Acceleration & Braking
S#7	Manage Maintenance Access to Track
S#8	Manage Traction Power Supply
S#9	Manage Emergencies – on Train
S#10	Manage Emergencies – on Platform
S#11	Manage Emergencies – on Trackside and others

However, there is a lot more that can be, and needs to be, said at this top level, before we immerse ourselves in the enormous complexities of the functional and logical design, and possibly lose sight of what we set out to achieve – an acceptably safe railway.

---

[1] In effect the track and the volume around it taken up by the kinematic envelope of the train

First of all, we need to describe the story of a typical normal "day in the life of a train", setting out the sequence of events and state-transitions that take place when everything happens as we would want, and expect, it to. For each stage of the journey, each event and each state of the railway, the narrative and supporting graphics must:

describe what Capabilities are provided by the (CBTC) system
describe how, when, and under what conditions the Capabilities are provided
show that all pre-existing hazards would be adequately mitigated
set Safety Objectives to capture each safety-related condition that must be satisfied in order for the consequences[1] of the pre-existing (and CBTC-system-generated), hazards to be mitigated.

This first scenario, so described and analysed, is called the reference Operational Scenario since it fits the description of an operational scenario (KPMG 2011) and provides a suitable reference from which to derive the full set of operational scenarios. These describe what happens under all other normal (i.e. variant) conditions and under all abnormal, degraded and emergency conditions.

We then analyse each of the additional scenarios, as above, to form the Service Model of Figure 4, including a full set of Safety Objectives which, if satisfied in the design, would ensure that a safe state is achieved for the railway under all the stated conditions – Arg 1 is thereby satisfied.

4.2.2 Functional Model

The Capabilities are naturally as generic as the Pre-existing Hazards, which they are intended to mitigate; however, the <u>means</u> by which they are delivered depends on (and is specific to) the functionality of the underlying system(s) – the question is where this functionality is derived from. In this example of an urban railway, we are fortunate to have available a very sound functional requirements specification (FRS) for a generic, CBTC-based, urban railway, in the form of the Urban Guided Transport Management System (UGTMS) specification (IEC/CENELEC 2011).

For a typical CBTC-based urban railway, of the order of 40 primary functions are needed to provide Capabilities S#1 to S#4 above and thereby mitigate the major pre-existing railway hazards – most of these functions are covered in the above UGTMS specification.

Although we must provide traceability of the FRS against the Capabilities, this would not give sufficient evidence that the Functional Model (FM) would deliver the required Capabilities under all the operational conditions that the CBTC system would encounter on the railway – i.e. the normal, abnormal, degraded and emergency conditions mentioned above. For this, we use tool-based Unified

---

[1] Causal analysis for the CBTC-system-generated hazards is do̶ine duri̶nge the next phase. |
Causes of the pre-existing hazards are, by definition, outside of our control

Modelling Language (UML) representations to capture the required functional and behavioural safety properties of the system at the (abstract) level of the FM[1], by means of, for example:

Activity Diagrams, to capture the very complex system safety functionality and interactions that are involved in, for example, each of Capabilities S#1 to S#4
Sequence Diagrams describing the required system behaviour for each of the *Operational Scenarios*.

Figure 5 is a simplified example of an activity diagram, for the Capability 'Ensure Safe Speed'. It shows, in general, how the initiation of a temporary speed restriction (TSR) interacts with the complex function F6.2 (which has its own activity diagram) to limit the speed of the train along the route and, in the special case of a 'Zero TSR' interacts with F4.2 and F4.1, respectively, to stop non-reporting and reporting trains immediately before the start of the segment in which train movement is prohibited.

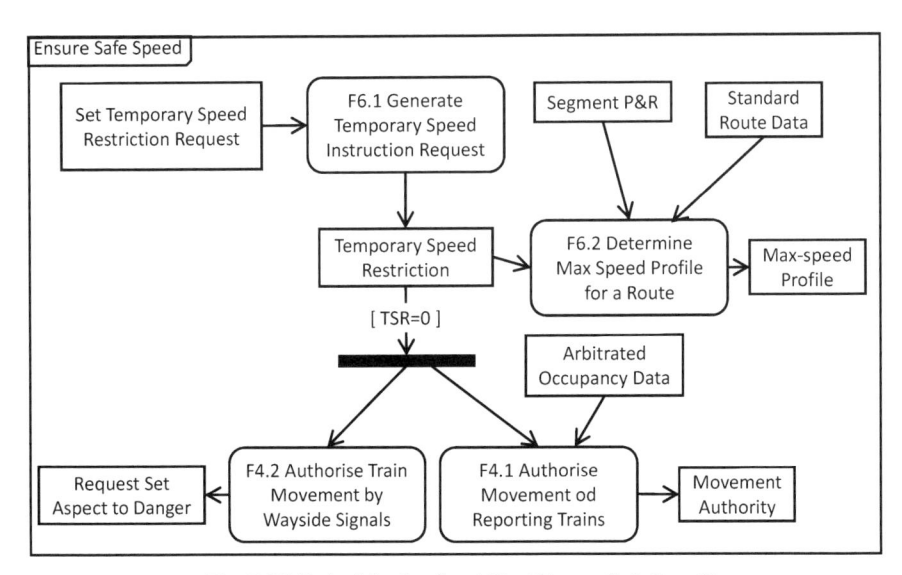

Fig. 5. UML Activity for Capability "Ensure Safe Speed"

A greatly simplified example of a Sequence Diagram, at this level, is shown in Figure 6.

The diagram is actually an overview of a protected train movement between two stations but, for the purposes of this paper, the detail is not important (and in any case would be difficult to reproduce here). What *is* important are the general characteristics of the diagram as follows:

---

[1] The IEC FRS (IEC/CENELEC 2011) is not safety specific not does it fully describe the required behaviour of the overall system – hence this needs to be done in the safety analysis in this Phase

Functional Safety by Design – Magic or Logic? 133

the actors that take part in the Operational Scenario are shown in the boxes along the top; in this simplified example (unlike in the Logical Model described below), there are only three actors – the whole CBTC-enable railway and two trains

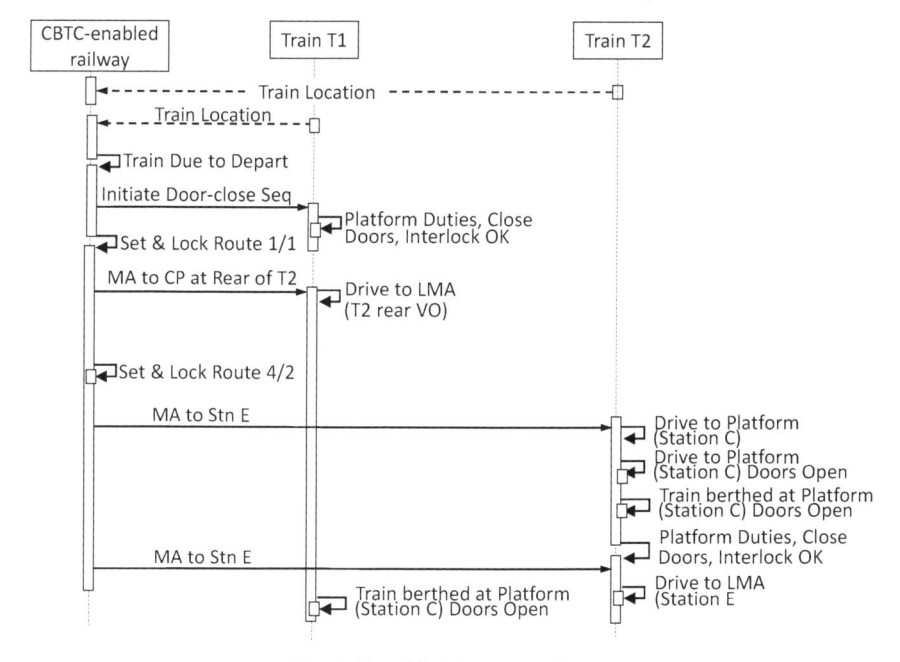

```
CBTC-enabled              Train T1                    Train T2
  railway
     ◄----------- Train Location ----------------◄
     ◄--- Train Location ---◄
     ◄Train Due to Depart
      Initiate Door-close Seq
                            ►Platform Duties, Close
     ◄Set & Lock Route 1/1   Doors, Interlock OK
      MA to CP at Rear of T2
                            ►Drive to LMA
                             (T2 rear VO)

     ◄Set & Lock Route 4/2
      MA to Stn E
                                                   ►Drive to Platform
                                                    (Station C)
                                                   ►Drive to Platform
                                                    (Station C) Doors Open
                                                   ►Train berthed at Platform
                                                    (Station C) Doors Open
                                                    Platform Duties, Close
      MA to Stn E                                  ◄Doors, Interlock OK
                                                   ►Drive to LMA
                            ►Train berthed at Platform (Station E
                             (Station C) Doors Open
```

Fig. 6. Simplified Sequence Diagram

the horizontal arrows represent the messages between the actors, which either simply provide information or, more usually, initiate a reaction by the receiving actor(s)
the vertical bars represent the response of an actor to each received message. These responses are defined by reference to a safety function (or Activity Diagram, in the case of multiple functions) and the specific safety requirements that apply to the response concerned
since every Operational Scenario has a defined start and end, there must be at least one continuous thread through the diagram.

To give some idea of the scale of the analysis task, it would not be unusual to find that around 100 Operational Scenarios were needed in order to define all the normal, abnormal, degraded and emergency modes of operation involved in a very complex urban railway, such as the older parts of the London Underground.

Because the Activity and Sequence diagrams describe behavioural attributes of the system, which are difficult to capture solely in textual descriptions of (or safety requirements for) individual safety functions, each diagram is deemed to be a safety requirement in its own right.

The process of producing the diagrams also proved to be a *very* effective way of checking the completeness and consistency of the textual safety requirements themselves.

At this point in the development process, we would have a largely solution-independent FM for our CBTC-based urban railway, comprising a set of functional descriptions and safety requirements, supported by a comprehensive set of UML products that capture:

the interactions between the functions, for each Capability
the required behaviour of the CBTC system for all the operational conditions that the system would encounter on the railway.

4.2.3 Logical Design

Whereas at the level of the FM, we expressed the CBTC system in abstract functional terms, we now need to continue the process of high-level design, with a logical architecture showing the main actors involved. A *greatly* simplified version is shown in Figure 7.

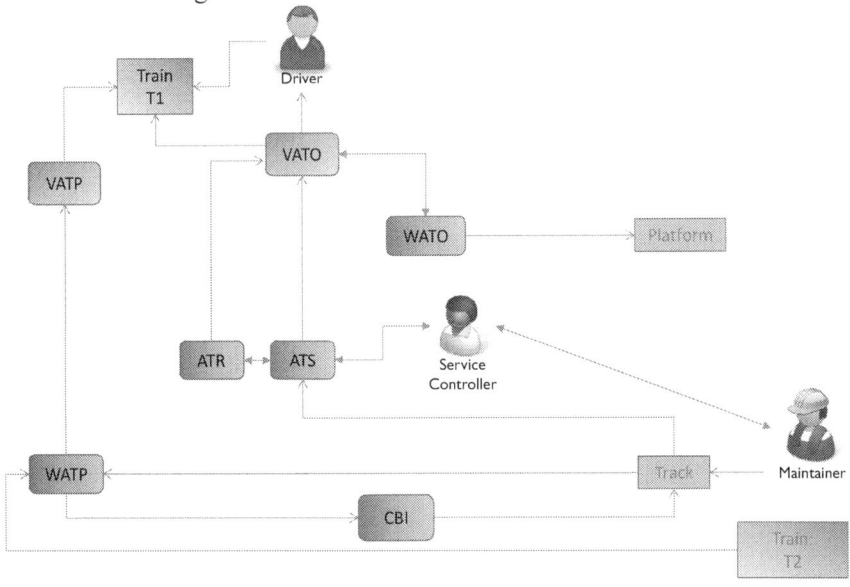

Fig. 7. Simplified Logical Model

In this example, the actors are as follows:

trains T1 and T2 – these are basic vehicles, without the CBTC system
three human operators – i.e. train driver, service controller, and (wayside) maintainer

Automatic Train Supervision (ATS): this interprets the timetable and delivers the planned service by applying the correct settings to infrastructure and signalling assets and by directing both the movement and movement authority for each train formed to deliver a timetabled service

Automatic Train Regulation (ATR): this continually monitors the progress of trains, detects when trains are running "off timetable", and regulates the progress of a train, or trains, to bring services back in line with the timetable.

Automatic Train Operation (ATO): this receives information from the signalling system regarding movement authority and required speed profile, and causes the train to proceed when in an automatic driving mode

Computer-based Interlocking (CBI): this controls all vital signalling functions in areas of converging or diverging routes, including signals, point mechanisms, and ATP facilities, such that no conflicting train movements occur, and protection is provided from following trains

Automatic Train Protection (ATP): this continuously detects the presence, or absence, of trains and transmits safety speed and distance data from the wayside so that train-borne equipment can continuously compare the actual train speed with the safety speed limit applicable at that time for the section occupied by the train.

ATO and ATP are both subdivided into wayside (WATO and WATP) and vehicle-borne (VATO and VATP) elements.

Having outlined a Logical Model, the next step is to allocate on to its constituent (machine-based and human) actors the abstract functions and safety requirements set out in the FM. This enables us to derive the tasks to be carried out by the human operators, the logical functions to be performed by equipment, and the associated safety requirements in each case.

What the LM does *not* do is to derive the required *behaviour* of the system at this level of representation. For this, we need to go back to the Capabilities and Operational Scenarios, used in the development, verification and validation of the FM, and carry out an equivalent analysis process for the Logical Model (LM).

The output of the process at this level would be sets of:

Safety requirements for each human and machine-based actor in the Logical Model

Sequence diagrams showing the required behaviour of the system, including the interactions between actors, for each identified Operational Scenario

Activity diagrams capturing complex functional interactions, in support of the Sequence Diagrams.

In parallel with this risk-reduction approach, an internal-failure analysis of the CBTC system is being carried out at the same level (i.e. Level 1) so that, by the end of this Phase, we should have collected sufficient evidence to support Arg 2, thus demonstrating that the Logical Design of the CBTC system is acceptably safe, in accordance with the safety criteria.

4.2.3 Implementation

Unlike the LM, the physical system is entirely about real-world entities, including hardware, software, operational procedures, training material, voice and data communication links, and HMIs.

The Implementation Phase is defined such that it comprises firstly the development of a Physical Model (PM) and then the realisation of that model in the built and integrated system. In making an argument for Implementation, we need to show that:

the properties of the PM satisfy the Safety Requirements for the LM
the causes or effects of any adverse, emergent safety properties (e.g. common-cause failures) or unwanted functionality have been mitigated in the PM such that they do not jeopardize the satisfaction of the Safety Requirements
the built system satisfies the Safety Requirements of the PM (i.e. verification)
the built and integrated system is consistent with the original qualitative Safety Objectives (i.e. validation).

In the physical design, we take the Safety Requirements from the LM and allocate them to the elements of the PM, as follows:

human tasks map on to, and provide the initial Safety Requirements for, skills, knowledge, procedures and training of those personnel filling the Driver, Service Controller and Maintainer roles.
machine-based functions map on to, and provide the initial Safety Requirements for, hardware and software design
human-machine interactions map on to, and provide the initial Safety Requirements for, HMI design.

These in turn lead to further levels of design, safety-requirements derivation and implementation for each of these elements and then to integration of the complete system – i.e. classic systems engineering.

Since the safety requirements for the LM include behavioural properties – in the form of, inter alia, Activity and Sequence diagrams – then we need to apply the same techniques to the PM, to show that it too exhibits the required behaviours for all the Operational Scenarios identified previously in the development phases. The outputs of these activities are also used to generate Use Cases for the subsequent testing of the elements of the physical system at all levels of integration.

The steps in the process follow, for example, the classical "V-model" of system development in which the safety engineer must ensure that the physical system as a whole (and its constituent parts) have sufficient reliability and integrity, and complete and correct functionality and performance, to satisfy the higher-level Safety Requirements.

It is a feature of such a hierarchical, systems-engineering approach that the lower the level of system representation the more remote, from the Operational Environment, the representations becomes. It is also inevitable that the demonstration of the safety of the end product – in this case, the working CBTC system – has to rely far more on evidence from verification than from validation.

4.3 Potential Application to Air Traffic Management

The need for a much broader approach to ATM safety assessment, than that set out in the Eurocontrol SAM, was recognised some years ago (Eurocontrol 2010) but, so far, has *not* led to the necessary further development of the SAM *nor* to extension of the scope of the corresponding European safety regulations[1].

This recognition led to the development of a new framework for the European SESAR) programme (Fowler et al 2011), an example of the application of which is described in detail in (Fowler 2012).

The generic framework, as further developed since 2012 and now set out in section 4.1, is currently being applied to a capacity-enhancement project at a major international airport – with considerable success to date (late 2014).

The airport has two, closely-spaced parallel runways that are employed in a segregated mode – the left-hand runway is used exclusively for arrivals and the right-hand runway for departures. The project involves a new operational concept under which, during peak-arrival-rate periods, a significant portion of arrivals will be paired (at a much closer spacing than today) and land on both runways, with departures interleaved with arrivals on the right-hand runway. The concept requires major changes to the airspace design, the arrival and departure procedures, and the surface traffic flows at the airport itself.

The safety-assessment task is based on a comprehensive, logical and rigorous Safety Argument, which is being used to drive the whole safety-management process, as detailed in the Project Safety Management Plan.

In line with the generic framework, the assessment itself started with the derivation of the following set of pre-existing hazards at, and in the airspace around, the airport, as relevant to the scope of the project.

Table 3. Pre-existing Hazards

ID	Hazard Description
Hp#1	Conflicts between pairs of aircraft trajectories
Hp#2	Controlled flight towards terrain or obstacle
Hp#3	Aircraft entry into unauthorized areas
Hp#4	Aircraft encounters severe weather conditions
Hp#5	Aircraft encounters wake turbulence
Hp#6	Runway incursion – i.e. conflict with another on landing or take-off

---

[1] It is understood that these regulations are finally in the process of being reviewed and revised.

ID	Hazard Description
Hp#7	Conflicts between taxiing aircraft and other aircraft on the ground, obstacles, or vehicle

The ATM Services to mitigate the above hazards were then derived, and are shown in Table 4.

Table 4. ATM Services

ID	Service Description
Sv#1	Maintain separation within the *same* arrival flow
Sv#2	Create & maintain separation *between* the arrival flows
Sv#3	Create & maintain spacing / separation between aircraft in *converging* arrival flows
Sv#4	Facilitate acquisition of the Final Approach path
Sv#5	Separate arrivals and departures from terrain/obstacles
Sv#6	Separate arrivals from departures, transit flights, over-flights and other arrivals - i.e. to other, nearby airports, and (non-segregated) UAVs
Sv#7	Separate departures from transit flights, overflights and other arrivals & departures - i.e. to/from other, nearby airports, and (non-segregated) UAVs
Sv#8	Separate arrivals and departures from restricted airspace (including the airspace which is considered restricted by local agreement) - both permanent and "temporary"
Sv#9	Prevent adverse-weather encounters
Sv#10	Prevent runway incursions, and collisions on the ground between aircraft and other aircraft, vehicles & obstacles

There are two operating modes under the concept. They involve identical pre-exiting hazards and ATM services, and share the same airspace design and the same procedures except for the last 12-15 NM before touchdown.

The differences, in the later stages of flight, are reflected in a separate Service-level Reference Operational Scenario description for each operating mode. To give some indication of the detail involved at this (Service) level, the following data apply to *each* operating mode:

the Scenario description occupies 20-30 pages of diagrams and narrative text.
there are up to 10 variants of the Reference Scenario, which reflect differences in, for example: aircraft pairings; runway direction; routes to the airport; manoeuvres to acquire the Final Approach path; and glidepath-intercept altitude; all of which have, or might have, specific implications for the safety assessment
a further 20 (approximately) scenarios, which represent *transitions* between significantly different operational configurations (system states), have been identified and analysed

a further 30-40 scenarios, which represent *abnormal*, undesired events in the Operational Environment, *outside* the boundary of the ATM system, have been identified and analysed

approximately 70 significantly different *failure* events, *inside* the boundary of the ATM system, have been identified and analysed

more than 70 high-level Safety Objectives, have been derived to cover the provision of the ATM services at each stage of the flight, for all the normal, abnormal and failure conditions.

The project is supported by collision-risk analysis, wake-turbulence risk analysis, and fast-time simulations – all of which have been carried out at the Service level, and have provided invaluable evidence for the Definition phase of the safety assessment.

The benefit of the above intensive, and extensive, safety analysis has been that most of the key issues and potential problems, underlying the two operational concepts, have been unearthed and resolved at a relatively early stage in the project.

The subsequent Logical and Physical levels involve few changes to the ATM equipment functionality. However, there are significant changes to the roles, workload and competencies of the Air Traffic Controllers, and to a lesser extent of the Flight Crew – all of these are being assessed through a Human Factors (HF) assessment, integrated into the safety-assessment programme, and supported by extensive real-time ('human-in-the-loop') simulations.

5 Conclusions

There are countless possible answers to the question: "*What's the difference between an electric toaster and a car airbag*"? From a logical, safety viewpoint, the most appropriate answer might be: "*Both can kill you but only the airbag can save your life*"!

Having explained this distinction in some detail, the paper has examined safety-assessment standards from the rail and ATM sectors and found that they are wholly inadequate in assessing the "life-saving" properties of systems, despite this being the basis of IEC 61508 – i.e. the concept of:

an *equipment under control* (EUC), as an inherently hazard-*creating* system, and control and/or protection systems, providing safety functions to mitigate these EUC-generated hazards, such that the resulting risk is *reduced* to an acceptable level.

It is recognised that failure of control & protection systems can have harmful side-effects – i.e. the introduction of new hazards / risks - and that the rail and ATM standards deal with such matters in depth. Even so, the full potential of such failure-based approaches is not always used (especially at the system level) in determining the rigour required of the system-development process itself.

The overall concern is that the vast detail in these standards can lead to a feeling that they provide a "magic" solution to every safety-assessment problem, resulting in illogical, and potentially misleading, conclusions.

In order to plug the major gap in such standards, the paper proposes a much more logical approach in the form of a generic framework for assessing the risk *reduction* capabilities of control & protection systems, alongside their risk-*inducing* properties, and provides an example of its application to a rail upgrade project and an airport-capacity-enhancement project.

Acknowledgments the author wishes to acknowledge the support and encouragement of countless colleagues – including Alasdair Graebner, Anthony Hall, Torkel Jensen, Chris Lowe, Ivan Lucic, Eric Perrin, Ron Pierce, and Steve Thomas - without whom the ideas presented in this paper would not have been developed.

References

Amey P (2001) Logic versus magic in critical systems. In proceedings of the 6th Ada-Europe International Conference, Leuven, Belgium May 2001.

Bieder C and Bourrier M (2013) Trapping safety into rules – how desirable or avoidable is proceduralization? Ashgate

CAA (2014), UK Civil Aviation Authority, CAP670, Air Traffic Services Safety Requirements

CENELEC (1999) EN 50126-1, Railway applications – the specification and demonstration of reliability, availability, maintainability and safety (RAMS), Part 1: Basic requirements and generic process.

CENELEC (2012) EN 50129, Railway applications —communication, signalling and processing systems — safety related electronic systems for signalling.

Eurocontrol (2012) Technical Review of Human Performance Models and Taxonomies of Human Error in ATM (HERA). Edition 1.0
http://www.eurocontrol.int/sites/default/files/content/documents/nm/safety/technical-review-of-human-performance-models-and-taxonomies-of-human-error-in-atm.pdf. Accessed 7 December 2014

Eurocontrol (2007) Safety assessment methodology https://www.eurocontrol.int/articles/safety-assessment-methodology-sam. Accessed 7 December 2014

Eurocontrol (2010) Safety assessment made easier, part 1, edition 1.0
http://www.eurocontrol.int/sites/default/files/field_tabs/content/documents/nm/safety/same_part_1_v1.0_released.pdf . Accessed 7 December 2014

Fowler D, Perrin E and Pierce R (2011) 2020 Foresight - a systems-engineering approach to assessing the safety of the SESAR operational concept. In ATC Quarterly, Vol. 19(4) pp 239-267, 2011

Fowler D (2012) Getting to the point: a safety assessment of arrival operations in terminal airspace. In ATC Quarterly, Vol 20(2), pp143-174, 2012

Fowler D, Pierce R (2012) A safety engineering perspective. In: Cogan B (ed) Systems engineering - *practice and theory*. InTech

Fowler D (2013) Proceduralization of safety assessment – a barrier to rational thinking? In:Bieder C, Bourrier M (eds) Trapping safety into rules – how desirable or avoidable is proceduralization? Ashgate

IEC (2010) IEC 61508 – Functional safety of electrical/electronic/programmable electronic safety related systems, V 2.0. International Electrotechnical Commission

IEC/CENELEC (2011), EN 62290-2, Railway Applications – Urban Guided Transport Management & Control Systems

IEEE (1992) IEEE Std. 100-1992, Standard dictionary of electrical and electronics terms.

KPMG (2011) Preparing for the unexpected – leading practice for operational risk scenarios http://www.kpmg.no/arch/_img/9697753.pdf . Accessed 7 December 2014

RTCA / EUROCAE (2012) DO-178C / ED-12C, Software Considerations in Airborne Systems and Equipment Certification, RTCA SC-205 / EUROCAE WG-12

SAE (1996) ARP 4761 - Guidelines and methods for conducting the safety assessment process on civil airborne systems and equipment. SAE International

SAE (2010) ARP 4754A - Guidelines for development of civil aircraft and systems. SAE International

Controlled Expression for Assurance Case Development

Katrina Attwood and Tim Kelly

University of York

York, UK

Abstract   *Guidance for developers of assurance arguments has generally focussed on issues concerning content, logical flow and structure. The use of natural language to express an argument can lead to problems with understanding the nature of the claims, with scope and with potentially obscure logical inferences. These problems can occur even if natural language is combined with the use of graphical notations to communicate the structure of an argument. In a supply chain, these problems are compounded by the involvement of numerous suppliers, each with his own "idiolect", which makes it difficult to integrate assurance data for components into the system argument, to evaluate it as evidence or to reuse it across projects. In this paper, we present work to develop controlled language and structured expressions to improve communication within and across domains and to provide some automated validation of assurance arguments.*

1 Introduction

Many industries require an assurance case for safety-critical services, systems and/or software. An assurance case typically comprises reasoned arguments justifying claims relating to the satisfaction of requirements concerning the safety, integrity and/or dependability of a system. These claims are supported by a body of evidence– analysis and test data, design information and process documentation. There is considerable literature providing guidance and methodologies for the development and presentation of assurance cases (e.g. Hawkins and Kelly 2010; Hawkins et al. 2011; Maguire 2006). Much of this guidance focuses on the desirability of presenting **structured assurance cases** (Object Modelling Group 2013), in which the relationships between different claims and between claims and evidence are made explicit. Graphical notations, such as GSN (GSN Community 2011) and CAE (Adelard 2014) have been developed in order to support the development and evaluation of structured assurance cases, although it should be noted that they are not an *a priori* requirement: there is no reason why a structured assurance case should not be presented using natural language alone, and indeed many are. However, guidance on the development of structured assurance cases to date has tended to focus on "macro-level" details of the structure and typical subject-matter of the arguments with relatively little attention paid to the "micro-level" issue of the language used to convey the argument. Issues such as the logical flow and overall readability of the argument – which can be readily summarized using graphical techniques – are foregrounded, while there is only limited guidance on how individual assertions and supporting statements should be phrased to ensure that the argument is correctly conveyed. In the GSN Community Standard (GSN Community 2011), for example, less than 10% of the total document is devoted to language issues.

Imprecise phrasing of claims and assertions can lead to a number of problems with the comprehension of assurance cases:

- **Semantic problems:** An assurance case typically has several authors, each of whom uses language in a distinctive, idiosyncratic way. This means that terms might be used with subtly different meanings at different points in an argument. Similarly, a reader or assessor might not share the writer's understanding of the terms used. This can lead to uncertainties in interpretation, particularly as to what the precise subject of a claim or assertion is or the scope within which the claim/assertion is valid. The author's intended meaning may differ from that conveyed to the audience. A related problem is that of inherent ambiguity, where a given term or phrase has more than one commonly accepted interpretation.
- **Syntactic problems:** Where an argument is carried by claims in 'freeform' text (such as is generally the case even where a graphical notation is used to structure the argument), it may be difficult for a reader to 'unravel' the structure of a sentence so as to establish the scope of the terms used, i.e. how they influence other terms beyond the single phrasal structure in which they occur (Lapore 2009). The relationships between elements under discussion might not be clear, making it difficult for a reader to identify the claims being made and the flow of the argument.

Although these problems may arise with any assurance case, the problems of multiple authorship and readership are compounded in current development practices, where safety-critical systems are increasingly developed by integrating multiple standalone components from a diffuse, multi-organisational, multi-national supply chain. Compositional approaches to certification have been developed for such environments. These require the collection of assurance data relating to discrete components ("Safety Elements Developed Out-of-Context" (ISO/FDIS 2011)). This data then needs to be matched and composed to form an integrated argument for the entire system. Aside from the difficulties inherent in aligning the intent and objectives of assurance requirements in different domains, there is a lower-level need for consistent use of domain- and system-specific terminology across the entire supply chain. This can help in communicating a shared understanding of the nature and limitations of the claims and evidence presented in the argument, and of assumptions made about the operational context in which component behaviour is guaranteed.

One means to address the problems associated with imprecise phrasing is to *constrain* the language used to write assurance cases. For example, a project dictionary can be used to address semantic issues. The dictionary declares the accepted meaning of terms that are inherently ambiguous or often misunderstood, or which may be commonly used with subtly different meanings in different usage contexts. There are several advantages to this approach, particularly in ensuring that the meaning intended by an author is that received by the reader, and also in making it easier to "match" terminology used in assurance data for discrete components, by making similarities and differences in meaning clear. However, there are considerable overheads in the initial development – and agreement – of the domain dictionary, and also in ensuring that it is adhered to. Constraining the syntax of assertions made in the assurance case – for example, by providing structured expressions – can help address uncertainties of linguistic scope and uncertainty. These are instantiatable "patterns" for sentence structures, which define the range of linguistic variables which can occur in particular relations. Although certain generic types of expression can be identified in the assurance case domain (see Attwood et al. forthcoming), for the most traction to be gained from this method specialised patterns for each domain are required. For this, we require a thorough understanding of the requirements for assurance in a particular domain, to ensure sufficient focus and relevance in the argument claims generated using the controlled expressions.

In this paper, we outline an approach to controlled language and expression for assurance cases which addresses the semantic and syntactic issues associated with natural language identified above. This approach has been developed on the EC-funded OPENCOSS project (OPENCOSS Consortium 2011), where it augments a model-based conceptualization of safety assurance which is used to address issues of cross-domain and cross-project reuse of safety assurance assets, including arguments. In principle, however, the language approach can be standalone, as it is presented here. Section 2 outlines the usage scenarios for controlled language identified in the OPENCOSS project. Section 3 presents conceptual metamodels to address each of these usage scenarios and encapsulate the proposed approach. Section 4 presents a brief worked example of the use of controlled vocabulary and structured expressions to develop an assurance argument in the automotive domain.

2 Usage Scenarios

In this section, we describe three basic usage scenarios for controlled language in the development of assurance cases. The general context for this work is a compositional certification context, where a system-level assurance case is being assembled by an integrator organisation from assurance data relating to freestanding, pre-existing components. In some cases, the components may have been developed within another safety-critical domain and are being reused in the integrated system (for example, a software module originally developed for an aerospace application is being reused in a rail system). Note that this paper does not address the complex technical implications of developing a compositional safety case, but is concerned simply with *linguistic aspects* of the project: principally, our concerns are to ensure that there is a clear, shared understanding of the meaning and scope of domain- and system-specific terminology throughout the assurance case, so that the technical aspects of integration of assurance case data can be assessed from a common basis. In particular, we are concerned that the differences in terminology across domains are clearly documented and the implications of reuse understood.

2.1 Development of Structured Vocabularies for Assurance

The development of vocabularies to provide constrained definitions of the terminology used in assurance cases is the prerequisite for the other usage scenarios outlined in this section. Section 3.1 below outlines a conceptual metamodel for the development of such vocabularies for each of the domains and projects relevant to the assurance case (e.g. for rail and avionics). The vocabularies are structured as concept hierarchies, along the lines of an English language Thesaurus (for example (Lloyd 1982)), with terms being grouped according to general concept types. A concept type is a general category, by which related concepts at lower levels of detail can be grouped. For example, in a vocabulary developed for the automotive domain using ISO 26262 (ISO/FDIS 2011) the terms "driver", "passenger" and "cyclist" are all grouped under the concept type "traffic participant", which is itself grouped under the general concept "agent". Definitions are provided in a semi-formal structured English, derived from Semantics of Business Vocabulary and Business Rules (SBVR) (Object Modelling Group 2008). SBVR is an OMG standard for the preparation of controlled or semi-controlled vocabularies for a given domain and for modelling the relationships between domain concepts in intuitive natural language. Where terms defined elsewhere in the vocabulary are used in the definitions, cross-references are provided. In order to facilitate comparison between terminology across domains, concept types are defined fairly generically across the domains, resulting in a commonality of general structure, even though the low-level terms grouped under each generic concept type might differ substantially between domains. Linguistic relations are used to indicate relationships between terms, such as "is-a" and "part-of". We develop vocabularies at two levels of abstraction: domain-specific vocabularies, which are derived from the terminology used in the relevant safety standards for the domain, and project-specific vocabularies, which specialise the domain vocabulary for a given project or organisation.

Having clear definitions of terminology in which concepts are related both vertically by type and sub-type relations and horizontally by being defined in terms of one another can help in ensuring consistency of reference across assurance cases. In particular, the terminology can be used to characterise the "interfaces" between concerns in an assurance argument (which may be represented explicitly in a modular assurance case), to ensure that the terms of reference are consistently understood across a supply chain.

2.2 Structured Expressions in Assurance Arguments

One important means of maintaining consistency in the natural language used to convey reasoning in an assurance argument is to regularize the structure of the statements used to express claims. Having a common syntactic structure makes it easier for a reader to parse claims, and avoids issues such as confusion over the scope of a given term in a sentence. It is quite common in "engineering prose" for there to be uncertainties over the interpretation of the scope of qualifiers, as in the phrase "failure mode and effect analysis", where "analysis" serves to qualify both "failure mode" and "effect" and where "failure" qualifies both "mode" and "effect": a non-specialist reader would be likely to look only for the most limited range, and associate "analysis" only with "effect" and "failure" only with "mode".

Structured expressions can be used to characterise the types of concepts which are discussed at a particular point in an argument and the relevant features which can be asserted about them. Typically, a structured expression comprises a "fixed" verb phrase, which carries the burden of the claim, while noun phrases, providing the subject and object over which the verb phrase ranges, are parameterisable. For example, a very simple structured expression in an assurance case claim might take the form "{systematic fault} is adequately mitigated by {fault mitigation technique}", where both "systematic fault" and "fault mitigation technique" are broad parameters. Simple expressions can be combined to form larger syntactic units, for example: "{systematic fault} is adequately mitigated by {fault mitigation technique} which addresses {hazard}".

In our approach, a series of generic structured expressions are defined, and used to refine the logical structures summarized in argument fragment templates captured in GSN patterns such as those in (Hawkins and Kelly 2013), by specifying the types of concepts which are in focus at particular points in the argument. The generic concept types which are used to structure the vocabularies (see section 2.1) are used to reference the variable parameters, e.g. 'fault", "fault mitigation technique", "hazard" are all high-level concept types common across the domain vocabularies. These expressions can then be instantiated to form claims in specific arguments, by supplying variables from the terms listed under the relevant concept types as appropriate. Since we have both domain- and project-specific vocabularies, we can instantiate claims at two levels of abstraction, meaning that we can use the concept types to specify detailed patterns at the domain level, for instantiation by a project. Automated support for this stage can be provided, with the concept types in the vocabulary models being used to present a series of potential instantiations of a given parameter from which an argument writer can choose.

2.3 Cross-Domain Vocabulary Mapping

As described above, the general context for our work involves the creation of a compositional assurance case from pre-existing components, some of which are being reused across domains. This cross-domain scenario intensifies the semantic problems concerning shared understanding of terminology observed in Section 1. There is no shared conceptualisation of assurance across safety-critical domains and the objectives and requirements of certification approaches differ fundamentally. Even where terms in different domains may appear similar, the concepts they represent may not be. As a motivating example, we consider the difficulties of reusing software developed according to IEC 61508 (IEC 2009) in an avionics context, where certification to DO-178B (RTCA 1992) is required. An assurance argument in the original context might assert that "software module Y is developed to safety integrity level SIL 4". In the avionics context, the manufacturer might wish to make a similar claim: "software component Y is developed to design assurance level DAL A". Since both the safety integrity level and the design assurance level are instantiations of the generic concept type "criticality level", it might be assumed that a direct replacement of one term for the other is allowable. Closer examination of the definitions of the terms, however, will reveal an important distinction between the two level descriptors. In IEC 61508, a SIL is directly associated with a (software) safety function which is modelled at the system level. In DO-178B, however, the DAL is associated with a software system or component and does not address the "function" concept at all. It is not possible to convert a SIL directly into a DAL without considering the extra process-related requirements that arise because of DO-178B's focus on the design of the system, rather than merely its functionality. What is required here is not a definition of individual concepts in isolation, but an appreciation of the interrelationships between the concepts, since these provide constraints on the reuse of the claim and associated assurance data.

We aim to provide a mechanism for informed "translation" between terminology used in assurance in different safety-critical domains, in such a way that the different understandings and scope of assurance concepts within the different domains are clearly documented and understood. Initially, we do this by providing guidance to a user in asserting mappings between concepts presented in the domain-specific vocabularies for the source and target domains. The metamodel underpinning this approach is presented in Section 3.2 below.

Three general classes of mapping are identified – exact map, partial map and no map. In the majority of cases, exact mappings of terms will not be possible across domains – instead, the mappings will need to be partial. The model allows for a distinction between "broad" and "narrow" partial mappings. Concepts in the vocabularies are organised hierarchically, using generic concept types. Initially, candidate mappings are presented to the user by simply displaying all of the elements in the vocabulary for the target domain which are categorized using the same concept type(s) as the source term. Although the source and target domain vocabularies are structured using a standard set of concept types, there may be a considerable number of candidate maps. The field is then narrowed by displaying the definitions for each of the candidate maps. The user can use these definitions to make an informed choice as to the most suitable replacement term, based on the definition, any relationships implied by the use of reserved terms in the definition and the degree of coverage of the source term in the chosen target. This information is recorded manually in a "map justification", so that the rationale for the reuse and any limitations on it are preserved.

3 Conceptual Metamodel

In this section, we present the conceptual metamodel which underpins our approach to structured language in assurance cases. The conceptual model has been heavily influenced by several existing models for concept definition and thesaurus development:

- **OMG Structured Assurance Case Metamodel (SACM)** (Object Modelling Group 2013) - an OMG standard which defines the treatment of argumentation and evidence in structured assurance cases and provides a common interchange format. The first version of SACM was published in 2013. Proposed revisions will be discussed in December 2014 with a view to a second version shortly thereafter.
- **Simple Knowledge Organization System (SKOS)** (W3 Consortium 2009) – a W3C recommendation for the representation of Thesauri and other classification schemes. SKOS was developed and refined in three EC-funded projects in the late 1990s and 2000s, and has some industry support. Its primary significance here is in the concepts it provides for the development of concept hierarchies linked by non-hierarchical relationships and in its introduction of the use of "labels" to support the recording of full synonyms.
- **ISO 25964:2011 Thesauri and Interoperability with Other Vocabularies** (ISO 2011) - This ISO standard provides guidance on mapping

between one vocabulary and another, as well as providing a series of relationships between concepts and terms (equivalence relationships, hierarchical relationships and associative relationships).
- **OMG Ontology Definition Metamodel** (Object Modelling Group 2009) - This document provides metamodels for three leading ontology representation paradigms – Common Logic, RDF and OWL, and explains their relationship with and divergences from UML.
- **OMG Semantics of Business Vocabulary and Business Rules (SBVR)** (Object Modelling Group 2008) - an OMG standard for the preparation of controlled or semi-controlled vocabularies for a given domain and form modelling the relationships between them in intuitive natural language. SBVR provides a mechanism for the presentation of "reserved terms" in semi-formal definitions of concepts and also for the super- and sub-typing of concepts in concept hierarchies. SBVR "fact types" provide a basis for some of the controlled expressions used in assurance arguments.

For ease of presentation, the metamodel is presented in three views. Section 3.1 describes basic vocabulary metamodel. Section 3.2 extends this model to include concepts required to support structured expressions in assurance arguments. Section 3.3 provides the mapping metamodel required to support cross-domain vocabulary "translation".

3.1 Vocabulary Metamodel

Figure 1 depicts the conceptual vocabulary metamodel developed in OPENCOSS. The notation used here and in Figures 2 and 3 below is Ecore, the semantics of which accord with the UML[1]. This model defines the concepts and relationships required for the development of structured vocabularies for safety-critical domains and projects, as outlined in Section 2.1. Textual definitions of the core elements of the model are presented below.

---

[1] Ecore is the foundational metamodel of the OMG's Meta-Object Facility standard for model-driven engineering (http://www.omg.org/mof/). The UML metamodel is defined in Ecore.

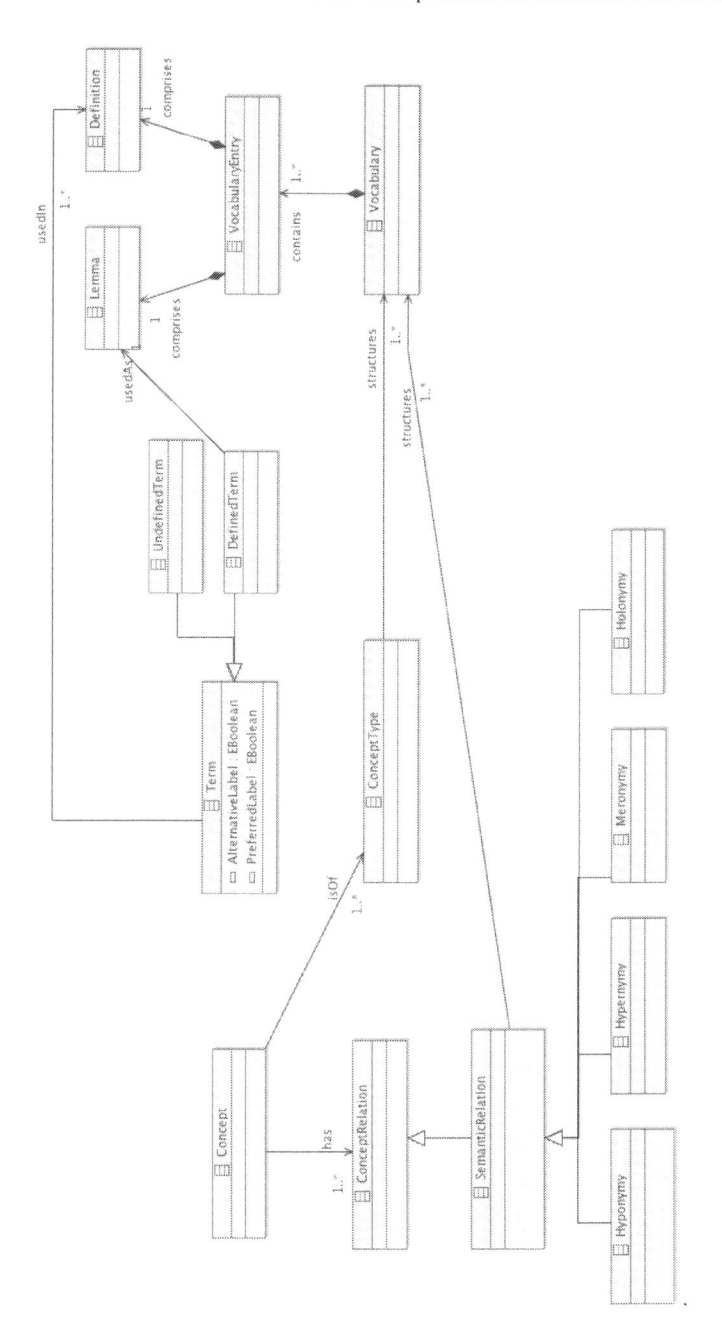

Fig. 1: Conceptual Vocabulary Metamodel

Concept: Any unit of meaning which can be described or defined by a unique combination of characteristics (Object Modelling Group 2008; W3 Consortium 2009). A Concept is represented/reified by the *Term* class and defined by the *Definition* class.

ConceptType: A container class which classifies things on the basis of their similarities. Each *Concept* belongs to one or more *ConceptType*. This is akin to the "ConceptType" relation in SBVR (Object Modelling Group 2008). The *ConceptTypes* provide the basic structure of the *Vocabulary* in that *Terms* are arranged according to the *ConceptTypes* of the *Concepts* they represent.

ConceptRelation: A container for the types of relation between *Concepts*, which are used to model the comparison between two *Concepts* in two different *Vocabularies* or to structure a *Vocabulary* in terms of the relationships between *Concepts* grouped within a *ConceptType*. Two types of *ConceptRelation* are identified: *MappingRelation* (discussed in Section 3.3) and *SemanticRelation*.

SemanticRelation: A container for the finer relationships between the meanings of *Concepts* within a *ConceptType*, used to structure *Terms* of the same *ConceptType* in a *Vocabulary*. *SemanticRelations* are kept separate from *MappingRelations* in this conceptual model, in order to highlight their linguistic basis, as opposed to the logical ones described by *MappingRelations*. *MappingRelations* are described in section 3.3 below.

Hyponymy: A *SemanticRelation* by which the meaning of one *Concept* is more specific than that of another *Concept*. In other words, a hyponym is used to designate a member of a general class. For example, 'systematic fault' and 'intermittent fault' are both hyponyms of 'fault'. This might be referred to as a "type-of" relationship.

Hypernymy: A *SemanticRelation* by which the meaning of one *Concept* is more general than another *Concept*. In other words, a hypernym designates the general category, of which its hyponyms are members or subdivisions. For example, 'fault' is a hypernym of 'systematic fault' and 'intermittent fault'. This might be referred to as a "supertype-of" relationship.

Meronymy: A *SemanticRelation* by which one *Concept* is a constituent part of a general whole captured in another *Concept*. For example, "wheel" is a meronym of "automobile". This might be referred to as a "part-of" relationship.

Holonymy: A *SemanticRelation* by which one *Concept* is an aggregation of other *Concepts*. For example, "automobile" is a holonym of "wheel", "chassis" etc. This might be referred to as a "contains" relationship.

Term: The word or phrase which represents (or reifies) a *Concept*, typically a noun or noun-phrase (ISO 2011). *Terms* are thus the basic domain vocabulary, and are stored in the *Vocabulary*. A *Term* may be thought of as providing a label for a *Concept*, an unambiguous means by which the *Concept* can be referenced. Each *Term* has two Boolean attributes, *PreferredLabel* and *AlternativeLabel*. If the *PreferredLabel* attribute is set true, then the *Term* serves as the primary means by which a *Concept* is referred to and the principal key to represent that *Concept* in the *Vocabulary*. If the *AlternativeLabel* is set true, the term serves as an alternate means to reference the *Concept*, there being a *PreferredLabel* elsewhere in the *Vocabulary*. This mechanism allows for the unambiguous recording of exact synonym relationships between terms.

DefinedTerm and UndefinedTerm: We identify two subtypes of *Term*. *DefinedTerms* are those which are included in the *Vocabulary* and which should be used with a single "reserved" meaning in project artefacts. Each *DefinedTerm* is represented in the *Vocabulary* by a *VocabularyEntry*. *UndefinedTerms* are those which have no "reserved" meaning in the project, and which are not therefore defined in the *Vocabulary*. *Terms* of all kinds can be used in *Definitions*.

Vocabulary: An aggregation of the *VocabularyEntries* which identify and define *DefinedTerms* for a particular domain. The broad structure of the *Vocabulary* is provided by the *ConceptTypes*, and relationships between the *Concepts* represented by *DefinedTerms* are captured by *SemanticRelations*.

VocabularyEntry: Each *DefinedTerm* has a *VocabularyEntry* which encapsulates the information stored concerning the *DefinedTerm* in the *Vocabulary*. Each *VocabularyEntry* comprises exactly one *Lemma* and exactly one *Definition*.

Lemma: A string representing the noun or noun-phrased used to designate a *DefinedTerm*. The *Lemma* serves as the key to identify the *DefinedTerm* in the *Vocabulary*.

Definition: A string which contains an unambiguous description of the *Concept* represented by a *DefinedTerm*. The nouns and noun-phrases used in *Definitions* may be either *DefinedTerms* or *UndefinedTerms*.

3.2 Metamodel for Structured Assertions

Figure 2 depicts the Ecore metamodel for structured assertions. This model derives from proposed modifications to the SACM (Object Modelling Group 2013), which have been made by the SACM Revision Task Force, and will be discussed at length at the OMG in December 2014[1]. This metamodel defines the use of controlled language in elements of assurance arguments. Two aspects of language constraint are addressed. Individual words and phrases are used with fixed definitions, as described in Section 3.1, to refer to concrete declared objects (the subjects of discussion in the assurance case). As described above, these terms and phrases are *DefinedTerms*, contained in a *Vocabulary*. Where structured assertions are used, these has a *SentenceStructure*, in which each term has a fixed *Role*, a clear function in the syntax and semantics of the sentence. *RoleBindings* link instances of these *Roles* to actual instances, which are used in *Assertions* in the argument. Textual definitions for the core elements of the model are given below.

---

[1] Note that this is an abbreviated version of the proposed model, in which only elements of interest to the present discussion have been included. Other subclasses of *ExpressionElement* and *ArgumentElement* occur in the SACM, but have been omitted here for clarity.

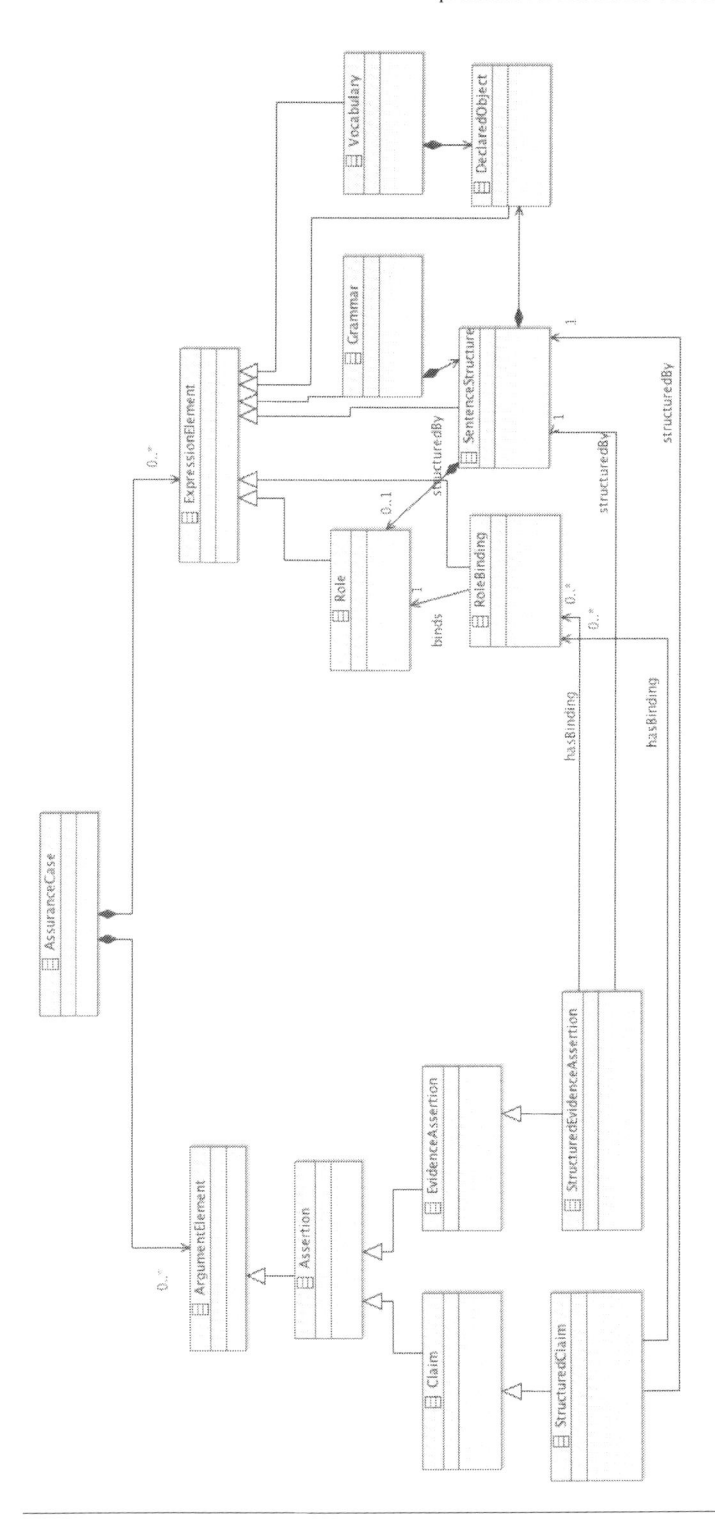

Fig. 2: Conceptual Structured Assertions Metamodel

AssuranceCase: A container class for elements comprising a structured assurance case: claims, argumentation and evidence which seek to establish that a given system satisfies relevant requirements.

ExpressionElement: A container class for those aspects of an *AssuranceCase* which are associated with the way in which the argument is expressed in natural language.

Vocabulary: An aggregation of *VocabularyEntries* (see section 3.1 above) relating to *DefinedTerms*.

DeclaredObject: A reference to an instance of a *Concept* to which a *DefinedTerm* refers. See section 3.1 above.

Grammar: An aggregation of *SentenceStructures*, representing the total of permissible structured expression types for propositions to be used in a given *AssuranceArgument*.

SentenceStructure: A pattern for a structured expression, which relates *DeclaredObjects* to one another in a proposition, using controlled syntactic structures in which the function of each *DeclaredObject* is clarified by the use of a *Role*.

Role: A description of the function played by a given *DeclaredObject* in the relationship captured in a *SentenceStructure*. Grammatically, a *DeclaredObject* will be either the subject or direct object of a verb. However, the *SentenceStructures* defined for *AssuranceCases* are more sophisticated, in that the verbs "dictate" the *DeclaredObjects* which can participate in the relationship. Hence, a *Role* is a clear description of the function of a *DeclaredObject* in the system or its assurance.

RoleBinding: The connection of a given *Role* to an actual variable, expression or concept which can be used to instantiate it.

ArgumentElement: A container class for those aspects of an *AssuranceCase* which are associated with the logical structure of the argumentation or the presentation of evidence.

Assertion: An abstract class to capture types of propositions which can be used in an argument.

Claim: A natural language statement which is used to record a proposition used as a premise or a conclusion in an argument, to express an opinion or judgment.

StructuredClaim: A *Claim* whose expression is an instantiation of a pattern captured in a *SentenceStructure*.

EvidenceAssertion: A natural language statement which is used to record a proposition concerning some quality or characteristic of an item of evidence or its relationship with some other element discussed in the argument.

StructuredEvidenceAssertion: An *EvidenceAssertion* whose expression is an instantiation of a pattern captured in a *SentenceStructure*.

3.3 Mappings Metamodel

Figure 3 depicts the conceptual mappings metamodel developed in OPENCOSS. This model defines the concepts and relationships required to support the technique to provide guidance for cross-domain vocabulary "translation", as described in Section 2.3. Subjective mappings of different degrees of exactness between concepts are identified, based on the degree of "overlap" between the meanings and usage of concepts in the different domains. In the OPENCOSS tooling, mappings are asserted manually, and a textual "Map Justification" is supplied, to describe the nature of the mapping, capturing the similarities and differences between the concepts. It is particularly important that *NoMap* relationships between concepts which might appear similar are recorded where "translation" is inappropriate. Textual definitions for the core elements of the model are given below.

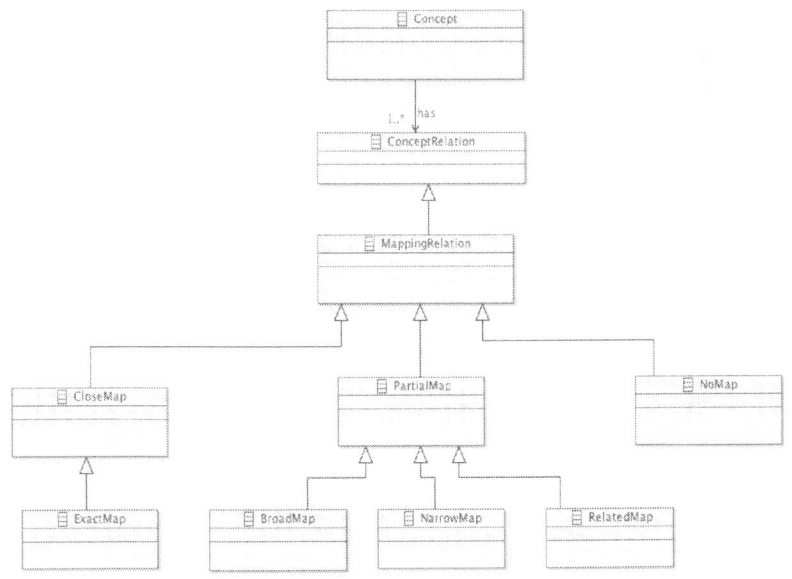

Fig. 3: Conceptual Mappings Metamodel

Concept: See section 3.1 above.

ConceptRelation: See section 3.1 above.

MappingRelation: A container for the relationships between the meanings and contexts of use of two *Concepts* represented by *Terms* in two different *Vocabularies* (across domains or projects).

CloseMap: A *MappingRelation* which indicates that there is a good degree of similarity between the meanings and contexts of use of two *Concepts* which are being compared across two *Vocabularies*.

ExactMap: A *CloseMap* which indicates that the relationship between two *Concepts* is such that the meaning and context of the target *Concept* is identical with that of the source *Concept*.

PartialMap: A *MappingRelation* which indicates that there is some similarity between the meanings and contexts of use of two *Concepts* which are being compared across two *Vocabularies*, but that the similarity is not sufficient to reach the threshold for a *CloseMap*.

BroadMap: A *PartialMap* in which the meanings and contexts of use of the two *Concepts* can be considered similar in terms either of important attributes or the proportion of overall similarity.

NarrowMap: A *PartialMap* in which the meanings and concepts of use of the two *Concepts* can be considered less similar in terms either of important attributes or the proportion of overall similarity than is the case for a *BroadMap*.

RelatedMap: A *PartialMap* which indicates that the meanings and concepts of use of the two *Concepts* can be considered to be related but there is no conceptual overlap between them.

NoMap: A *MappingRelation* which asserts that there is no similarity or overlap between the meanings and contexts of use of two *Concepts* which are being compared across two *Vocabularies*, with respect to features considered relevant for asserting a mapping.

4 Example

In this section, we present a simple example to illustrate the ways in which structured expressions using controlled vocabulary can be exploited to instantiate claims in an assurance argument. The example is based on a simplified, fictitious automotive anti-lock braking system (ABS), which is developed to ISO 26262 (ISO/FDIS 2011). Correct operation of the ABS allows the wheels to maintain contact with the road surface during hard braking, preventing the wheels from locking and avoiding an uncontrolled skid. The system comprises a software controller, four wheel sensors (one for each wheel) and two hydraulic valves (one for each axle). The system has two basic operational scenarios. The software monitors the speed at which the wheels rotate, measured via the wheel sensors. If it detects that one wheel is rotating at a slower speed than the others, the controller actuates the hydraulic valves to reduce hydraulic pressure to the brake, thus reducing braking force on that wheel and allowing it to turn faster. Alternatively, if the software detects that one wheel is turning significantly faster than the others, the valves are operated to increase hydraulic pressure to that wheel, thus increasing braking force to that wheel and slowing down its rotation. The software controller contains a critical function to calculate the hydraulic pressure demand value from the wheel speed sensor inputs. Failure of this function results in the incorrect braking force being applied to the wheel, which could result in a skid.

The assurance argument for the ABS software controller needs to address the issue of potential faults in the hydraulic pressure demand calculation function. In this example, that issue will be addressed as part of a top-down argument concerning the mitigation of the "uncontrolled skid" hazard by the software. An argument of this type can be structured using the high-level software safety argument pattern in (Hawkins and Kelly 2013), which is presented in Figure 4, using the GSN (GSN Community 2011; Kelly 1998). High-level patterns for the expression of the structured assertions are captured in this pattern. In terms of the metamodels presented in sections 3.1 and 3.2 above, *Roles* are captured as parameterisable noun-phrases contained in {}, while the *SentenceStructure* is captured as the entirety of the phraseology contained within a given GSN claim. The totality of *SentenceStructures* in the diagram represents the *Grammar* for this particular pattern.

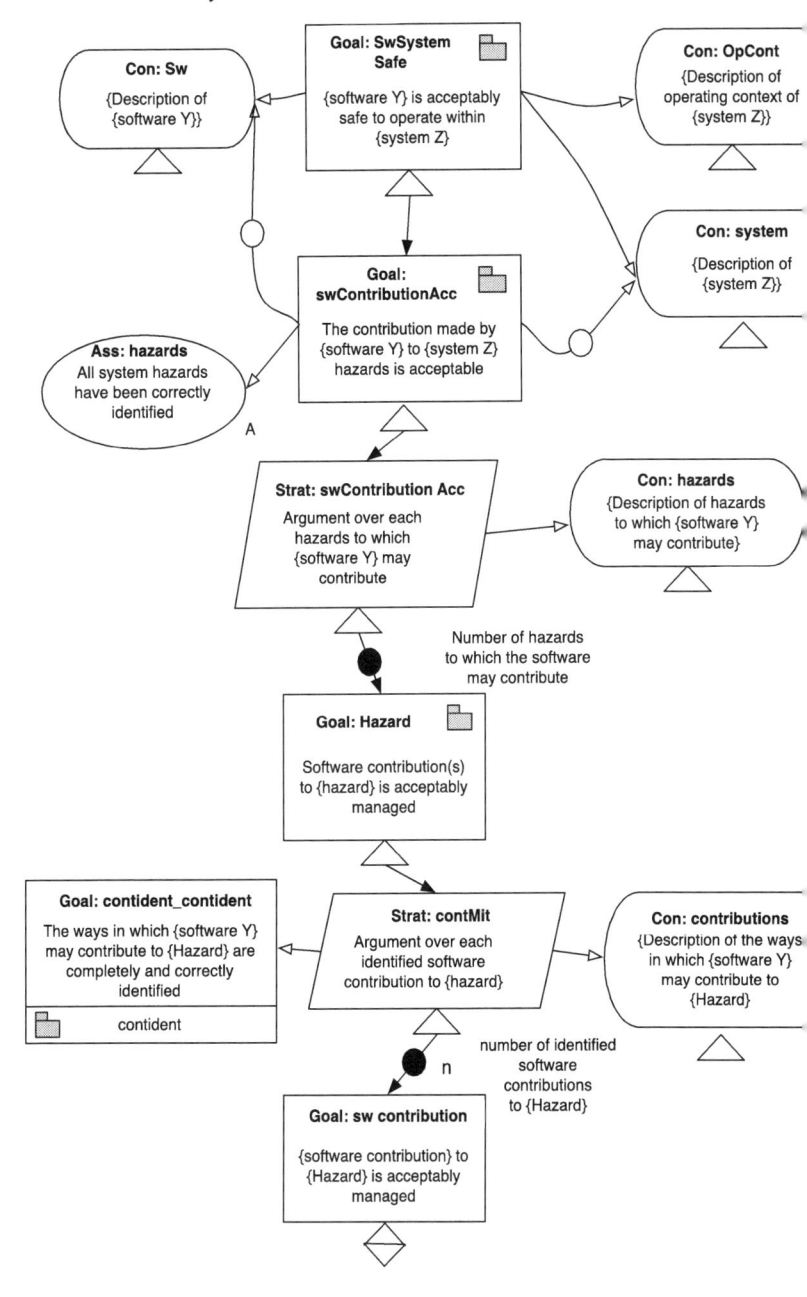

Fig. 4 High-Level software safety argument pattern (from (Hawkins and Kelly 2013))

Our discussion draws on the lower part of the pattern in Figure 4, the claim in Goal:Hazard that the software's contribution to a particular Hazard is adequately managed and the subsequent argument addressing each potential way in which the software could contribute to the hazard.

The example requires two distinct *Vocabularies* containing *VocabularyEntries* by which *DefinedTerms* are grouped initially as abstract ConceptTypes and then as concrete instances of these types relating to particular *DeclaredObjects* in the ABS project documentation. Firstly, the ABS System is represented in a vocabulary, terms in which are drawn from the organisational vocabulary for the system as a whole and which define concepts in the deployment context of the ABS software. Concepts relating to the structure and functionality of the ABS software are also represented in a dedicated, project-level, vocabulary.

Figure 5 shows a partial instantiation of the template pattern presented in Figure 3, as an assurance argument for the ABS Software. The underlined terms here ("ABS software", "uncontrolled skid hazard", "safety requirement 123", "fault tree analysis") are *DefinedTerms* in the vocabulary for the ABS System (populated from project documents at the system level, such as system descriptions, requirements documents, system safety analysis), and relate to *DeclaredObjects* which fulfil the *Roles* indicated in the *SentenceStructures* in Figure 4.

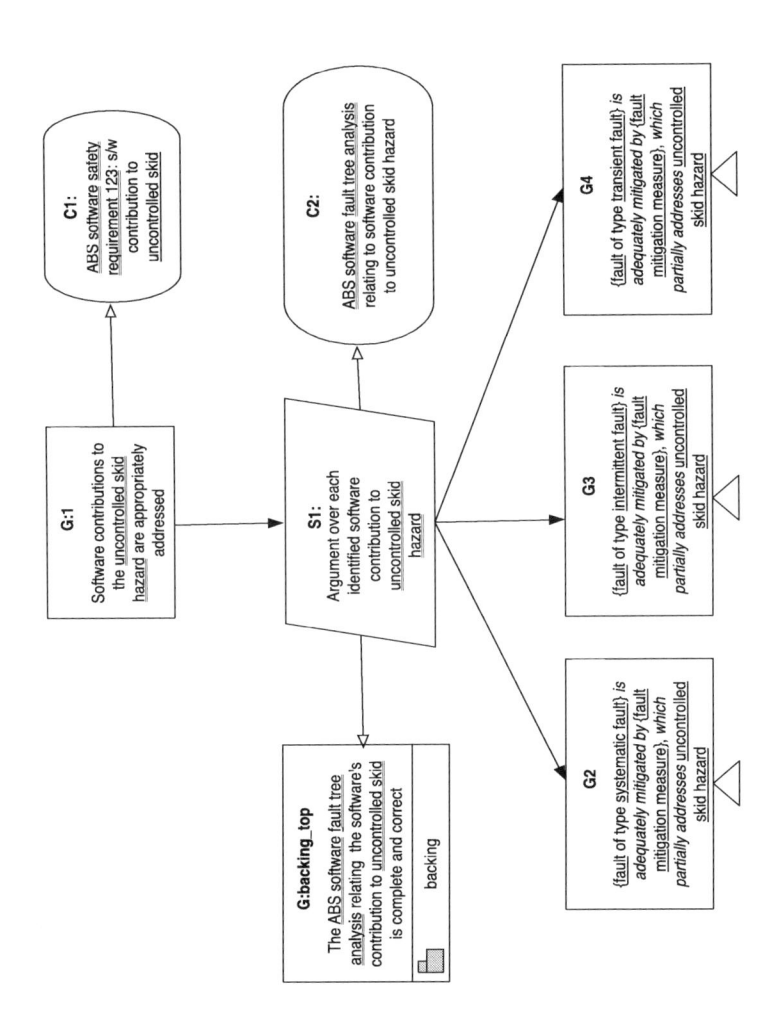

Fig. 5 Restatement of lower portion of software safety argument pattern

It will be clear that the first part of the claims in Goals G2, G3 and G4 are represented here as a less abstract *SentenceStructure* than in Figure 4, but that the *RoleBinding* to a definite *DeclaredObject* is not yet complete. Instead, they assert the relationship which is modelled between two *ConceptTypes* in the *Vocabulary*: "fault" and "safety measure". This relationship is captured in the *SentenceStructure* safety measure *mitigates* fault. The second part of the claim is generated directly from a *RoleBinding* linking a specific *DeclaredObject* to the ConceptType {Hazard} presented in Figure 4. Further traversals of the *ConceptTypes* presented in the *Vocabulary* are required to instantiate the *RoleBinding* for the first part of the claim, as shown in the partial instantiation of Goal G2 as two parallel goals addressing systematic faults:

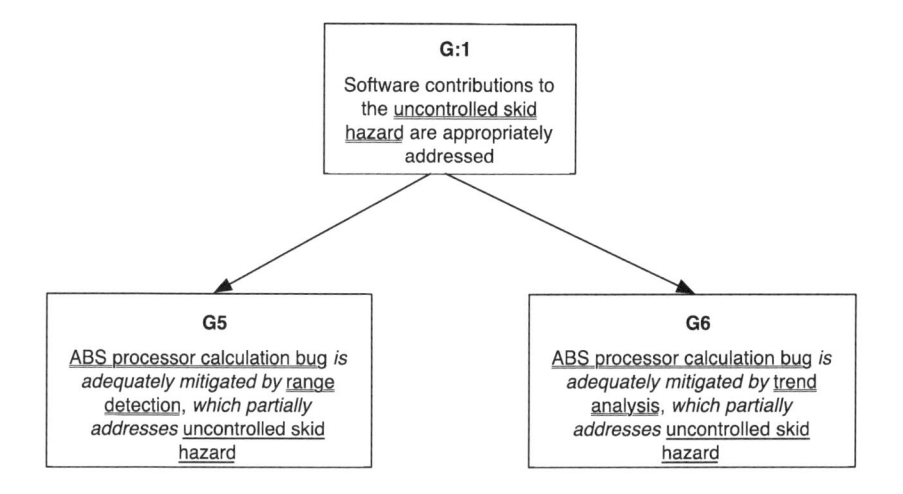

```
                    G:1
          Software contributions to
             the uncontrolled skid
          hazard are appropriately
                  addressed
```

```
         G5                              G6
ABS processor calculation bug is   ABS processor calculation bug is
adequately mitigated by range      adequately mitigated by trend
detection, which partially         analysis, which partially
addresses uncontrolled skid        addresses uncontrolled skid
hazard                             hazard
```

Fig. 6 Partial instantiation of Goal G2

Here, Goal G2 from Figure 4 has been instantiated twice, populated using instances of the "systematic fault" and "fault mitigation measures" *ConceptTypes* (a synonym for "safety measure") from the *Vocabulary* for the actual ABS system as *DeclaredObjects* fulfilling the *Roles* indicated in Figures 4 and 5. Note that the intention here is to show the population of the generic claim type using *DefinedTerms* from the vocabulary, rather than to present a complete argument.

5 Conclusion and Further Work

This paper has demonstrated the potential use of controlled vocabulary and structured expressions to add rigour to the language used to convey assurance arguments for safety-critical systems. We have presented metamodels for the creation of structured vocabularies, for controlled expressions and for cross-domain comparison of terminology, and have demonstrated how controlled expressions can be instantiated to inform the reuse of assurance assets. Work to develop this method and to provide tooling is currently at a relatively early stage. Further work is required to define structured vocabularies for the OPENCOSS target domains (rail,, automotive and aerospace) and to agree a common hierarchy of *ConceptTypes* which can be used to structure and compare them. In principle, assuming that there is a degree of consistency in the terminology used for *ConceptTypes* and *Definitions* across the safety-critical domains, automated support could be provided for the cross-domain "translation" of terminology. This could be achieved by providing information about the relative positions of source and candidate target terms in the concept hierarchy provided by the *ConceptTypes* (i.e. the conceptual distance between a *Term* and a generic concept and the nature of the mappings between interim terms) and the relative similarity of their *Definitions*. In practice, it is unlikely that such a technique could ever replace manually-asserted mappings between vocabulary terms: at most, they could provide a means for narrowing and prioritizing the list of candidate mappings. The technique could also be deployed to provide informed "translation" of generic assurance argument patterns into domain-specific versions, by the tailoring of vocabulary to suit the application context.

Acknowledgments The work presented here was carried out as part of the OPENCOSS Project, No:289011, which is funded by the European Commission under the FP7-ICR Framework. For further details, see the Project website: http://www.opencoss-project.eu.

References

Adelard (2014) Claims Argument Evidence Notation. http://www.adelard.com/asce/choosing-asce/cae.html. Accessed 30 September 2014
Attwood K, Conmy P, Kelly T (forthcoming) The use of controlled vocabularies and structured expressions in the assurance of CPS. Ada Users' Journal 2014
GSN Community (2011) Goal Structuring Notation Community Standard. Issue 1. http://www.goalstructuringnotation.info. Accessed 1 October 2014
Hawkins R, Kelly T (2010) A systematic approach to developing software safety cases. J System Safety 46: 25-33
Hawkins R, Kelly T (2013) A software safety argument pattern catalogue. University of York. ftp://ftp.cs.york.ac.uk/reports/2013/YCS/482/YCS-2013-482.pdf, Accessed 25 September 2014
Hawkins R, Kelly T, Knight J, Graydon P (2011), A new approach for creating clear safety arguments. In Dale C, Anderson T (eds) Advances in systems safety. Springer
IEC (2009) IEC 61508: International standard – functional safety of electrical/electronic/programmable electronic safety-related systems
ISO (2011) ISO 25964 International standard for thesauri and interoperability with other vocabularies.
ISO/FDIS (2011) ISO/FDIS 26262 International standard – road vehicles, functional safety
Kelly T (1998) Arguing safety – a systematic approach to managing safety cases. DPhil Thesis. University of York.
Lapore E (2009) Meaning and argument. John Wiley and Sons

Lloyd S (ed) (1982), Roget's Thesaurus of English words and phrases. Penguin
Maguire R (2006), Safety cases and safety reports: meaning, motivation and management. Ashgate.
Object Modelling Group (2008) Semantics of Business Vocabulary and Business Rules (SBVR). Version 1. http://www.omg.org/spec/SBVR/1.0. Accessed 25 September 2014
Object Modelling Group (2009) Ontology Definition Metamodel. Version 1. http://www.omg.org/spec/ODM/1.0. Accessed 24 June 2014
Object Modelling Group (2013) Structured Assurance Case Metamodel (SACM). Version 1. http://www.omg.org/specs/SACM. Accessed 3 October 2014
OPENCOSS Consortium (2011) http://www.opencoss-project.eu. Accessed 3 October 2014
RTCA (1992) RTCA/DO-178B: Software considerations in airborne systems and equipment certification
W3 Consortium (2009) Simple Knowledge Organization System (SKOS) Reference. http://www.w3.org/TR/2009/REC-skos-reference-20090818. Accessed 24 June 2014

Systematically Self-Reflecting Safety-Arguments: *Introduced, Illustrated and Commended*

Stephen E. Paynter

MBDA (UK) Ltd,

Bristol, UK

Abstract *It is contended that safety-arguments can be improved by being made "systematically self-reflecting" (SSR). A safety argument should systematically address its own quality at each point in its argumentation. This can be done by addressing specific questions at each argument node, without changing the structure of the argument. The approach has been applied to software safety-cases expressed using the Goal Structuring Notation (GSN), and lessons are drawn from this industrial experience. A comparison is made between SSR-arguments and Kelly's criteria for assessing assurance-cases, the 4+1 principles for software safety-cases and the confidence-arguments of Hawkins et al.*

1 Introduction

The ideas in this paper arose out of a reflection that an argument, A say, for some course of action, is neither as helpful nor as compelling as the same argument enriched with another argument, B say, that attempts to show that argument A is a sufficient argument to commend the course of action in question. This combined argument, $A + B$, can be called a ***self-reflecting (SR)*** argument.

A moment's reflection on the nature of argument B leads one to realise that argument B has to refer to argument A in order to comment upon it. There appear to be three different options as to how B can do this:
1. B may refer to A in an *ad hoc* manner, according to the structure of its own argument;
2. B may replicate the structure of A, in order to comment upon it;
3. Argument B may be distributed throughout argument A, commentating upon the quality of the argument at each point.

The first option suffers from the problem of being *ad hoc*, and hence of making it difficult for a reader or reviewer to assess whether every aspect of argument A has been established to be "adequate" by argument B. Furthermore, both the first two options are open to the problem of having to repeat argument A within them, creating an unwelcome problem of ensuring the two representations of argument A are consistent. This is a task that becomes increasingly onerous the more A undergoes modification after B has been developed. Therefore it is the third option which is pursued here, and it is the argument $A+B$ with such a structure which is called a *systematically self-reflecting (SSR)* argument.

An obvious application area for such insights is in structuring safety-case arguments. A safety-case argument marshals the reasons why a given product is safe to deploy. It presents a series of claims about the product in question, and ultimately points to the evidence which supports those claims. An SSR safety-case argument will have the same structure, but at each point will recursively reflect upon its own quality and the quality of the evidence that it rests upon.

This paper explains how the above insights were developed into a practical method when applied to safety-arguments expressed in the *Goal Structuring Notation* (GSN), (Wilson et al. 1995; Kelly 1998; and Kelly 2004), and it illustrates how they looked when applied to a safety-case for an embedded software system. An attempt is made to assess the experience of developing an SSR safety-argument, and to compare the SSR approach with other work on assessing or arguing the quality of safety or assurance-arguments.

2 Developing the idea of SSR Safety-Arguments

When safety-arguments are structured using the ideas from GSN, the argument consists of three principal kinds of nodes: *claims* (which are sometimes known as "goals"); *strategies*; and *evidences* (solutions), perhaps supported by nodes which define *context*, document *assumptions* and provide *justifications*. Typically, a claim will be either directly supported by evidence, or decomposed into sub-claims, which are combined according to a strategy to meet that claim, and where these sub-claims are established either by evidence or further argumentation, which only ultimately calls upon supporting evidence.[1]

Different issues impact the "quality" of an argument depending upon whether it is a node (claim) established by evidence or argument. The three basic issues identified in this work are:

1. Counter evidence to the argument's claim or claims;
2. The "assurance deficit" between what is claimed and what has been established;
3. Whether more could reasonably have been done to reduce the assurance deficit, if a deficit has been identified.

---

[1] The GSN considered here is used in a stylised way, where a claim is never established by a single claim (as this amounts to a mere re-wording of the higher-level claim), A claim is assumed to only ever be established directly by evidence or by multiple claims combined by a (single) strategy. These are considered to be sensible well-formedness constraints upon good GSN arguments.

The "assurance deficit" of a node not only depends upon the quality of the evidence it rests on, but also upon the logical validity of the argument itself. Another consideration is whether the assurance deficits of the sub-claims that support the argument are acceptable in combination. Clearly, in some circumstances, if every sub-claim is uncertain, too much uncertainty for the higher-level argument might be introduced, even if individually the assurance deficits would be tolerable.

These insights combine to raise a small number of questions for claims directly supported by evidence, and a similar but slightly different set of questions for claims established by an argument.

2.1 Quality Questions for Claims Directly Supported by Evidence

Consider a claim in a safety-argument that is directly supported by evidence, as illustrated in the Goal Structuring Notation (GSN) in Fig. 1. The argument is simply a claim (made in the rectangle) that the software in question has some property "P", purportedly established by some evidence (represented by a circle).

Claim N

The software has property P.

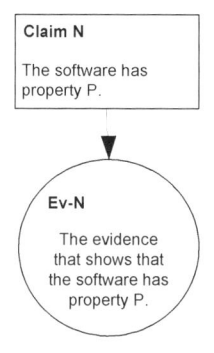

Ev-N

The evidence that shows that the software has property P.

Fig.1. A Basic "Evidence Node" Fragment in GSN

In an SSR-argument each node of this kind should be supplemented with evidence that addresses the following questions:
- Has counter-evidence been sought? Has any been found? If counter-evidence has been found, is it compelling? How has the discounting of this counter-evidence been justified?
- Has the assurance deficit of this evidence been considered? Is it tolerable?
- If there is a significant assurance deficit, have possible extra measures been considered? If such measures were identified, how have they been discounted, and is that discounting adequate?

How this information contributes to the overall argument is shown graphically in Fig. 2. The original argument is still present, but it is now enriched with supplementary claims and a higher-level claim which integrates them.

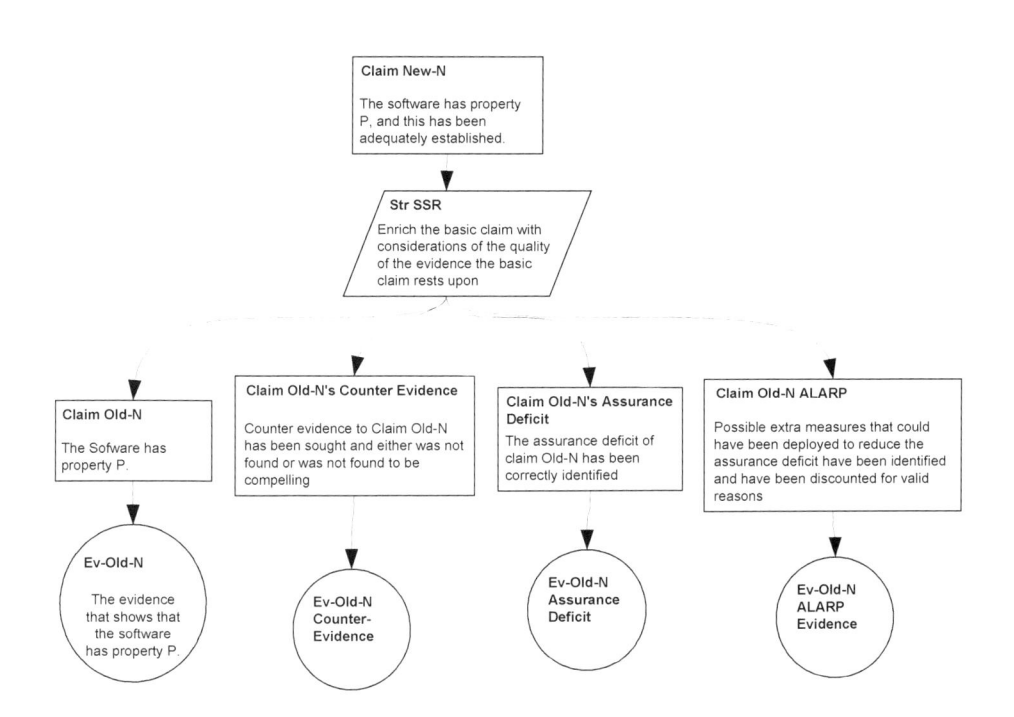

Fig. 2. An Enriched Self-Reflecting "Evidence-Node" Fragment in GSN

However, whether one actually enriches an argument in GSN in this way is a matter of style. When the ideas were applied to an industrial safety-case, the decision was made to leave the GSN unchanged, and simply to add three sub-sections to the safety-case report for each section documenting this kind of evidence-node. These sub-sections assessed the evidence, and, when appropriate, pointed to the external evidence that justified the discussion. This kept the safety-argument and the argument about the quality of the safety-argument separate, and stopped the quality-argument from swamping and obscuring the safety-argument.

2.2 Quality Questions for Claims Supported by an Argument

Similarly, consider a typical GSN argument node in a software safety-argument, such as is depicted in Fig. 3. Here the higher-level claim (claim N) is established by two (presumably) more specific claims, (sub-claims N1 and N2), combined using a strategy (represented in GSN by a parallelogram). The need for further argumentation to establish a claim is indicated in GSN by a diamond under the rectangle.

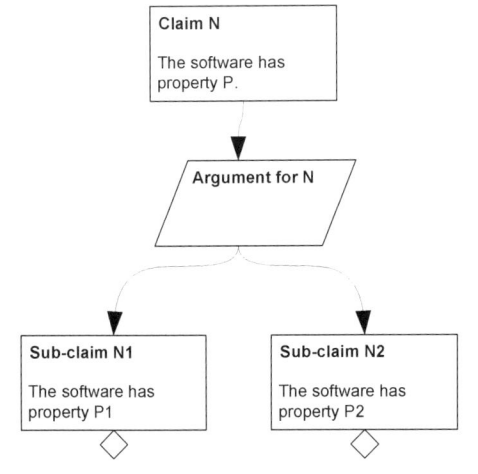

Fig. 3. A Basic "Argument" Fragment in GSN

In an SSR argument, each argument fragment of this kind should be supplemented with evidence which addresses the following questions:
- Has counter evidence been sought? Has any been found? If counter-evidence has been found, is it compelling?
- Are the sub-claims adequate to establish the higher-level claim?
- Are the assurance deficits (if any) for the sub-claims acceptable in combination?
- Are the sub-claims themselves adequately supported?

How this information contributes to the argument can be presented graphically in GSN, as shown in Fig. 4. Again, the original argument is still present, but it is now enriched with supplementary claims and a higher-level claim which integrates them, and the sub-claims are themselves re-cast (recursively) to make them self-reflecting.

As before, when these ideas were applied to an existing safety-argument, the original GSN was left unchanged and sub-sections were added to the safety-case report for each section documenting this kind of node. These sub-sections assessed the adequacy of the evidence and, when appropriate, referenced external evidence that justified this assessment.

3 Experience with an SSR Software Safety-Case

The ideas expressed in this paper have been applied by the author to a software safety-case for an embedded software system, which was used to control high-performance and semi-autonomous equipment with hard real-time constraints and safety-related functionality. A safety-argument had already been written for earlier versions of the system when these ideas about systematically self-reflecting arguments were developed. The first application of them, therefore, was to an existing argument. Subsequently, they have been applied by colleagues in MBDA to software intensive products with no existing safety-argument.

The following discussion is informed by this industrial experience. The aim has been to illustrate the approach being advocated, rather than to present the safety-arguments for public scrutiny.[1]

3.1 The Safety-Case Report Context

The software safety-case arguments were captured in reports. After introductory sections, placing the safety-related software in context (i.e. describing the system design; the safety requirements; and the software design), these reports had a section containing the safety-argument. This was divided into sub-sections, each dealing with a claim or sub-claim. For each such claim the relevant fragment of GSN was included. As described above, each of these sub-sections were further sub-divided into sections which addressed the "systematically self-reflecting" claims for that argument fragment. For consistency and ease of navigation:

---

[1] Details of these systems and their safety-cases are not available in the public domain.

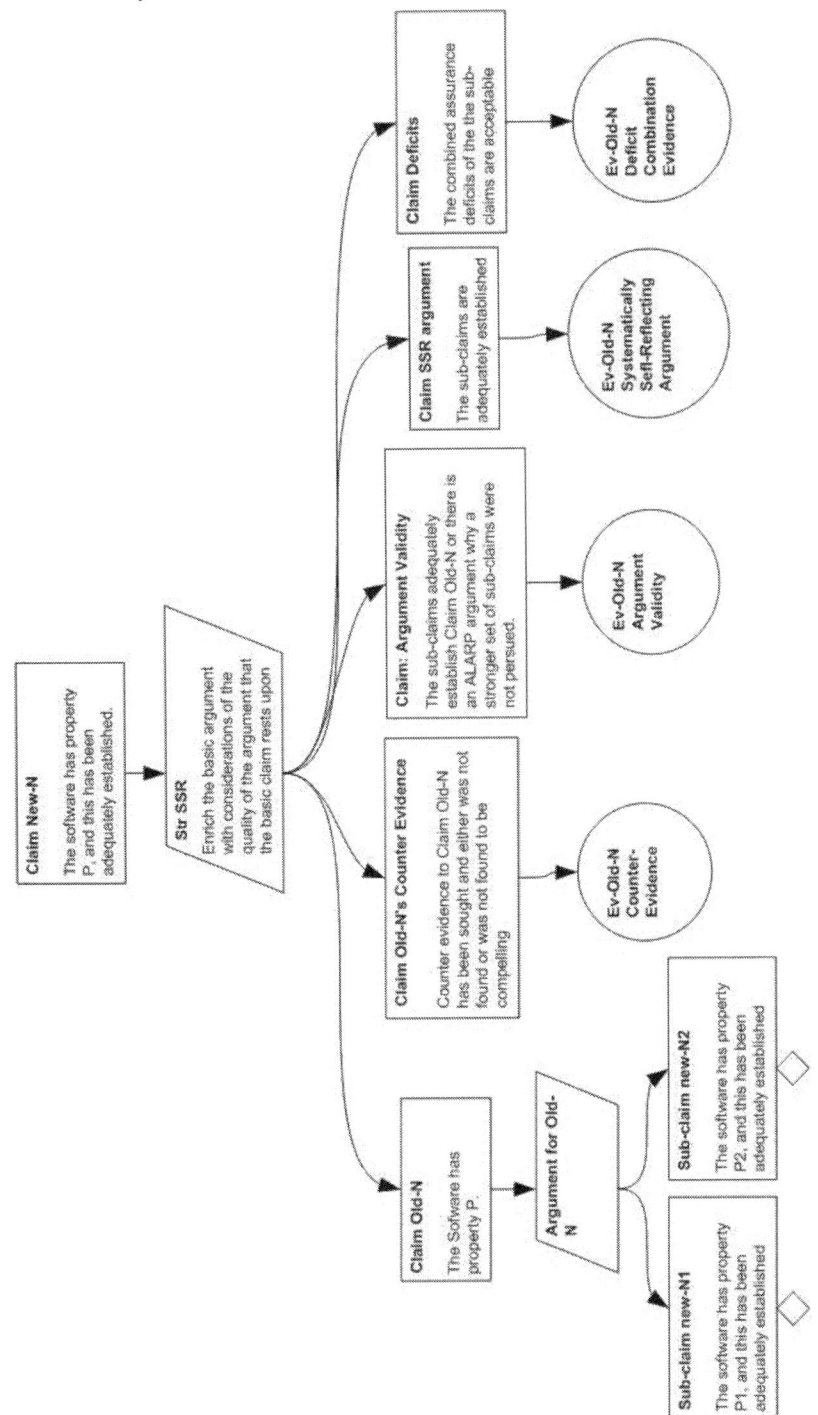

Fig. 4. An Enriched Self-Reflecting "Argument" Fragment in GSN

- the report adopted a depth-first, left-to-right, presentation order for the claims;
- Each section dealing with a different claim was made to start on a different page; and
- Cross-references to the section dealing with each claim were included in the section dealing with the higher-level argument fragment.

3.2 Counter-Evidence

It is important that counter-evidence to any claim be sought, for safety-arguments can all too easily degrade into the selective presentation of evidence and arguments to support a predetermined course of action (the deployment of the potentially dangerous system) without consideration of whether there are good reasons not to deploy the system (as it currently stands). Furthermore, it is not adequate to claim that '*no counter-evidence is known*', for it might be the case that none was sought and collected. It is necessary to show that counter-evidence was actually sought.

It was found that for numerous terminal nodes of a safety-case this question was addressed by the very nature of task of developing the evidence which supports the claim. For example, when the claim is '*No significant anomalies were detected in the software by testing*' and the evidence is the Test Report, the very nature of a Test Report, being compiled from the results of testing the software, is that it would involve a check for anomalous test results. In such a case, the subsection dealing with counter-evidence ends up containing a sentence to the effect that: '*The collection of the evidence to support the claim necessarily involved looking for counter-evidence, and given that the evidence supports the claim, no compelling counter-evidence was found.*'

When counter-evidence has been found, its presence is, of course, evidence that it has been sought. It is then necessary to argue that the counter-evidence is not compelling. The form of this argument will crucially depend upon the details of the claim and the counter-evidence in question. For example, if it is the case that static analysis of the software has revealed certain anomalies in the code, it might also be true that subsequent analysis of these static analysis results has shown that the anomalies identified are justifiable, and do not compromise the safety-functionality of the software.

At certain nodes the evidence that counter-evidence has been sought for the claim may be that counter-evidence has been sought for each of the sub-claims that the argument asserts establishes the higher-level claim. This can be indicated by a sentence to the effect that: '*The checks for counter-evidence to the sub-claims are checks for counter-evidence to this claim.*'

However, on other occasions, it may be possible to seek direct evidence that could undermine a claim. An example is the claim '*all the software safety requirements have been satisfied in the software solution*', where the sub-claims address whether:
- every software safety requirement is traceable to code via the relevant design element or elements;
- the software design architecture supports the software safety requirements, and has been correctly implemented;
- the safety-related software cannot be credibly interfered with by other (non safety-related) software.

Although each of these sub-claims should be supported by a claim that counter-evidence has been sought for it specifically, direct counter-evidence to the higher claim may also be sought. For example, if a project has a process for capturing problems and queries (P&Qs) against the requirements or design, which can be raised at any stage in the development, these could be checked to ensure that no outstanding P&Qs involving software safety-requirements or design elements remain outstanding.

It is observed that the presentation of counter-evidence can be marshalled into an adversarial counter-argument, as the author has contended for elsewhere, (Armstrong and Paynter 2004). While, no doubt, that has merit, it goes beyond what is being advocated here.

3.3 Logical Strength

The logical strength or validity of an argument is important to assess, as it is easy to construct arguments where the sub-claims in combination do not amount to a justification of the higher-level claim. This may be because the higher-level claim asserts more than is needed for the overall argument, or it may be because more strands of argument (or stronger sub-claims) are needed to warrant the needed higher-level claim.

It is a matter of style whether one attempts to support a demanding higher-level claim with a set of fully warranted sub-claims, which in total do not amount to a complete case for the higher claim, or whether one has stronger sub-claims, which together satisfy the higher-level claim, but where some or all of the sub-claims are not themselves fully warranted by the evidence. In other words, where some or all of the sub-claims have significant assurance deficits associated with them.

As a general principle, it is often clearer for the reader if either one weakens the top-level claim or one adopts stronger sub-claims, and subsequently one makes explicit the assurance deficits in the evidence that supports them, than it is to have an incomplete argument. Generally, weaknesses in arguments can be harder to appreciate and understand, even when pointed out. However, in cases where the available technologies are not able to amass enough evidence it is sometimes necessary to use a weak argument, and point out the weakness of the argument. This typically occurs when the higher-level claim is abstract, and does not have a precisely defined meaning.

Consider the top-level claim that is sometimes made: '*The software is adequately safe*'. However one argues this, there will almost certainly be an assurance deficit in the argument, and serious consideration should be given to whether the overall system safety-case could rely instead on the weaker claim, '*all the software safety requirements are satisfied by the software*'. The higher levels of the (system safety-case) argument might have to be extended to capture statements about the adequacy of the software safety requirements, but the resulting argument is more explicit and is therefore likely to be more plausible.

An example of the option of keeping an inadequate argument, but explicitly pointing out the weakness of it, is when one wants to claim that the software construction process is sound. One might point to the quality of the design to software mapping, the fact that the mapping has been consistently followed; the fact that the design tools used to generate software have been qualified and extensively tested; and that the hand-written software is unable to break the design. However adequately these sub-claims are supported, it will remain an open question whether these (or some other set of) sub-claims are adequate to justify the claim that the software construction process is 'sound'. However, the explicit discussion of the assurance deficit provides an opportunity for the author of the safety-case to articulate the rationale for the approach taken.

3.4 Assurance Deficits

The idea of *assurance deficits* in safety-cases has gained currency in the safety engineering community. Notably, work on safety-case patterns that come out of the MoD funded Software Systems Engineering Initiative (SSEI) at the University of York, emphasised the need to address "assurance deficits", (Hawkins 2009) and (Menon et al. 2009).

When a claim in an argument is supposedly directly supported by evidence, it is necessary to reflect upon whether the evidence referenced is strong enough to establish the claim, and if it is not, to gain some understanding of the "assurance deficit" that remains. The terminology of "assurance deficits", however, should not lead one to expect a numerical value for the deficit, as such deficits are rarely quantifiable and are often subject to engineering judgment. Furthermore, one will not have a calculus for combining disparate assurance deficits and assessing their combined acceptability. The subjective nature of answers to these questions does not mean that they should not be considered. Reflecting on such deficiencies is likely to drive engineering advancement and excellence.

An interesting example is the claim that '*the object-code has been adequately tested*'. Part of the meaning of what "adequately" means may be that '*a high coverage measure of the object code has been achieved by the testing*'. However, sometimes a direct measure of coverage is not possible, perhaps because of the real-time nature of the software means that the distortion due to the inserted code to log the coverage will be significant, or because the non-intrusive collection of logging data is not available for the embedded target environment being used. In that case, one might have to rely on indirect evidence, such as the coverage achieved in source code testing, the fact that the same tests are applied on the target, and the fact that the compiler's object-code (for the constructs used in the source code) has been scrutinised, and has not been observed to introduce unexpected paths.

Clearly, this set of claims cannot support any particular coverage claim for the object-code. There is an "assurance deficit" with this evidence. However, there is no way to quantify this deficit, and the significance of the shortfall is a matter of engineering judgement. This is not uncommon with assurance deficits.

It is noted that the approach to assurance deficits advocated in the SSEI standard of best practice, (Menon et al 2009, page 174), is problematic. It argues that assurance deficits be characterised as *intolerable*, *broadly acceptable*, or *tolerable*, and provides a GSN argument pattern for recording the extra work that was done to turn an intolerable assurance deficit into one of the others. Here it is noted that safety-case arguments do not need to focus on some arbitrary point earlier in the development of the system and discuss the subsequent adoption of extra development tasks which produced more evidence of the product's safety. Rather the safety-case is responsible for recording the evidence that there is for the safety of the system that is to be deployed, and it should identify any remaining assurance deficits with that evidence. Almost by its very nature, a safety-case argument that is arguing that the system can be safely deployed must be claiming that the assurance deficits are at worst tolerable.

In the project to which these ideas were initially applied, the following sentence was consistently used when the evidence supported the claim without any deficit: '*There does not appear to be any assurance deficit with this evidence.*' This was typically the case when the claim was focused and factual, and often this was when the claim concerned the fact that some analysis or development task had been performed or reviewed, and the output from the task was proof that it had been performed. When there was no apparent deficit, but the nature of the claim and evidence was conceptually less straightforward, the slightly weaker sentence, '*There does not appear to be any significant assurance deficit with this evidence*' was systematically adopted instead. No claim is made that this wording is ideal, but the practice of making a systematic distinction between these cases is commended.

3.5 Mini-ALARP Arguments

Arguments that the safety risk has been reduced "*as low as reasonably practicable*" have an important role in safety engineering in the United Kingdom, (HSE 2001) and (Redmill 2010). The aim of this section is not to defend the validity or coherence of ALARP arguments,[1] but to point to the fact that SSR safety-arguments encourage ALARP arguments to be embedded in a safety-case at each step of the argument. In other words, rather than one safety-argument, and a global ALARP argument about the engineering approach taken and the perceived risk remaining, SSR-arguments foster the introduction of ALARP considerations at each point in the argument.

Arguably this is exactly how it should be. The question of whether a risk has been reduced as low as is practicable is not a global question about the system and its development, but questions about a myriad different design and development decisions which have gone into the system's development.

For example, the assurance deficit in object-code test coverage discussed above, calls for a discussion of the issues involved in generating a more direct measure of object-code coverage, and an assessment of whether such approaches are technically feasible (given the project and product environment) and a determination of the effort, cost and timescales involved in overcoming any technical limitations, and an assessment of how significant a reduction in the assurance deficit would result, and whether that is justified, given the nature of the hazards one is trying to protect against.

As should be clear, the approach of SSR-arguments is not restricted to any particular approach to arguing ALARP properties. One's favoured ALARP argument pattern may be instantiated and used as appropriate.

3.6 The +1 Principle

It has recently been argued that a software safety-case argument should be based on what have been called the "4+1 principles", (Hawkins et al. 2013a) and (Hawkins et al. 2013b). The first four principles call for such arguments to address the ways that the software might impact the safety of the product. In particular, they call for the following issues to be addressed:

1. that software safety requirements address the software contribution to system hazards;
2. that the intent of the safety requirements is maintained throughout their decomposition;
3. that it is shown that the software satisfies its safety requirements; and
4. that any hazardous behaviours are identified and mitigated.

The +1 principle differs in kind from the first four (this is why it is not called a fifth principle). It calls for the other principles *to have been established in a way which generates confidence commensurate with the system risk.*

It would seem that making a safety-argument systematically self-reflecting is a way of ensuring and demonstrating that this +1 principle has been consistently

---

[1] With J.M. Armstrong, the author attempted a brief deconstruction of the concept of ALARP in (Armstrong and Paynter, 2003).

applied to the safety-argument in question. The mere adoption of an SSR approach does not address the questions concerning how risk is to be characterised, but it suggests a way of considering and documenting the issue, whatever approach is adopted.

4 Consideration of an Objection

It might be argued that it is the reviewers or readers of a safety-argument who are responsible for determining for themselves whether or not the argument in question is adequate, not for the writers of the safety-argument to decide the matter for themselves.

That, of course, remains the case. A systematically self-reflecting safety-argument, however, assists in making explicit the issues that a reviewer ought to consider and does so systematically, which even a diligent reader may fail to do. Furthermore, it is an aid to the developer of the safety-case to have a framework which encourages the systematic consideration of the quality of the argument and evidence being presented. The experience is that this will improve the quality of the safety-argument and evidence being presented, as it will encourage the developer of the argument to modify it until it can be justified.

5 Comparison with Kelly's Criteria for Reviewing Assurance Cases

T.P. Kelly has written independently on the criteria for reviewing assurance cases, (Kelly 2007). It is illuminating to reflect on the relationship between them.

Firstly, Kelly advocated early review (by the certification authorities): '*If there are problems with the arguments and evidence being offered up by the assurance case it is desirable to find this out as early as possible in the lifecycle.*' SSR-arguments encourage the consideration of such concerns as the argument is developed, although not necessarily by external authorities.

Secondly, Kelly advocates a staged review process, including argument (1) comprehension; (2) well-formedness; (3) expressive sufficiency; and (4) criticism and defeat checks. The first three are basically covered if GSN is used well. Kelly identifies six attributes which are pertinent to the fourth, argument criticism. These are: *coverage; dependency; definition; directness; relevance;* and *robustness*.

Coverage concerns the extent to which the argument and evidence "covers" the conclusion. This is addressed in SSR-arguments by consideration of the logical strength of argument-nodes, and the assurance deficits of evidence-nodes.

Dependency concerns whether a lack of independency undermines the evidence or argument. This issue is not directly addressed by SSR-arguments, but it would be natural to deal with it when consideration is given to the logical strength of argument-nodes which rely upon independency (for not all do.)

Definition concerns overly constrained arguments or evidence. This is addressed only implicitly in SSR-arguments, where consideration is given to the logical strength of arguments and the assurance deficit of evidence. In a GSN argument where context-nodes are present, this issue should also draw attention to the adequacy of their definition.

Directness concerns whether evidence establishes a goal directly or only indirectly, and *relevance* concerns the pertinence of the evidence to the goal. To the extent that these issues are addressed in SSR-arguments, these are issues relating to the assurance deficit associated with evidence-nodes.

Finally, *robustness* concerns the fragility of the argument to possible changes in evidence and consequent claims. This issue has not been to the fore in the development of SSR-arguments so far, but such a review could be added to the SSR-argument schema, if required.

Kelly proceeds to discuss auditing the evidential basis of the argument. In particular, he identifies issues of the "buggy-ness" of the evidence; the level of review it has been subject to; and either the experience of the personal creating the evidence, or the assurance that there is in the tools used to create the evidence. These issues would naturally be addressed in assessing the assurance deficit associated with evidence-nodes in SSR-arguments.

Kelly ends by discussing the rebuttal and undercutting of an argument which leads to its "defeat". This issue is systematically addressed in SSR-arguments through the checks for logical soundness and counter-evidence at each argument node.

There are two issues that SSR-arguments encourage to be treated, which are not explicitly addressed in (Kelly 2012). The first is consideration of whether the assurance deficits of nodes are acceptable in combination; and the second is consideration of whether more evidence or tasks could have realistically been done to strengthen the argument.[1]

Arguably, a synergistic combination of Kelly's criteria and an SSR-argument structure should work together to produce a strong argument that guides a reviewer through a systematic assessment of it.

6 Comparison with the "Assured Safety-Arguments" of Hawkins *et al.*

There have been other attempts to apply to safety the idea that an argument is more compelling if it is supported by another which argues that the first is sufficient. In independent work, (Hawkins et al. 2011), have argued that to achieve

---

[1] Arguably, this last issue is implicitly addressed in Kelly's "coverage" attribute. However, it is not described in terms of localised ALARP argumentation, as in SSR-arguments.

assured safety-arguments, one needs to have safety-arguments which are supported by *confidence-arguments*. Assured safety-arguments have many similarities with the SSR-arguments described here, but with intriguing differences.

(Hawkins et al. 2011) advances a number of reasons as to why confidence-arguments should be developed separately from the safety-argument, including: preventing the combined argument from becoming too large and unwieldy; poor structure for both arguments, and possible omissions and weaknesses overlooked, due to the arguments mutually obscuring each other; loss of focus; inclusion of unnecessary material; and the difficulty of reviewing the arguments, due to their size and lack of focus.

In reply, it is felt that SSR-arguments escape many of these pitfalls because there is a clear distinction between the safety and confidence parts of the argument (especially when the safety argument is captured in GSN, and the `self-reflecting' part is captured in report sub-sections relating to each node). The result is neither much larger than the safety-argument by itself, nor too unwieldy; the structure of the safety-argument is not obscured, and the self-reflecting arguments are treated systematically and clearly, making any omissions and weaknesses noticeable. Furthermore, it is not felt that the danger of loss of focus or the inclusion of unnecessary material is elevated, and that, on the contrary, SSR safety-arguments are easier to review than safety-arguments by themselves.

Confidence-arguments are linked with their corresponding safety-argument in (Hawkins et al. 2011) by reference to the "assurance claim points" (ACPs) of the safety-argument. This allows the safety-argument to be clearly referenced in the confidence case, but requires each reader to convince themselves that the two arguments are in-step.

An important insight of (Hawkins et al. 2011) is that there are ACPs associated with *context* nodes, and, although not mentioned, one may assume with *assumption* and *justification* nodes too. Ideally, in SSRs these issues would be addressed when considering the logical strength of argument-nodes which have such nodes. However, nothing prevents one extending the SSR approach to address explicitly the adequacy and correctness of such ancillary nodes.

Confidence-arguments in (Hawkins et al. 2011) are based on qualitative *assurance deficits*, as are SSR-arguments, and they also encourage the consideration of ALARP-like arguments regarding what are called in (Hawkins et al. 2011) "residual assurance deficits", and here just "assurance deficits". (Kelly 2007) also addresses the search for counter-evidence, but associates that with the residual assurance deficits, rather than with the claims in the safety-argument.

The overall structure of confidence-arguments advocated in (Hawkins et al. 2011) has three legs[1], arguing confidence in all inferences; all contexts; and all evidences. Clearly, a complete confidence-argument ought to argue all these things, but in structuring the argument around different kinds of nodes, important inter-related issues become separated. In contrast, a SSR-argument addresses the

---

[1] However, if ACPs associated with *assumption* and *justification* nodes are also recognised, perhaps there should be *five* legs to such arguments.

same scope through a systematic consideration of each node, and, by being integrated into the structure of the safety-argument, keeps inter-related issues together.

In spite of these specific differences, there remain much commonality between *assured safety-arguments* and *systematically self-reflecting safety-arguments*, and (Hawkins et al. 2011) contains much wisdom which can be directly applied to SSR-arguments, as indeed does (Ayoub et al. 2012) which builds upon (Hawkins et al. 2011).

7 Conclusions

Systematically self-reflecting arguments have been introduced, and it has been argued that they should be applied to safety-arguments. Drawing upon an application of the ideas to software safety-cases, it has been concluded that:
1. Constructing an SSR-argument was not too much work to be practical.
2. Turning an unreflecting safety-argument into an SSR one is practical, and does not necessarily lead to changes in the structure of the original safety-argument (except, of course, where the extra reflection reveals the original argument to be weak and in need of modification or supplementation.)
3. SSR-arguments help to focus ALARP considerations, and properly treat ALARP as concerning a lot of focused issues about the technologies adopted and the design practices used, at the appropriate point where the technologies in question impact the safety-argument.
4. SSR-arguments encourage the producer of the safety-argument systematically to ask the questions that a reviewer should be asking, and that therefore, they encourage deficiencies to be removed before the reviewers are presented with the case. In some cases, it might reveal weaknesses in the proposed engineering practices being deployed, and so encourage more rigorous techniques or more extensive or focused evidence to be gathered about the product and how it was developed.
5. SSR-arguments encourage the explicit search for counter-evidence and the documentation of the outcome of this search.
6. SSR-arguments should be easier for a reviewer to assess because they should be able to see that the right questions have been asked and addressed, and should the answers to these questions be weak, this fact is likely to be more obvious as they will have had to have been made explicit in order to be articulated.

This positive assessment of making safety-arguments systematically self-reflecting is predicated upon the idea that the role of a safety-case is to make as clear as possible the argument why a product should be deployed, and to draw attention to the ways that the argument might be weak. In this way, the parties involved are able to make informed decisions about whether or not to accept the argument, deploy the system, and accept any residual risk.

Finally, it is inevitable when discussing "self-reflection" that self-aware meta-considerations be addressed: in particular, recognising that a SSR-argument might itself be the subject for systematic self-reflection, should SSR-arguments be systematically reflected upon? However, this way lies an infinite regress, and engineering pragmatism and the law of diminishing returns must hold sway. Nevertheless, this paper attempts to justify the systematic self-reflecting approach being proposed. As you have read, its conclusion is that it has some benefits over an unreflecting approach and that, therefore, future safety-arguments should be systematically self-reflecting. As a reader, you are left with the task of assessing this claim and the case for it.

Acknowledgments This work was funded by MBDA (UK) Ltd., although the opinions expressed are the author's. The help of various colleagues are acknowledged in developing the ideas of this paper, including: Jim Armstrong; Bob Born; Dave Nuttall; Alex Matthews and James Doulton. Thanks are due to Tim Kelly for pointers to the literature on *assured safety-arguments*.

References

Armstrong J, Paynter S (2003) Safe Systems: Construction, Destruction and Deconstruction. In: Redmill F, Anderson T (eds) Proceedings of the 11th Safety-critical Systems Symposium (SSS'03), Bristol, Springer, pages 63-76.

Armstrong, J, Paynter S (2004) The Deconstruction of Safety Arguments through Adversarial Counter-Argument. In: Heisel M, Liggessmeyer P, Wittmann S (eds), Computer Safety, Reliability and Security, 23rd International Conference, SAFECOMP 2004, Springer, LNCS 3219, pages 3-16.

Ayoub A, Kim B, Lee I, Sokolsky O (2012) A Systematic Approach to Justifying Sufficient Confidence in Software Safety Arguments. In: Proceedings of the 31st International Conference on Computer Safety, Reliability and Security, SAFECOMP 2012, LNCS 7612, Springer

Hawkins R (2009) Software Systems Engineering Initiative: A Systematic Approach to Software Safety Argument Construction. SSEI-TR0000023, January

Hawkins R, Habli I, Kelly T (2013a) The Principles of Software Safety Assurance. In: 31st International System Safety Conference, Boston, Massachusetts, August

Hawkins R, Habli I, Kelly T (2013b) Principled Construction of Software Safety Cases. In: 2nd Workshop on Next Generation of System Assurance Approaches for Safety Critical Systems (SASSUR), Toulouse, France, September

Hawkins R, Kelly T, Knight J, Graydon P (2011) A New Approach to Creating Clear Safety Arguments. In: Proceedings of the 20th Annual Safety-critical Systems Symposium (SSS'11), Southampton

HSE (2001), Reducing Risks, Protecting People. Health and Safety Executive, HSE Books

Kelly T (1998) Arguing Safety - A Systematic Approach to Managing Safety Cases. DPhil Thesis, Department of Computer Science, University of York, September

Kelly T (2004) A Systematic Approach to Safety Case Management. In: Proceedings of SAE 2004 World Congress, Detroit, March

Kelly T (2007) Reviewing Assurance Arguments - A Step-by-Step Approach. In: Proceedings of Workshop on Assurance Cases for Security - The Metrics Challenge, Dependable Systems and Networks (DSN), July

Menon, C, Hawkins R, McDermid J (2009) Software Systems Engineering Initiative: Interim Standard of Best Practice on Software in the Context of DS 00-56, SSEI-BP000001, Issue 1, August

Redmill F (2010) ALARP Explored, Technical Report. CS-TR-1197, University of Newcastle, March

Wilson S, Kelly T, McDermid J (1995) Safety Case Development: Current Practice, Future Prospects. In: Proceedings of the 12th CSR Workshop, Springer-Verlag

A Case Study of Security Case Development

Benjamin D. Rodes, John C. Knight, Anh Nguyen-Tuong, Jason D. Hiser, Michele Co, and Jack W. Davidson

University of Virginia

Charlottesville, VA, USA

Abstract   *Security concerns that arise in safety-critical domains, such as air-traffic control and energy system management, might be analyzed using rigorous security cases. Such analysis has been explored minimally. We present a case study of the application of rigorous security arguments to a novel approach to thwarting command-injection attacks based on transformation of the binary form of a program (no source code). The case study illustrates an approach to security argument structure, the construction process for the argument, and organizes and reveals the defensive capabilities of the technique and its limitations thereby demonstrating the power of security argument.*

1 Introduction

Safety cases have been studied extensively for many years, but the application of assurance case technology to security has been explored far less thoroughly (Alexander 2011). To gain insight into the practical application of security cases, we conducted a case study in which we developed a prototype security case for a particular security technology designed to defend against a particular type of attack. Rather than develop a security case for a specific system, we developed a prototype security case for the defense technology. This activity was an exploratory exercise, but the prototype could be further developed for any system employing that technology.

Despite extensive research and the deployment of new tools and techniques, successful security attacks against high-value targets continue to take place with alarming frequency. A major security concern in modern software engineering is the reliance of systems on software for which minimal information is available about the development and quality of the software (Goertzel et al. 2007). Such software is often referred to as *Software Of Unknown* (or *Uncertain*) *Provenance* (or *Pedigree*), or *SOUP* (Redmill 2001). Explicit and complete attention to security properties is not necessarily guaranteed for SOUP, yet SOUP is used frequently because of the associated economic benefits (Goertzel et al. 2007).

In response to uncertainties about the security properties of SOUP, researchers have developed numerous SOUP modification techniques to enhance SOUP security (Hiser et al. 2012, Kc et al. 2003, Pietraszek and Berghe 2006, Rodes et al. 2013). Ostensibly, every software security enhancement is accompanied by a rationale for the technique and some form of evidence to indicate that the technique is useful, although typically the rationale is ad hoc, incomplete, and informal.

While relying upon this type of rationale might provide a general justification that a given security enhancement has some utility, users wishing to apply the technique lack a reasoning structure for determining how useful the technique is for their specific goals. A security case provides that structure. In this paper, we examine the application of security cases to a security-enhancing SOUP modification called *Software DNA Shotgun Sequencing*, or S^3. S^3 is designed to thwart command-injection attacks by transforming a binary program based only on information from that binary program (Nguyen-Tuong et al. 2014).

Using the development of the security case for S^3, we describe how the security case is structured, document the benefits and weaknesses of the S^3 modification technique that are revealed by the security case, and introduce a minor enhancement to argument semantics to allow additional flexibility in argument structures.

2 Concept of Security Arguments

When engineering software systems, developers have the burden of demonstrating assurance that the software establishes properties and characteristics desired by the system's stakeholders. For example, stakeholders might require assurance that the software is adequately safe, secure or reliable for use in the intended operating environment. The challenge, therefore, is to demonstrate that the system has the property or properties desired by the stakeholders.

The complexity and size of modern software systems makes providing definitive, complete, and irrefutable proof that any particular software system has any significant property impractical for all but the most basic and trivial software. In practice, developers rely on available evidence, possibly derived from the application of a standard, to infer the existence of the desired properties.

A software assurance argument documents the rationale for belief in a crucial claim about a software system, referred to as a *top-level* goal. In a goal-structured assurance argument, the top-level goal is decomposed into sub-goals recursively until each goal is supported directly by evidence. The premise of an assurance argument is that the goal structure combined with the evidence provides a valid and compelling rationale for belief that the system has the property specified in the top-level goal. To date, assurance arguments have been applied most widely in the form of *safety cases*. The special case of a top-level goal stating a security property is a *security case*.

4 Target of Evaluation: S^3

Software DNA Shotgun Sequencing (Nguyen-Tuong et al. 2014) is a novel approach to thwarting command-injection security attacks. S^3 is designed to operate with no information about the software other than the binary form. The concept is to make transformations to the binary program to prevent exploitation of vulnerabilities that might be present but are unknown.

Vulnerabilities that can result in *operating system* (OS) command injection attacks (a special case of command injection attacks) are the second entry in MITRE's 2011 CWE/SANS list of top 25 most dangerous software errors (Mitre and SANS 2011). Operating system commands are the means whereby programs make requests for operating system services, including file and network operations. If an attacker can compromise an application and issue arbitrary commands to the underlying operating system, the damages could be extensive. A network-facing server running with high privileges could be attacked, for example, with the potential for the loss or leakage of extensive sensitive data.

The OS commands of interest are those that contain parameters derived from outside the software. For example, an application can issue an OS command to output the contents of a file identified by the user:

```
char[MAX_LEN] cmd = "/usr/bin/cat";
fgets(file, BUFFERSIZE, stdin);
...
strcat(cmd,file);
system(cmd);
```

If instead of providing a file name, the user provides "; rm –rf /", cat will execute with no file, and consequently fail, followed by the rm command. The rm command would proceed to delete the entire root partition of the file system if the software were running with administrator privileges.[1]

Clearly, the solution cannot be to deny all OS commands. Instead, many modern defenses rely upon *taint analysis*, i.e., determination of whether information to be used in an OS command can be trusted. Prior to issuing a security-sensitive OS command, the command is first checked against its taint markings to ensure that critical parts of the command are not tainted.

Taint analysis can be based upon either (1) *positive* taint in which trusted data is analyzed or (2) *negative* taint in which untrusted data is analyzed. Analysis is performed either by (1) *tracking* the flow of data (and hence taint) from an external source as the data propagates through a program to a security-sensitive operation, or by (2) *inference* in which the taint is inferred in some way.

---

[1] We note that the dangerous and common SQL injection attacks in which malicious parameters from outside the software are included in a command to an SQL database are non-OS command injection attacks. The techniques described here have been adapted for SQL injection attacks.

Applying existing taint tracking techniques to binary programs is problematic, because the execution-time overhead can be prohibitive (Bosman et al. 2011). The goal for S^3 is to provide effective, low overhead taint analysis that can be applied to binary programs. The approach used by S^3 is positive taint via inference. Prototype implementations of S^3 have been developed for binary programs running on Intel's X86 architecture for both OS command injection attacks and SQL injection attacks. In this work, we focus on the application of S^3 for OS command injections specifically.

4.2 The S^3 Approach

The S^3 approach is summarized here. Full details are available from Nguyen-Tuong et al (Nguyen-Tuong et al. 2014).

S^3 is structured as five major components. The *DNA Fragment Extraction* component extracts string literals, i.e., DNA fragments, from the binary program and the associated libraries. This analysis is done once, prior to program execution, and the analysis time is not part of the execution-time overhead. The *Command Interception* component intercepts security-critical commands generated by the subject program during execution so that the commands can be examined.

By matching the command against the extracted DNA string fragments, the *Positive Taint Inference* component determines which characters in the intercepted command should be trusted. Any unmatched character is deemed untrusted. Using DNA fragments native to the software to infer taint is a novel form of taint inference and one of the key contributions of the S^3 architecture.

The *Command Parsing* component parses an intercepted command to identify critical tokens and keywords (i.e., components of the command that can be used to achieve a malicious purpose). The *Attack Detection* component combines the output of the Positive Taint Inference and Command Parsing components to determine whether an attack has occurred. A command is deemed an attack if a critical token or keyword is not marked as trusted (i.e., the critical token or keyword is not found within the extracted set of fragments).

If S^3 detects an attack, the command is either rejected outright or altered before being passed to the operating system. The current prototype implementation simulates a failed command invocation by substituting an error code in place of the actual command.

S^3 meets the goal of operating with low overhead, but what about S^3's efficacy? The detection of malicious commands is based on inference, and two types of failure are possible: (1) a benign command could be inferred to be an attack, i.e., a *false positive*, or (2) a malicious command could be inferred to be benign, i.e., a *false negative*.

The inference in S^3 is based on *speculation* (Rodes and Knight 2014). The core of S^3 is that positive taint inference based on the set of strings recovered from the subject binary program is correct. The two types of failure noted above arise because the speculation in S^3 is imperfect. The rate of false positives and false negatives depends upon the algorithms used in S^3 and on the specific program to which they apply. The crucial importance of the security case for S^3 (and any technique that relies upon speculation) is to provide a framework for judgment about the *accuracy* of the speculation. There is no way to guarantee the absence of either type of failure. Thus the security case for S^3 documents the rationale for belief that S^3's efficacy is adequate to meet the needs of the stakeholders of a system to which S^3 has been applied.

5 Security Case Development for S^3

In this section, we present key aspects of the S^3 security case. Recall that the target of the security case is not the application of S^3 to a particular information system. Rather, we developed a prototype security case for the application of S^3 to a generic information system. By doing so, the resulting security case reveals the complete spectrum of issues associated with S^3 and provides a framework for developing instantiations of S^3 for specific systems.

In the next subsection (5.1), we review the top-level goal in the argument. The following subsection (5.2) discusses the overall structure of the argument. The remaining subsections discuss details of two significant fragments of the argument – those associated with the goals for false-negatives (5.3) and false-positive failures (5.4).

5.1 S^3 Top Level Goal

The "obvious" top-level goal for S^3 is that a system to which the transformation has been applied provides adequate protection against command-injection attacks where the stakeholders define "adequate". Unfortunately, as well as providing security protection, S^3 can have a variety of effects on the subject system, because S^3 makes changes to the software. For example, the transformation could actually change the functionality of the software, because the transformation is necessarily speculative.

Along with security protection, the factors that S^3 could affect include: (a) *functionality* – the functionality seen by the user might change, (b) *execution time* – the operation of the subject program might be slowed perhaps by an unacceptable amount, (c) *execution space* – the memory used by the subject program might be increased perhaps by an unacceptable amount, and (d) *vulnerabilities* – new vulnerabilities might be introduced such as vulnerability to denial of service attacks. With these factors in mind, stakeholders must determine the degree of change (if any) that is acceptable in each issue area.

Combining these points, the top-level goal in the S^3 security argument is that the system modified by S^3 is *fit for purpose*. Fit for purpose is an amalgam of the various issues in which each is considered and is appropriately weighted in the final definition of fit for purpose.

5.2 Security Argument Structure

The first challenge in developing the argument is to determine how to structure the argument. In previous work (Rodes and Knight 2013), we described a general argument structure for software modifications that are designed to enhance the security of SOUP. The structure we described is:

- **Level 1** – Fit for Purpose: The top-level argument structure organizes the assets that need to be protected, the security requirements for the subject system, i.e., the security attacks that are within the threat model, and specifies the pragmatic and other stakeholder-defined constraints and restrictions.
- **Level 2** – Attack Classes: The next level of the argument structure argues over classes of vulnerabilities for each security requirement. We refer to each class of vulnerability as an attack class. Each attack class defines a general vulnerability that malicious adversaries could exploit to achieve a successful attack. To the extent possible, we derive the content and details of each attack class from a taxonomy of classes of vulnerabilities.
- **Level 3** – Decomposed Attack Classes: Level 3 further decomposes and refines attack classes. Decomposed attack classes are necessary to bridge the gap between general security requirements and limitations of selected mitigation techniques.
- **Level 4** – Attack Class Mitigation: Arguments demonstrating how the threats associated with each vulnerability are mitigated.

We have adapted and extended this earlier structure for this study. The arguments used in the S^3 security case are all documented in the Goal Structuring Notation (GSN) (Kelly and Weaver 2004). Here we summarize the top levels of the argument in text form because of space limitations. Fragments of the GSN arguments are included in the later subsections.

The fit-for-purpose level of the security argument argues over the security requirements for the subject system and the four issue areas of interest discussed in Section 5.1 within the various argument contexts. The sub-arguments in the four issue areas are not discussed further, because these sub-arguments are, for the most part, typical of software assurance arguments.

In the case of S^3, there is only one attack class of interest, OS command injections, and so level 1 of the argument is simplified to that effect. The taxonomy used to determine the known attack details was the Common Weakness Elaboration (MITRE 2011) (CWE). For OS command injections, the CWE of interest is CWE 78.

Level 3 of the argument structure (decomposed attack classes) separates the arguments for statically- and dynamically-linked binaries. For purposes of security analysis, these two cases are distinct, and the vulnerabilities are completely different because of implementation details specific to S^3. In modern information systems, dynamic linking is much the more common, and so the prototype S^3 argument is developed for this case only.

Recall that the basic taint inference in S^3 assumes that strings within the binary can be trusted and that OS command injection attacks will originate from externally supplied strings. Nevertheless, attacks could originate from internal strings, e.g., strings that were corrupted by a previous *preparatory* attack. Level 3 of the S^3 argument argues over these two cases although only the former is developed. Finally, the level 4 sub-argument for this element of the S^3 argument begins with the claim:

> All maliciously crafted command strings are adequately detected and remediated prior to execution of the OS command.

The argument structure used for this goal contains two sub-goals:

- Malicious commands are adequately detected prior to command execution.
- All detected malicious commands are rendered inert.

The argument structure used for the first of these two sub-goals is itself subdivided into two sub-goals:

- Malicious commands are detected prior to execution with sufficiently low rates of false negative failures.
- Malicious commands are detected prior to execution with sufficiently low rates of false positive failures.

We discuss the details of the sub-arguments for each these two goals in the following subsections.

5.3 False Negative Failures

Recall that, for S^3, a false negative is a failure to recognize a malicious OS command when a binary program protected by S^3 is executing – an extremely serious situation.

5.3.1 ALARP

Ideally, the false negative argument in the S^3 security case would compel belief that either (a) a false negative failure is not possible, or (b) that the probability of a false negative failure occurring is below some prescribed threshold. Clearly false

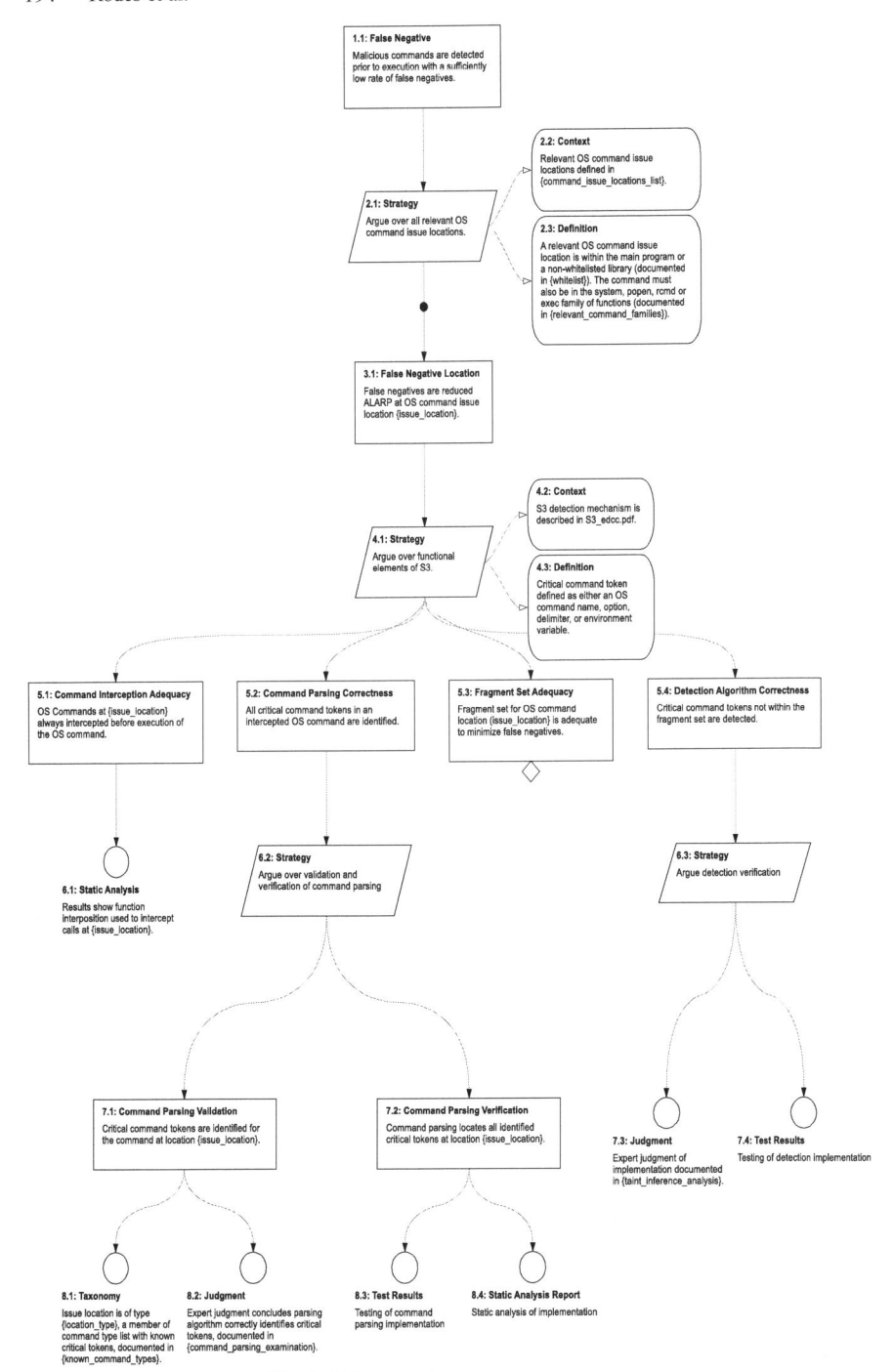

Fig. 1: False negative argument summary

negatives can occur with S^3, and so we cannot argue that a false negative failure cannot occur.

The S^3 technology prohibits comprehensive probability quantification, because several elements of the technology rely upon distributions that are completely unknown. We therefore can only argue that the risk associated with false negatives is acceptably low by arguing that risks are reduced As Low As Reasonable Practicable (ALARP) (Redmill 2010). Determination of whether the level achieved is acceptable is then the responsibility of the stakeholders.

The false negative ALARP argument in the S^3 security case is summarized in Figure 1. Arguments in this paper are only summarized because the complete arguments are quite large.

The false negative argument structure depends extensively on terminology specific to S^3. The original publication by Nguyen Tuong et al. (Nguyen Tuong et al. 2014) presents all of the S^3 terminology. Some of the key S^3 concepts are:

- **Critical Token:** An OS command contains one or more tokens (in the sense of a language definition) that are considered critical, because they could be used maliciously. More specifically, S3 defines command names, options, delimiters, and the setting of environment variables as critical tokens.
- **Fragment Set:** The fragment set is a set of string literals, referred to as fragments, that are considered trusted if found within an intercepted OS command.

S^3 infers taint by matching tokens in an intercepted OS command to fragments. If the OS command contains critical tokens that are not matched to a fragment, the OS command is considered tainted. And tainted commands are interpreted as attacks.

5.3.2 Argument Granularity

The first strategy in the argument for false negative failures argues over all locations from which OS commands are issued. An argument about false negatives must demonstrate that, for *all* such locations within the application, the rate of false negatives have been reduced ALARP. This element of the argument is both important and subtle.

The essence of the issue is that the mechanics of an exploit and the associated cost of an OS command injection attack depend upon the memory location from which the OS command is issued, i.e., the circumstances of the command. For example:

How a command is parsed might differ at different locations, i.e., what is considered a critical token might differ depending on the particular OS call.

The contents of the fragment set should be customized to each location. For example, the files to which a call might need access might depend on the location from which the OS command is issued.

If the strategy were to argue about all locations that issue OS commands at once, the argument would assume the same command parser and fragment set in all instances. The lack of specificity presents more opportunities for false negatives. For example, fragments that might never be legitimately used at one location would have to be part of the fragment set if the fragment serves a legitimate purpose at any other location.

In principle, arguing about *all* locations from which OS commands are issued at once might be possible (or at least considered), but to do so would require that one determine that the associated argument is, in fact, identical for each location. Prior to completing the S^3 argument, we cannot be certain that a single argument is applicable to all locations, and that determination is program dependent. Thus, the S^3 false negative argument argues over all locations that issue OS commands, each with an individual false negative goal. If the argument were found to be identical for some subset (or indeed all) locations, the argument could be repaired later. In practice (as described below), we concluded that, for the most part, the argument at each location should differ.

The strategy that argues over each site that issues an OS command has to be in the form of a GSN pattern. We do not know prior to applying S^3 how many OS commands will be issued or their locations in any given application. Instead, we must use structural and entity abstractions typically used within GSN argument patterns (GSN Standard 2011) and within product line arguments (Habli and Kelly 2010). In the S^3 argument, a black dot is used to indicate a repetition of the argument structure, and text within curly braces refers to entities that are dependent on the specific application of S^3.

Arguing over OS command locations requires the definition of *relevant* locations. Stakeholders might consider applying S^3 to all OS commands in an application, or they might trust some entities such as libraries like libc. In the current prototype implementation of S^3, the system trusts some libraries, and they are identified using a *white list*. In principle, there might be entities that stakeholders always prefer to trust across all applications, and this possibility is provided for within the S^3 argument as a context item (see Figure 1). Similarly, all locations from which OS commands are issued within the target software must be identified, and this set is also referenced within a context.

For each of these contexts, strong confidence must be demonstrated that the context is appropriate, sufficient, and trustworthy. In the S^3 argument, separate confidence arguments are assumed (Rodes et al. 2014, Hawkins et al. 2011). Confidence in the white list might be supported by evidence in the form of expert judgment. Confidence in locating all locations issuing OS commands might include some form of static analysis of the application.

5.3.3 False Negative Functional Hazards

At the next level of the argument, we argue that, for each injection location (of which there could be many), the associated rate of false negative failures is

ALARP by arguing over four goals derived from the basic S^3 mechanism. These four goals are:

- **Command Interception Adequacy:** OS commands at a given location are intercepted. If not, any associated attack(s) might not be detected.
- **Command Parsing Correctness:** Given an intercepted OS command, command parsing identifies all critical tokens correctly. Since detection of attacks is based on analysis of critical tokens, failure to identify a critical token could allow an attacker to inject a malicious command.
- **Fragment Set Adequacy:** The fragment set must have specific characteristics in order to properly imply trust. Fragments within the fragment set are compared to critical tokens in intercepted OS commands, and the parts of an OS command that match fragments are considered trusted.
- **Detection Algorithm Correctness**: As with the previous goal, the fragment set and the parsed critical tokens are used to infer taint, and so the associated algorithm must work correctly.

The sub-arguments for these four goals are summarized in Figure 1. Compelling evidence has not been obtained for the prototype implementation of S^3, and so we hypothesize feasible types of evidence. The first and fourth goals could be solved by verification evidence. For example, the adequacy of command interception might be solved by evidence from static analysis of the transformation of the binary and the associated insertion of the necessary probes. Similarly, algorithm correctness might be shown by evidence from testing, static analysis, proof, or some combination.

The second goal, Command Parsing Correctness, is decomposed in the argument into two sub-goals: (1) Command Parsing Validation, and (2) Command Parsing Verification. The first sub-goal refers to the need to identify the necessary set of tokens and the second to the need to identify the critical tokens in a specific intercepted OS command. In general, the evidence for the solution of each of these goals could derive from expert judgment, testing of similar programs, taxonomies developed separately, and so on.

The third goal is the most difficult of the three for which to develop a compelling sub-argument because of the speculative basis of the fragment set. The goal in its entirety is "Fragment set for OS command location {issue_location} is adequate to minimize false negatives." where *adequate* is a practical manifestation of *accurate*. The sub-argument associated with this goal is discussed in the next subsection.

5.3.4 Fragment Set Adequacy

Many different approaches to the sub-argument for the fragment set adequacy goal could be developed. The approach we use in our S^3 security case is summarized in Figure 2. We argue the following two sub-goals:

1.1: Fragment Set Adequacy
Fragment set for OS command location {issue_location} is adequate to reduce false negatives ALARP

1.2: Definition
A constituent fragment is a string containing a sequence of one or more critical command tokens for the given location in an acceptable order. Acceptable order defined as: (1) order that the tokens appear in a legitimate command, or (2) order deemed acceptable by expert judgment.

1.3: Definition
Minimal fragment set contains no excess fragments, i.e, fragments that will either not support attack detection or will cause excessive number of attacks to be missed

2.1: Strategy
Argue over elements of fragment set adequacy.

3.1: Constituent Fragment
All fragments within the fragment set for command location {issue_location} are constituent fragments.

3.2: Minimal Fragment Set
Fragment set at command location {issue_location} is adequately minimal.

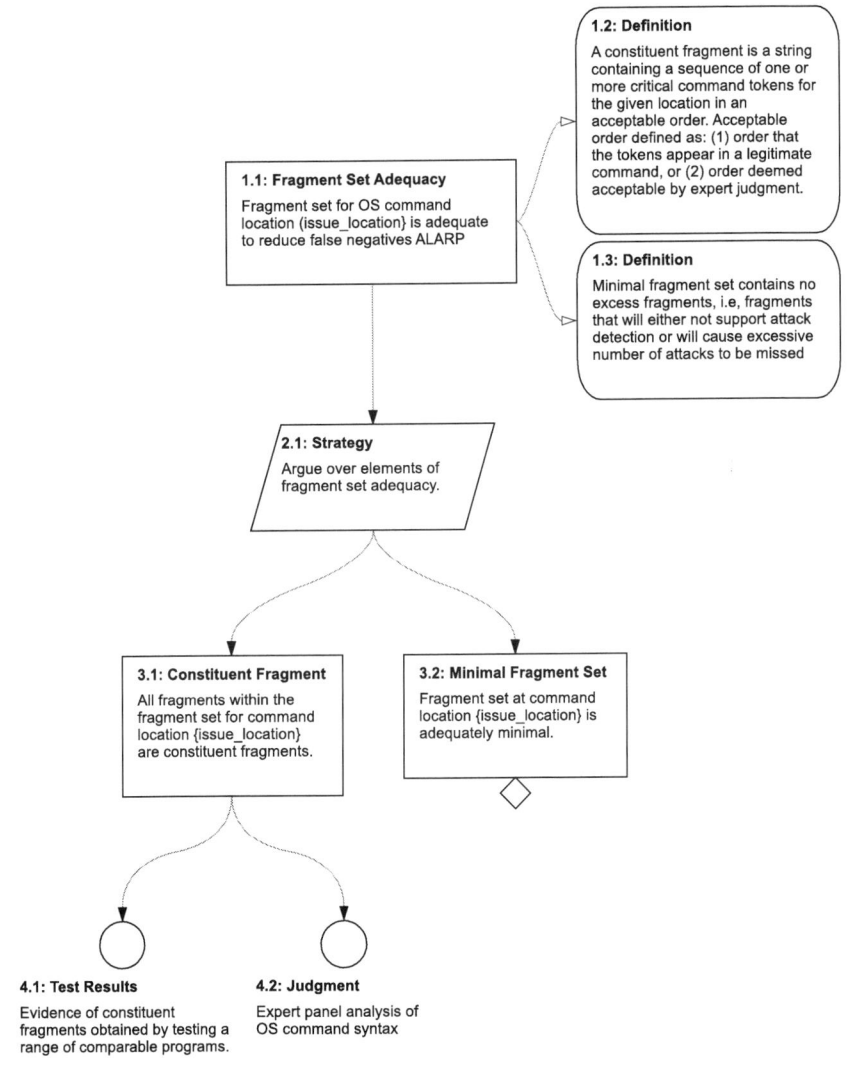

4.1: Test Results
Evidence of constituent fragments obtained by testing a range of comparable programs.

4.2: Judgment
Expert panel analysis of OS command syntax

Fig. 2: Fragment set adequacy sub-argument summary

- **The Constituent Fragment Goal**. The fragment set should only containing constituent fragments. A constituent fragment is a string containing a sequence of one or more critical command tokens for the given location in an acceptable order.
- **The Minimal Fragment Set Goal**. Even if the fragment set is shown to contain only constituent fragments, some fragments might not be effective, single characters for example. Such fragments can be thought of as "junk" DNA. In the limit, if there were too many such fragments, few or no attacks would be

detected. We therefore stipulate that the fragment set must be adequately "minimal". How best to identify ineffective fragments is not presently known.

Evidence for these goals could take many forms. First, we note that *acceptable order* in the definition of constituent fragment might be defined to be one or more of the following:

- Any order.
- The order that the tokens appear in a legitimate command.
- An order deemed acceptable by expert judgment.
- An order determined by analysis of strings used in practice by similar programs.
- An order that could occur as a result of the control flow of the program that leads to the generation of an OS command.

Thus, the evidence used for determination of the constituent fragments might derive from a process as simple as referring to a taxonomy of tokens to a process as complex as complete control and data flow analysis of the subject program.

The evidence for the fragment set being of minimal size depends upon many factors. Again, expert judgment could be elicited, but testing a wide variety of programs and sampling the OS command contents of the programs when operated in a benign environment might provide a more compelling body of evidence.

5.4 False Positive Failures

A compelling false *negative* argument could be constructed for a security defense that identified *all* OS commands as malicious. Every malicious command would be detected, but the rates of false positive failures would be unacceptable.

Thus, the S^3 defense attempts to determine the difference between malicious and benign OS commands. The false negative argument discussed above provides the rationale for belief that attacks are detected. Here we discuss the argument that benign OS commands are not identified as attacks.

5.4.1 Similarities to the False Negative Argument

In developing the false *positive* argument, we use the same initial strategy that we used for the false negative argument, i.e., argue over all relevant OS command locations that false positives failures are reduced ALARP.

For each relevant OS command location, false positives are demonstrated to be reduced ALARP using the same general goals that were used in the false negative argument: (1) Command Interception Adequacy, (2) Command Parsing Correctness, (3) Fragment Set Adequacy, and (4) Detection Algorithm Correctness.

Each goal within the false positive argument is used to *balance* the corresponding goal within the false negative argument. For example, to reduce false negatives, each relevant OS command location must be intercepted. To reduce false positives, we must demonstrate that only relevant OS command locations are intercepted when an OS command is actually executed.

Most of the false positive goals could be supported by evidence in the form of expert judgment. For example, experts could conclude that command interposition is not known to intercept functions spontaneously when the function is not actually called. We note that such evidence for false positives does not rely on the location of the OS command as heavily as false negatives do. Many claims might therefore be argued generally over all locations in a single argument.

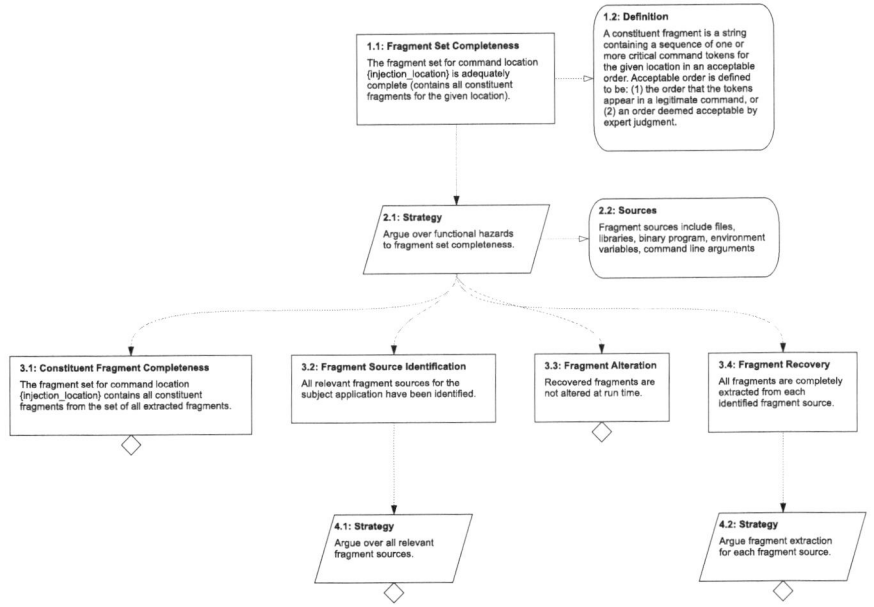

Fig 3: Fragment set completeness sub-argument summary

5.4.2 Fragment Sources

An important difference between the false positive and the false negative arguments is that the false positive argument depends heavily on characteristics of the particular piece of software that is modified, especially within the fragment set completeness sub-argument. For the fragment set to be complete (see Figure 3), all possible *sources* of fragments must be identified, and then fragments must be adequately extracted from each source.

The present prototype implementation of S^3 relies on the source of fragments being the binary program and relevant libraries. There is an assumption that fragments either do not originate from other locations or can be easily extracted from other sources. For the purposes of building a strong argument, we cannot rely on this assumption. Instead, we must demonstrate how application of S^3 accounts for the possibility of fragments in alternative locations. In principle, fragments might exist in any of the following:

- The binary program.
- Libraries references by the binary program.
- Files used for configuration of the programs OS commands.
- Environment variables.
- Command line arguments.

The argument must therefore show that all fragment sources have been identified and properly considered in the argument. Each of the above sources is an example of a fragment source that could be overlooked easily.

The success of identifying fragment sources and the success of fragment extraction are therefore dependent on the specific characteristics of the subject application. In our S^3 security case, we include a goal to that effect but note here that the associated evidence would probably be limited to control and data flow analysis of the binary program to detect possible input sources, expert judgment, or some kind of evolving taxonomy.

5.4.3 Fragment Alteration

The prototype implementation of S^3 relies on the assumption that fragments, once extracted, are never altered. Since fragment extraction occurs only once, if the fragments were altered after extraction, all fragment sets might be invalidated.

Some users might be willing to accept the assumption that fragment alteration does not occur, but a generalized fragment set completeness argument (shown in Figure 3) requires a goal to this effect together with the associated sub-argument. Evidence in support of this goal might include data flow analysis of fragments, instruction analysis of string manipulation operations, or expert judgment.

The use of analyses or expert judgment might depend on how well experts or specific analyses can assess a given instance of software. For example, static analysis might be the preferred source of evidence, but if the analysis cannot generate sufficient data for a given program, expert opinion will be tolerated instead. Static analysis might fail for a variety of reasons, for example: (1) the analysis might use excessive resources, (2) the analysis might be unable to complete the required analysis, or (3) the analysis might not be applicable to the subject software.

If alternatives exist for structure of the S^3 argument, product line argumentation and argument pattern techniques could be adopted (as was used to argue over each OS command location). A key difference with traditional uses of product line argumentation is the need to be more explicit about the preferences between alternatives and the conditions that could invalidate any given argument alternative.

An enhancement to argument structures that we propose is the inclusion of *decision models* and *guarding constraints* within the argument. Decision models express how a selection is made between alternative argument options, referenced with obligation elements. Decision models capture a preference between argument alternatives, but a preference might be invalidated by the characteristics of the software.

Guards express constraints about software specific data that restricts (i.e., guards) the validity of an argument. We express the concept of an argument guard as a GSN constraint element. Guards highlight key evidence that is not known prior to applying S^3 that must be generated and assessed based on the constraints specified within the guard.

5.4.4 Constituent Fragment Set Completeness

The final goal in support of fragment set completeness demonstrates that of all extracted fragments, the constituent fragment set for the specific OS command location is *complete*. In the prototype implementation of S^3, only one fragment set was developed for the entire program. The fragment set was considered complete if all fragments were extracted. For our argument, however, since fragment sets might differ for each OS command issue location, we must reevaluate the concept of a complete fragment set.

To produce a complete fragment set, we could also use all extracted fragments as the fragment set for each OS command location; however, such a policy conflicts with the false negative argument. In the false negative argument, our goal was to narrow the fragment set as much as possible. Specifically, we required that a fragment set at an OS command issue location must contain "constituent" fragments (defined in Section 4). Based on the concept of a constituent fragment, for a given fragment set to be complete, the argument must (1) show that all possible fragments have been extracted (discussed above) and (2) show that a given fragment set has all the fragments that are also constituent fragments (a subset of all extracted fragments).

As with the false negative argument, evidence used for determination of the constituent fragments could be derived from a process as simple as referring to a taxonomy of tokens or a process as complex as complete control and data flow analysis of the subject program.

6 Conclusion

In this paper, we have described some of the key elements of a security case for a security enhancement technique, *Software DNA Shotgun Sequencing*. This technique is designed to modify binary programs to prevent OS command injection attacks.

Developing the S^3 security case revealed a range of assumptions and weaknesses in the defense that S^3 provides. The development team knew about many of these, but, importantly, reasoning about the security case yielded all of the known and some unknown weaknesses.

In the authors' opinion, some of the security argument structures developed for S^3 will not be familiar to developers of safety cases. For example, the need to argue over all possible locations from which OS commands might originate has no common analogy in typical safety arguments. Based on our experience with S^3, we conclude that the subtlety of security properties and the complexity of malicious attacks will lead to such argument strategies frequently.

Overall, we have concluded that development of a security case for security technologies that operate on SOUP, such as S^3, is both feasible and practical. With one exception, all of the basic concepts of rigorous argument, evidence generation, and case development were applied using the same processes and procedures that are typical in safety case development.

The exception to the above is the need to work with S^3 (and similar techniques) with no specific target, because the S^3 concept is to apply the transformation to a wide variety of target systems, i.e., a single and static assurance case cannot be constructed for these technologies. Instead, an assurance case for security modifications must be able to express a *range* of acceptable modifications (i.e., a solution space) and how this solution space is navigated.

To accommodate the notion of a solution space, the form of the S^3 security case required the integration of ideas from argument patterns and the extensions that have been proposed to support safety cases for software product lines. Since there are elements of the argument that require argument structure selection that is dependent on the final target system, we conclude that a security case for SOUP modifications must include: (a) a *fit-for-purpose* argument which argues that a generated modification is acceptable, and (b) a *success* argument (Graydon and Knight 2008) which argues that the implied process for modifying the particular system of interest will succeed in finding valid alternatives for each argument guard and will not exceed stakeholder-defined development limitations on resource consumption.

Acknowledgments This research is supported by National Science Foundation (NSF) grant CNS-0811689, the Army Research Office (ARO) grant W911-10-0131, the Air Force Research Laboratory (AFRL) contracts FA8650-10-C-7025 and FA8750-13-2-0096, and DoD AFOSR MURI grant FA9550-07-1-0532. The views and conclusions contained herein are those of the authors and should not be interpreted as necessarily representing the official policies or endorsements, either expressed or implied, of the NSF, AFRL, ARO, DoD, or the U.S. Government.

References

Alexander, R., Hawkins, R., and Kelly, T. (2011). Security Assurance Cases: Motivation and the State of the Art. University of York, Tech. Rep. CESG/TR/2011.

Bosman E., Slowinska, A., and Bos, E. (2011) Minemu: The World's Fastest Taint Tracker. In proceedings of the 14th International Conference Recent Advances in Intrusion Detection (RAID), pp. 1-20. Springer Berlin Heidelberg.

Nguyen-Tuong, A., Hiser, J., Co, M., Kennedy, N., Melski, D., Ella, W., Hyde, D., Davidson, J., and Knight, J. (2014) To B or not to B: Blessing OS Commands with Software DNA Shotgun

Sequencing. In Proceedings of the 10th European Dependable Computing Conference (EDCC), pp. 238–249. IEEE.

Graydon, P., and Knight, J. (2008). Success Arguments: Establishing Confidence in Software Development. University of Virginia, Tech. Rep. CS-2008-10.

GSN Standard (2011) GSN Community Standard Version 1. http:// http://www.goalstructuringnotation.info/. Accessed 22 September 2014.

Goertzel, K.,Winograd, T., McKinley, H., Oh, L., Colon, M., McGibbon, T., Fedchak, E., and Vienneau, R. (2007) Software Security Assurance: A State-of-the-Art Report (SOAR). Information Assurance Technology Analysis Center (IATAC) Herndon, VA.

Habli, I., and Kelly, T. (2010). A Safety Case Approach to Assuring Configurable Architectures of Safety-Critical Product Lines. In Architecting Critical Systems, pp. 142-160. Springer Berlin Heidelberg.

Hawkins, R., Kelly, T., Knight, J., and Graydon, P. (2011). A New Approach to Creating Clear Safety Arguments. In Advances in systems safety, pp. 3-23. Springer London.

Hiser, J., Nguyen-Tuong, A., Co, M., Hall, M., and Davidson, J. (2012). ILR: Where'd My Gadgets Go?. In Security and Privacy (SP), pp. 571-585. IEEE.

Kc, G., Keromytis, A., and Prevelakis, V. (2003). Countering Code-Injection Attacks with Instruction-Set Randomization. In Proceedings of the 10th ACM conference on Computer and communications security, pp. 272-280. ACM.

Kelly, T., and Weaver, R. (2004). The Goal Structuring Notation–A Safety Argument Notation. In Proceedings of the Dependable Systems and Networks (DSN) workshop on assurance cases.

Mitre (2011). Common Weakness Enumeration (CWE). http://cwe.mitre.org/. Accessed 22 September 2014.

Mitre and SANS (2011). CWE/SANS Top 25 Most Dangerous Software Errors. http://cwe.mitre.org/top25/. Accessed 22 September 2014.

Pietraszek, T., and Berghe, C. (2006). Defending Against Injection Attacks Through Context-Sensitive String Evaluation. In Recent Advances in Intrusion Detection (RAID), pp. 124-145. Springer Berlin Heidelberg.

Redmill, F. (2010). ALARP Explored. University of Newcastle upon Tyne, Computing Science.

Redmill, F. (2001). The COTS Debate in Perspective. In Computer Safety, Reliability and Security, pp. 119-129. Springer Berlin Heidelberg.

Rodes, B., and Knight, J. (2013). Reasoning About Software Security Enhancements Using Security Cases. In The First International Workshop on Assurance for Argument and Agreement (AAA).

Rodes, B., Nguyen-Tuong, A., Hiser, J., Knight, J., and Davidson, J. (2013). Defense Against Stack-Based Attacks Using Speculative Stack Layout Transformation. In Runtime Verification, pp. 308-313. Springer Berlin Heidelberg.

Rodes, B., Knight, J., and Wasson, K.(2014). A Security Metric Based on Security Arguments. In Proceedings of the 5th International Workshop on Emerging Trends in Software Metrics (WETSoM), pp. 66-72. ACM.

Rodes, B., and Knight, J. (2014). Speculative Software Modification and its Use in Securing SOUP. In Dependable Computing Conference (EDCC), 2014 Tenth European, pp. 210-221. IEEE.

Explicate '78: Uncovering the Implicit Assurance Case in DO–178C

C. Michael Holloway

NASA Langley Research Center

Hampton, VA, U.S.A.

Abstract *For about two decades, compliance with Software Considerations in Airborne Systems and Equipment Certification (DO–178B/ED–12B) has been the primary means for receiving regulatory approval for using software on commercial airplanes. A new edition of the standard, DO–178C/ED–12C, was published in December 2011, and recognized by regulatory bodies in 2013. The purpose remains unchanged: to provide guidance 'for the production of software for airborne systems and equipment that performs its intended function with a level of confidence in safety that complies with airworthiness requirements.' The text of the guidance does not directly explain how its collection of objectives contributes to achieving this purpose; thus, the assurance case for the document is implicit. This paper presents an explicit assurance case developed as part of research jointly sponsored by the Federal Aviation Administration and the National Aeronautics and Space Administration.*

1 Introduction

Software Considerations in Airborne Systems and Equipment Certification (DO–178B) (RTCA 1992)[1] was published in 1992. Compliance with this document has been the primary means for receiving regulatory approval for using software on commercial airplanes ever since. Despite criticisms of the DO–178B from various quarters, the empirical evidence suggests strongly that it has been successful, or at worst, has not prevented successful deployment of software systems on aircraft. Not only has no fatal commercial aircraft accident been attributed to a software

---

[1] The European Organisation for Civil Aviation Equipment (EUROCAE) uses a different document numbering scheme, but the content of the documents is equivalent. For example, DO–178C is equivalent to ED–12C. For simplicity, only the DO numbering scheme is used in the body of this paper. Also, please note that although once upon a time RTCA was an abbreviation for Radio Technical Commission for Aeronautics, since 1991 the four letters have been the freestanding name of the organization.

This is a work of the U.S. Government and is not subject to copyright protection in the United States. Published by the Safety-Critical Systems Club.

failure, many of the technological improvements that have been credited with significantly reducing the accident rate have relied heavily on software. For example, controlled flight into terrain—once one of the most common accident categories—has been nearly eliminated by software-intensive Enhanced Ground Proximity Warning Systems (Rushby 2011).

A new edition of the standard, DO–178C, was published by the issuing bodies in late 2011 (RTCA 2011a). New editions of two existing associated documents and four entirely new guidance documents were also published at the same time. More information about these documents is provided later in this paper. The relevant documents received official regulatory authority recognition in 2013 (Federal Aviation Administration 2013b, European Aviation Safety Agency 2013).

The stated purpose of DO–178C remains essentially unchanged from its predecessor: to provide guidance *'for the production of software for airborne systems and equipment that performs its intended function with a level of confidence in safety that complies with airworthiness requirements.'* The text of the guidance provides little or no rationale for how it achieves this purpose. A new section in the revised edition of DO–248C (RTCA 2011b), 'Rationale for DO–178C / DO–278A', contains brief discussions of the reasons behind some specific objectives and collection of objectives; nevertheless, the overall assurance case for why DO–178C achieves its purpose is almost entirely implicit.

Although empirical evidence suggests that this implicit assurance case has been adequate so far, its implicitness makes determining the reasons for this adequacy quite difficult. Without knowing the reasons for past success, accurately predicting whether this success will continue into the future is problematic, particularly as the complexity and autonomy of software systems increases. Equally problematic is deciding whether proposed alternate approaches to DO–178C are likely to provide an equivalent level of confidence in safety.

As a potential way forward for addressing these problems, the Federal Aviation Administration (FAA) and the National Aeronautics and Space Administration (NASA) are jointly sponsoring an effort, called the Explicate '78 project within NASA, to uncover and articulate explicitly (that is, explicate) DO–178C's implicit assurance case. Early work in this effort was described in (Holloway 2012, Holloway 2013).

This paper describes the current status of the research, and is organized as follows. Section 2 provides background material. Section 3 presents the key concepts underlying, and several excerpts from, the explicit assurance case developed to date. Section 4 discusses the next steps in the research and makes concluding remarks.

2 Background

Fully understanding this paper requires at least a passing familiarity with DO–178B/C and the assurance case concept. This section provides background information on these two subjects for readers who do not already possess the requisite knowledge. This section also provides a brief discussion of prior related published work.

Although some excerpts from the assurance case are expressed using the Goal Structuring Notation (GSN), background material about GSN is not provided because of space limitations. Readers unfamiliar with GSN should consult (GSN Committee 2011).

2.1 About DO–178C

The information in this section is based on Appendix A in DO–178C, which contains a summary of the history of the DO–178 series of documents. The initial document in the 178 series was published in 1982, with revision A following in 1985. Work on revision B began in the fall of 1989; the completed document, which was a complete rewrite of the guidance from revision A, was published in December 1992. Among many other changes, the B version expanded the number of different software levels based on the worst possible effect that anomalous software behaviour could have on an aircraft. Level A denoted the highest level of criticality (for which satisfying the most rigorous objectives was required), and Level E denoted the lowest level (which was objective-free). The B version also introduced annex tables to summarize the required objectives by software level.

Twelve years after the adoption of DO–178B, RTCA and EUROCAE moved to update the document by approving the creation of a joint special committee / working group in December 2004 (SC-205/WG-71). This group started meeting in March 2005, and completed its work in November 2011. The terms of reference for the group called for (among other things) maintaining the 'objective-based approach for software assurance' and the 'technology independent nature' of the objectives. SC-205/WG-71 was also directed to seek to maintain 'backward compatibility with DO–178B' except where doing so would fail to 'adequately address the current states of the art and practice in software development in support of system safety', 'to address emerging trends', or 'to allow change with technology.'

Ultimately the effort produced seven documents. In addition to DO–178C, new editions were written of two existing, associated documents: DO–278A: Software Integrity Assurance Considerations for Communication, Navigation, Surveillance and Air Traffic Management (CNS/ATM) Systems (RTCA 2011c), and DO–248C: Supporting Information for DO–178C and DO–278A (RTCA 2011b). The former is very similar to DO–178C, but addresses software in certain ground-based systems, which operate within a different regulatory scheme from airborne systems. The latter provides answers to various questions and concerns raised over the years by both industry and regulatory authorities. It contains 84 frequently asked questions, 21 discussion papers, and, as noted above, a brief rationale.

Four new guidance documents were also published to address specific issues and techniques: DO–330: Software Tool Qualification Considerations (RTCA 2011d); DO–331: Model-Based Development and Verification Supplement to DO–178C and DO–278A (RTCA 2011e); DO–332: Object-Oriented Technology and Related Techniques Supplement to DO–178C and DO–278A (RTCA 2011f); and DO–333: Formal Methods Supplement to DO–178C and DO–278A (RTCA 2011g). The subject matter of these documents is evident from their titles.

As a result of the terms of reference and operating instructions under which DO–178C was developed, the document is only an update to, as opposed to a rewrite or substantial revision of, DO–178B. Differences between the B and C versions include corrections of known errors and inconsistencies, changes in wording intended for clarification and consistency, an added emphasis on the importance of the full body of the document, a change in qualification criteria for tools and the related creation of a separate document for tool qualification, modification of the discussion of system aspects related to software development, closing of some perceived gaps in guidance, and the creation of the technology-specific supplements enumerated above for formal methods, object-oriented technology, and model-based design and verification.

2.2 About assurance cases

The concept of an assurance case is a generalization of the safety case concept. A common definition of a safety case is 'a structured argument, supported by a body of evidence that provides a compelling, comprehensible and valid case that a system is safe for a given application in a given operating environment' (UK Ministry of Defence 2007). Claims are made concerning the achievement of an acceptable level of safety, and arguments and evidence are focused on providing justified confidence that those safety claims are satisfied. A more general assurance case is concerned about providing justified confidence that claims are satisfied about other desired attributes such as correctness, functionality, performance, or security.

Claims, arguments, evidence[1], context, and assumptions constitute five components of a well-structured assurance case (Knight 2012). *Claims* are statements about desired attributes. Other names that are used for the same concept include *goals*, *propositions*, and *conclusions*. In a full assurance case, there will likely be many claims that must be shown to hold, at varying levels of generality. An example of a high-level claim is **The software performs its intended function at an acceptable level of safety** (**bold Arial font** is used throughout the

---

[1] The claims, argument, evidence distinction (perhaps using slightly different words) is well established within the literature. A strong case can be made that *argument* is more properly thought of as a broad term, of which *claims* and *evidence* are components; however, this particular paper is not the place to try to clean up the terminology, so the standard terms and distinctions are maintained.

paper to denote assurance case text). Examples of claims with an increasing level of specificity are as follows: **High-level requirements are a satisfactory refinement of system requirements**; **Adequate configuration management is in place**; and **Configuration items are identified**.

An *argument* explains how a stated claim is supported by, or justifiably inferred from, the evidence and associated sub-claims. Other terms sometimes used for the same concept include *strategies*, *warrants* (Toulmin 2003), and *reasons*. Just as a system nearly always consists of multiple sub-systems, an argument nearly always consists of multiple sub-arguments; but the term sub-argument is almost never used.

Evidence refers to the available body of known facts related to system properties or the system development processes. *Data*, *facts*, and *solutions* are synonymous terms. Examples of evidence include hazard logs, testing results, and mathematical theorems.

Context generally refers to any information that is needed to provide definitions or descriptions of terms, or to constrain the applicability of the assurance case to a particular environment or set of conditions. As example, the context for the claim **The software performs its intended function with a level of confidence in safety that complies with airworthiness requirements** would likely include the applicable airworthiness requirements (Federal Aviation Administration 2013a), a description of the intended function of the software, and any constraints on the environment in which the software is expected to be used. Some recent research defines context more strictly than has been done previously (Graydon 2014).

Assumptions are statements on which the claims and arguments rely, but which are not elaborated or shown to be true in the assurance case. As an example, an argument concerning safety that shows that all identified hazards have been eliminated may rely on the assumption **All credible hazards have been identified**.

Claims, arguments, evidence, context, and assumptions are all present implicitly in the collective minds of the developers of any successful engineered system. An assurance case simply provides a means for ensuring that this implicit knowledge is documented explicitly in a form that can be examined carefully and critically, not only by the developers, but also by others. An active research community is exploring how to best create, express, analyze, improve, and maintain assurance cases. Examples include (Matsuno 2014, Ayoub et al. 2013, Denney et al. 2013, Hawkins et al. 2013, Rushby 2013, Goodenough et al. 2012, Yuan and Kelly 2011, Bloomfield and Bishop 2010, Hawkins and Kelly 2009, Holloway 2008).

2.3 Previous work

No published work was found that has attempted to accomplish the same goals as the current effort, but two previous projects did address related aspects concerning DO–178B and assurance cases.

The MITRE Corporation tried to map three different standards into an assurance case framework (Ankrum and Kromholz 2005). The primary purpose of this effort was to explore two primary hypotheses: all assurance cases have similar components, and an assurance standard implies the structure. One of the three standards used in the study was DO–178B. The created assurance case was structured rigidly around the DO–178B chapters. For example, the top-level claim was **DO–178B Software Considerations are taken into account**. Sub-claims were given for each of the DO–178B chapters 2 – 9; for example: **2.0 System Aspects are taken into account**; **5.0 Software Development Process is executed as planned**; and **9.0 Certification Liaison process is properly established & executed**.

The effort appears to have concentrated on translating the textual and tabular form of DO–178B into a graphical form with as little interpretation or abstraction as possible. This differs substantially from the current research, which is concentrating on discovering the underlying implicit assurance case, not rigidly translating one form of concrete expression into another form.

Researchers at the University of York and QinetiQ in the United Kingdom conducted the other related previous work (Galloway et al. 2005). The primary goal of this research was to explore ways to justify substitution of one technology for another. In particular, a major emphasis was placed on developing arguments showing that the evidence produced by replacements for testing (such as formal proof) could be at least as convincing as the evidence produced by testing. As part of this research, certain aspects of the testing-related objectives of DO–178B were explored and GSN representations were produced. Unpublished results from the research were submitted to SC–205/WG–71, and considered by the Formal Methods sub-group, which wrote the document that eventually become DO–333. This material was also considered during the process of developing the assurance case for DO–178C that will be discussed in the next section.

3 The explicit case

The first version of a complete, explicit assurance case in the Explicate '78 project was completed and expressed in GSN at the end of 2013. It was structured in a modular fashion, with separate arguments for each of the four main software levels A-D. To the extent consistent with the 178C text, arguments from lower software levels were referenced directly in the arguments for higher software levels.

This version was reviewed in varying levels of detail and rigor by a handful of FAA personnel and other interested parties over a period of six months.

Revisions based on the review yielded a version (called e78-1.5) that was substantially similar in overall structure to the original, but which differed in some subtle ways and in several specific details. This version also introduced generic primary and confidence arguments, which were not strictly necessary, but which served to illustrate a consistent argument structure across levels. A lengthy presentation describing e78-1.5 was delivered to over 100 people at the FAA-sponsored 2014 National Systems, Software, and Airborne Electronic Hardware Conference in September 2014. Comments received at the conference prompted several minor modifications to the GSN structures, and the creation of textual representations of portions of the case, yielding version e78-1.6, which is the version described here.

The section is organized as follows:
1) Four fundamental concepts that greatly influence the structure and content of the e78-1.6 assurance case are discussed.
2) Salient characteristics about the case itself are provided.
3) Five excerpts from the case are presented.

3.1 Fundamental concepts

The following four concepts provide the foundation on which the explicit assurance case is built: transforming safety into correctness, allowing life cycle flexibility, using confidence arguments, and explicating before evaluating. The first two of these concepts permeate the DO–178C guidance itself. The latter two concepts arose as solutions to difficulties encountered in the early days of trying to structure an explicit assurance case. All four are discussed below.

3.1.1 Transforming safety into correctness

A fundamental assumption of DO–178C is discernable only through inferences from the text; it involves the relationship between safety and correctness. Although in the general case, these two concepts are not equivalent (Knight 2012), DO–178C rests implicitly on the assumption that within the constraints established by the guidance, establishing justifiable confidence in the correctness of the software with respect to its requirements *is* sufficient to establish justifiable confidence that the software does not contribute to unsafe conditions.

The validity of this assumption rests on the further assumption that adequate system safety processes have been followed in determining the requirements placed on the software and its associated criticality level. As stated in the Rationale: 'Software/assurance levels and allocated system requirements are a result of the system development and safety assessment processes' (RTCA 2011b, p. 9).

The system requirements allocated to software are further assumed by DO–178C to include all of the requirements that must be satisfied by the software to ensure an adequate level of safety is maintained. DO–178C is not concerned with determining or analysing these safety requirements, but only in satisfying them. Hence it is strictly true, as is often asserted, that the standard is not a safety standard. Conducting system safety analysis is intentionally outside the scope of the guidance. Guidance for it is expected from other documents (SAE International 1996, SAE International 2010).

Any new requirements that arise during software development must be passed back to the system processes, including system safety processes, for analysis of (among other things) potential safety implications. Such requirements were called *derived requirements* in DO–178B; the term is retained in 178C. (This choice of terminology has been a frequent source of confusion, because the phrase *derived requirements* is not commonly used in the broader software engineering community. When encountering the term for the first time, many people assume that it means requirements derived from higher level requirements, as opposed to new requirements that are explicitly *not derived* from higher level ones. Some members of SC–205/WG–71 tried, but failed, to change the terminology.)

With these assumptions understood, DO–178's emphasis on software correctness is consistent with its stated purpose. Given that all the requirements necessary for ensuring adequate safety are eventually specified, then developing software that is correct with respect to those requirements is sufficient to ensure that the software does not negatively affect safety. Transforming safety into correctness is valid in this particular case.

As will be shown below, the e78-1.6 assurance case makes the transformation explicit. It also highlights the special role played by derived requirements.

3.1.2 Allowing life cycle flexibility

Another foundational concept of DO–178C may come as a surprise to people whose only exposure to the guidance and its ancestors comes through criticisms by academics: developers are permitted wide flexibility in choosing how to develop their software. Neither specific development methods nor life cycles are prescribed by the guidance. As stated in the Rationale:

> The committee wanted to avoid prescribing any specific development methodology. [The guidance] allows for a software life cycle to be defined with any suitable life cycle model(s) to be chosen for software development. This is further supported by the introduction of "transition criteria". Specific transition criteria between one process and the next are not prescribed, rather [the guidance] states that transition criteria should be defined and adhered to throughout the development life cycle(s) selected.' (RTCA 2011b, p. 126)

The guidance does include detailed descriptions of specific activities that may be performed in order to satisfy particular objectives. References to the text of these activities are even included in the Annex A tables in 178C. However, the guidance also explicitly states that the activities themselves may be changed:

> The applicant should plan a set of activities that satisfy the objectives. This document describes activities for achieving those objectives. The applicant may plan and, subject to approval of the certification authority, adopt alternative activities to those described in this document. The applicant may also plan and conduct additional activities that are determined to be necessary. (RTCA 2011a, p. 3).

To emphasize the flexibility allowed by the guidance, the e78-1.6 assurance case does not explicitly include accomplishing any activities as goals that must be satisfied. Activities are only referenced within contextual items in the case.

3.1.3 Using confidence arguments

Researchers from the University of York and the University of Virginia (Hawkins et al. 2011) introduced the idea of a confidence argument to accompany a primary safety argument. The primary safety argument documents the arguments related to direct claims of safety; the confidence argument documents the arguments related to the sufficiency of confidence in the primary argument.

This separation into two different argument structures differs from the prevailing practice of intermixing concerns of safety and confidence in a single unified argument, and offers the potential promise of eliminating or mitigating some of the difficulties recognized in the prevailing approach (Haddon-Cave 2009). Although the original research concentrated on safety arguments, the general concept applies equally to any property of interest.

Even a cursory reading of DO–178C reveals that the guidance contains a mixture of objectives about the desired properties of the final software product, objectives related to intermediate products, and objectives concerning the processes used to develop the product. A more careful reading, keeping the notion of separating primary and confidence arguments in mind, suggests that some of these objectives naturally fit well into a primary argument about properties of the final software, and some naturally fit well into a confidence argument that affects the degree of belief in the sufficiency of the primary argument. Only a comparatively few objectives are difficult to classify.

These observations make using confidence arguments a foundational concept for the explicit assurance case. Reviewers of previous versions of the case commented favorably on this approach.

3.1.4 Explicating before evaluating

The fourth foundational concept is that accurately articulating the implicit case contained in DO–178C must precede trying to evaluate the sufficiency of the case. Evaluation is an important eventual goal of the research, but unless agreement can be reached about what the guidance really says, reaching agreement on whether it says the right thing is impossible.

The e78-1.6 assurance case is intended to properly capture what 178C says. Great effort was made to represent accurately the implicit arguments in the guidance, without trying to correct any perceived deficiencies. The coherence and cogency of this explicit case should be neither greater nor lesser than that of the guidance itself.

3.2 Characteristics of the case

The e78-1.6 assurance case expression in GSN consists of a primary argument module and a confidence argument module for each software level (A, B, C, D), generic primary and confidence argument modules, and a simple primary argument for software level E. Additional modules support the Level A-D primary and confidence arguments as follows:

Level D

- Primary argument: five supporting modules
- Confidence argument: five support modules

Level C:

- Primary argument: two unique and two directly referenced level D supporting modules
- Confidence argument: eight unique and five directly referenced level D supporting modules

Level B:

- Primary argument: one unique supporting module and a direct reference to the level C primary argument
- Confidence argument: three unique, four directly referenced level C, and one directly referenced level D supporting modules

Level A:

- Primary argument: no unique supporting modules and a direct reference to the level B primary argument

- Confidence argument: two unique, two directly referenced level B, two directly referenced level C, and one directly referenced level D supporting modules

Overall the 34 GSN modules for Levels A-D comprise 131 goals, 42 strategies, 176 context items, 34 justifications and assumptions, and 161 references to evidence. In some instances, the style of the GSN representation used in the project may rightly displease purists. Strict adherence to standard practices has been sacrificed in places under the belief that the sacrifice better achieves visual simplicity and enhances readability for the primary intended audience of the work, few of whom are experts in the notation.

Also, for the benefit of the intended audience, textual representations have been manually created for 15 of the GSN modules, with more in the works. For two of the five examples presented in the next section, a textual representation accompanies the GSN structure.

3.3 Excerpts from the case

Obviously the full case is too large to reproduce in this paper. Five representative excerpts are presented in this section: a simple version of the general primary argument, and one example each from the four main software levels.

3.3.1 Simple generic primary argument

Figure 1 shows a GSN representation of a very simple generic primary argument that captures the essence of the safety to correctness transformation, which, as noted above, constitutes the heart of the DO–178C implicit assurance case. It intentionally omits context, justifications, and assumptions for the sake of initial simplicity. These missing items do appear in the instantiation of the Level D primary argument shown in the next section.

```
                    ┌─────────────────────────────┐
                    │ 1.1: SwAcceptable{Level X}  │
                    │ Software performs its       │
                    │ intended function at        │
                    │ acceptable level of         │
                    │ safety for {Level X}        │
                    │                           1 │
                    └─────────────────────────────┘
                                   │
                                   ▼
                    ┌─────────────────────────────┐
                    │ 2.1: ArgByCorrectness       │
                    │ Argument by correctness of  │
                    │ the software relative to    │
                    │ allocated system requirements│
                    │ and derived requirements    │
                    │                           2 │
                    └─────────────────────────────┘
                          /                \
        ┌─────────────────────┐    ┌─────────────────────┐
        │ 3.1: HLRSat{LevelX} │    │ 3.2: EOCSat{Level X}│
        │ High-level          │    │ Executable Object   │
        │ requirements are a  │    │ Code is a           │
        │ satisfactory for    │    │ satisfactory for    │
        │ {Level X} refinement│    │ {Level X} refinement│
        │ of the allocated    │    │ of the high-level   │
        │ system requirements │    │ requirements        │
        │                   3 │    │                   4 │
        └─────────────────────┘    └─────────────────────┘
```

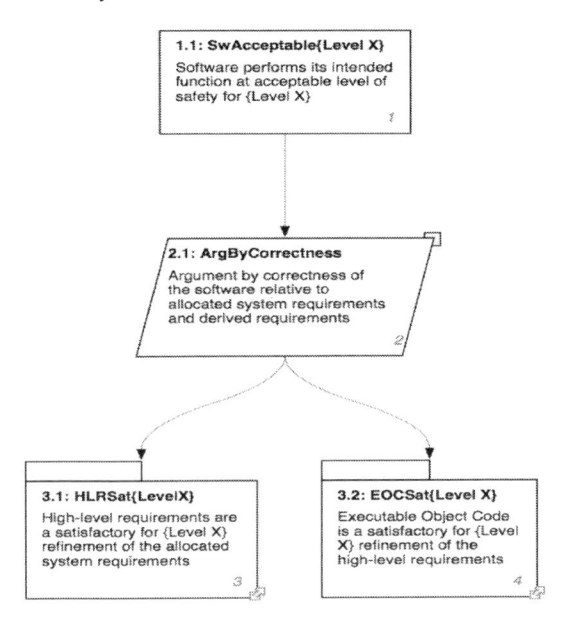

Fig. 1. Simplified generic primary argument in GSN

Two aspects of the figure may be unclear to anyone unfamiliar with the particular tool set used in the project[1]. The number in the lower right hand corner of each element is a tool-generated unique identifier. It permits easy reference to a particular GSN element across an entire collection of arguments. The small appendage on the upper right corner of the **ArgByCorrectness** strategy element indicates a link to an associated confidence argument, which is contained in a separate GSN module.

A top-level primary argument for each software level D, C, B, and A could be expressed using an instantiation of this generic argument. In the e78-1.6 assurance case, the primary arguments for levels D (shown below) and C (not shown) are expressed in this way. The primary arguments for levels B and A are not, because using a different structure that highlights the specific ways these levels differ from the lower levels seemed more enlightening.

Using the structured textual format developed for the project, the simple generic argument may also be expressed as shown in Figure 2. Note that the text contained in item C within the 'if' clause corresponds to the top-level goal of the associated confidence argument, which is not shown here.

The conclusion
    Software performs its intended function at acceptable level of safety for {level X}
is justified by an argument

---

[1] The GSN structures were produced using tools created by Dependable Computing, Inc. Use of these tools does not imply an endorsement of them by the U.S. Government.

by correctness of the software relative to allocated system requirements and derived requirements

if
- A. High-level requirements are a satisfactory for {level X} refinement of the allocated system requirements; **and**
- B. Executable Object Code is a satisfactory for {level X} refinement of the high-level requirements; **and**
- C. The evidence provided is adequate for justifying confidence that the correctness of the software has been demonstrated to the extent needed for {Level X}

Fig. 2. Simplified generic primary argument in structured text

3.3.2 Level D primary argument

A GSN expression of the primary argument for software Level D is shown in Figure 3.

Text contained within double quotation marks is quoted directly from either DO–178C if no document is specified, or from the specified document otherwise. The location of the quotation is given in parentheses. For example, the text in **MeaningAnomBeh** comes from page 109 in the Glossary of DO–178C, and the text in **HLRDev** comes from Annex A table 2 row 1 of 178C. The text in 3.1 References comes from bullet 6 in section 5.4 of DO–248C. To keep the size of some elements reasonably small, quotations are not always given, but instead references to document locations are listed.

The Level D primary argument follows the structure illustrated in the previous section, but with appropriate context and assumptions added. Five salient points about the argument are as follows.

(1) The five context elements attached to the top-level claim in the GSN representation emphasize that the meaning of the claim can only be understood within an environment containing a description of the intended function of the software and definitions for acceptable level of safety, Level D, and anomalous behavior. Also, the top-level claim is relevant only for software that has been assigned to level D. In the textual representation, these pieces of information are identified as 'givens' when considering whether the desired conclusion holds.

1.1: SwAcceptableLevD
Software performs its intended function at acceptable level of safety for level D
50

1.2: IntFun
Description of intended function of the software
36

1.3: DefAwReg
Definition of acceptable level of safety from airworthiness regulations
37

1.4: LevelDAssign
The software has been assigned to level D
40

2.1: MeaningLevelD
Description of the meaning of level D. "Software whose anomalous behavior, as shown by the system assessment process, would cause or contribute to a failure of system function resulting in a minor failure for the aircraft." (2.3.3.d).
38

2.2: MeaningAnomBeh
"Anomalous behavior: Behavior that is inconsistent with specified requirements" (Glossary, p. 108.)
39

2.3: ArgByCorrectness
Argue by correctness of the software relative to allocated system requirements and derived requirements
49

2.4: ReqAllocValidSuff
System requirements allocated to software augmented by any derived requirements are valid and sufficient to define intended function and ensure acceptable level of safety
42 A

2.5: HLRDev
"High-level requirements are developed" (A-2.1)
44

2.6: DerHLProv
"Derived high-level requirements are defined and provided to the system processes, including the system safety assessment process" (A-2.2)
46

3.1: References
"The relationship of the requirements development process to the safety process [is] defined to ensure that the safety analysis [is] not compromised by either the improper implementation of safety-related requirements or the introduction of new behavior (that is, derived requirements) that was not envisioned in the original safety analysis" DO-248C 5.4 bullet 6
41

3.2: References
5.1.1.a; Activities 5.1.2.a, 5.1.2.b, 5.1.2.c, 5.1.2.d, 5.1.2.e, 5.1.2.f, 5.1.2.g, 5.1.2), 5.5a; Glossary; 248C 5.5.1
43

3.3: References
5.1.1.b; Activities 5.1.2.h, 5.1.2.i; Glossary; DO-248C 5.5.1
45

3.4: Module HLRSatLevD
High-level requirements are a satisfactory (for level D) refinement of the allocated system requirements
47

3.5: Module EOCSatLevD
Executable Object Code is a satisfactory (for level D) refinement of the high-level requirements
48

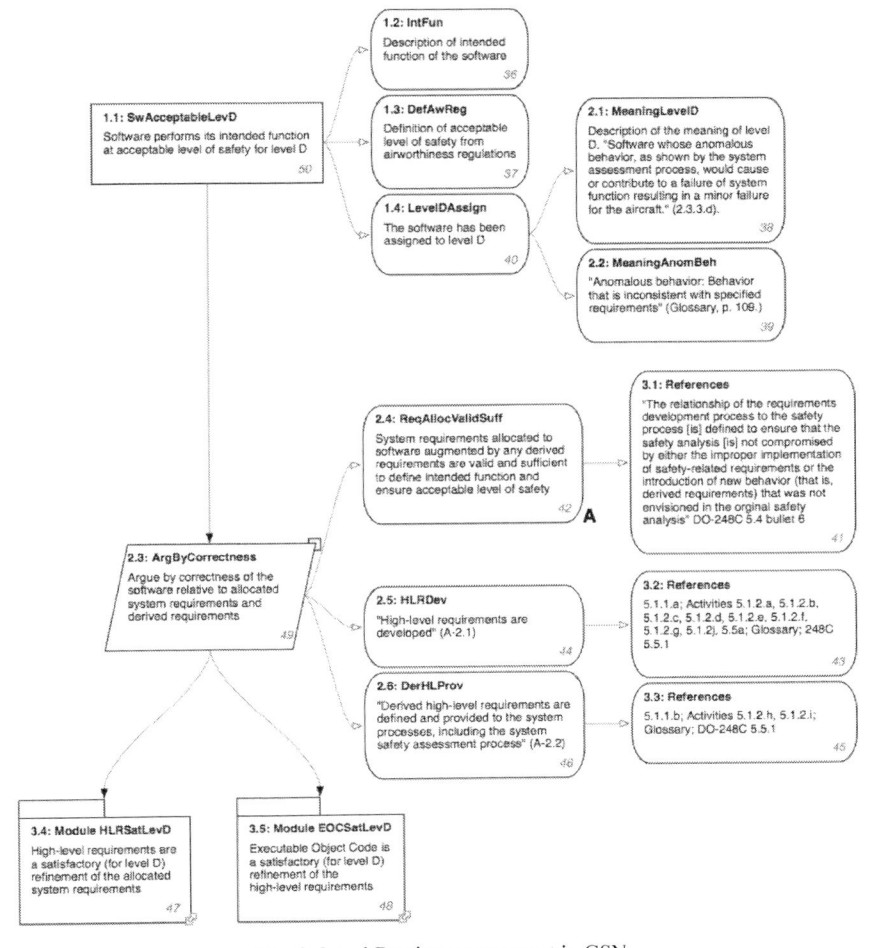

Fig. 3. Level D primary argument in GSN

(2) The assumption **ReAllocValidSuff** explicitly identifies an essential part of the implicit assurance case within 178C. As discussed in section 3.1.1, the guidance is grounded in the belief that the requirements to which the software is built are sufficient to both fully define the intended function of the software and ensure achievement of an acceptable level of safety. The guidance itself does not directly justify this belief, but it does include objectives intended to ensure that the safety analysis processes are provided with adequate information to conduct a proper assessment.

(3) **HLRDev** and **DerHLProv** both refer to specific objectives in 178C. From the vantage point of the assurance case, these objectives seem more properly to establish the context in which the implicit correctness argument makes sense and satisfies the **ReAllocValidSuff** assumption than to identify propositions that must be shown to be true as part of the argument.

(4) **HLRSatLevD** and **EOCSatLevD** are the two prongs of the correctness argument. If the high-level requirements are a satisfactory refinement of the system requirements, and the executable object code is in turn a satisfactory refinement of these high-level requirements then the software can be said to be correct with respect to the allocated system requirements. By the safety to correctness transformation previously discussed, the software can therefore be said to perform its intended function at an acceptable level of safety for Level D.

(5) The associated confidence argument is not shown here, but its goal is identified in the textual representation as **The evidence provided is adequate for justifying confidence that the correctness of the software has been demonstrated to the extent needed for level D**.

Figure 4 presents an equivalent structured text representation of the same argument.

The conclusion
    Software performs its intended function at acceptable level of safety for Level D
given
    A. Description of intended function of the software
    B. Definition of acceptable level of safety from airworthiness regulations
    C. The software has been assigned to Level D
    D. Description of the meaning of Level D: "Software whose anomalous behavior, as shown by the system assessment process, would cause or contribute to a failure of system function resulting in a minor failure condition for the aircraft for the aircraft." (2.3.3.d)
    E. "Anomalous behavior: behavior that is inconsistent with specified requirements" (Glossary, p. 109.)
is justified by an argument
    by correctness of the software relative to allocated system requirements and derived requirements
if
    A. High-level requirements are a satisfactory for Level D refinement of the allocated system requirements; and
    B. Executable Object Code is a satisfactory for Level D refinement of the high-level requirements; and
    C. The evidence provided is adequate for justifying confidence that the correctness of the software has been demonstrated to the extent needed for Level D
The argument assumes
    A. System requirements allocated to software augmented by any derived requirements are valid and sufficient to define intended function and ensure acceptable level of safety [see DO–248C 5.4 bullet 6]
    B. "High-level requirements are developed" (A-2.1) [see 5.1.1.a; Activities 5.1.2.a, 5.1.2.b, 5.1.2.c, 5.1.2.d, 5.1.2.e, 5.1.2.f, 5.1.2.g, 5.1.2j, 5.5a; Glossary; 248C 5.5.1]
    C. "Derived high-level requirements are defined and provided to the system processes, including the system safety assessment process" (A-2.2) [see 5.1.1.b; Activities 5.1.2.h, 5.1.2.i; Glossary; DO–248C 5.5.1]

Fig. 4. Level D primary argument in structured text

3.3.3 Level C confidence argument

Thus far, confidence arguments have been mentioned several times, but none have been shown. Figure 5 remedies the situation by showing a GSN expression of the confidence argument for Level C software, slightly simplified to allow legible display on paper.

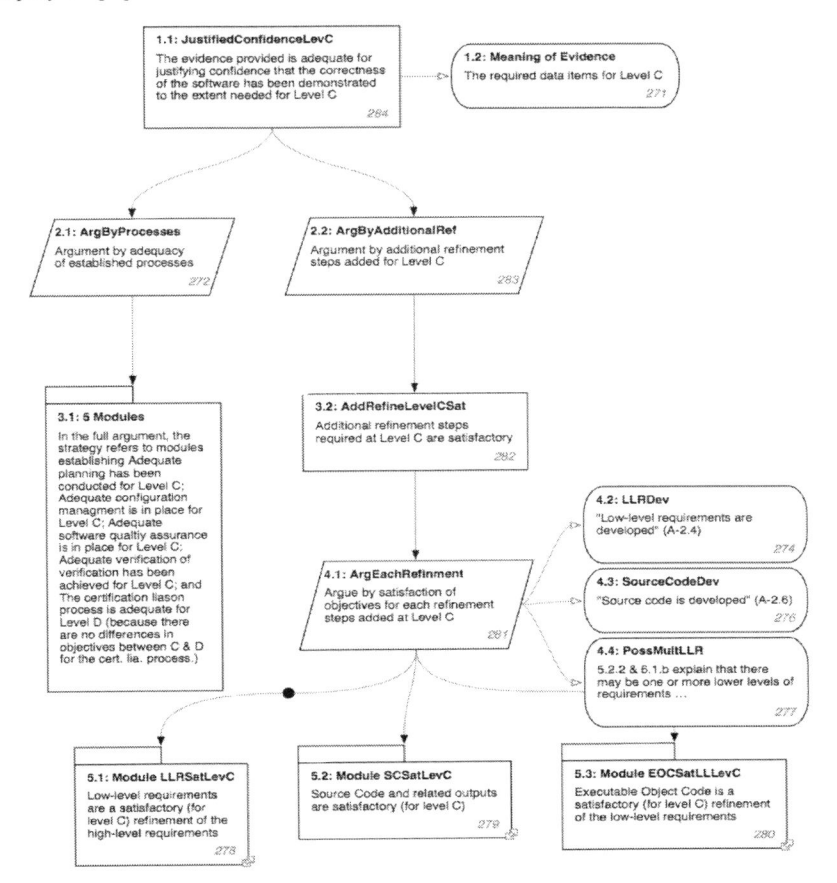

Fig. 5. Level C confidence argument in GSN

The goal of the confidence argument is to establish that the evidence used in the primary argument is adequate to justify believing that software correctness has been established. DO-178C's guidance related to showing the adequacy of processes for planning, configuration management, software quality assurance, verification of verification, and certification liaison all support gaining sufficient confidence. To enable Figure 5 to fit on the page, all of these are summarized in 3.1 5 Modules. In the full assurance case, separate modules exist related to each of the five processes.

To enhance confidence in the sufficiency of the two-level refinement process (system requirements to high-level requirements to executable object code), for Level C software, DO–178C introduces additional refinement steps, of which the guidance requires at least one (high-level to low-level), but allows for multiple in which 'the successive levels of requirements are developed such that each successively lower level satisfies its higher level requirements' (6.1.b). The possibility of multiple iterations of low-level requirements is denoted in the figure by the black circle on the arc from **ArgEachRefinement** to **Module LLRSatLevC**. The full assurance case includes details for each of the indicated modules.

3.3.4 Level B adequate planning argument

As an example from the Level B part of the e78-1.6 assurance case, Figure 6 shows the adequate planning component of the confidence argument.

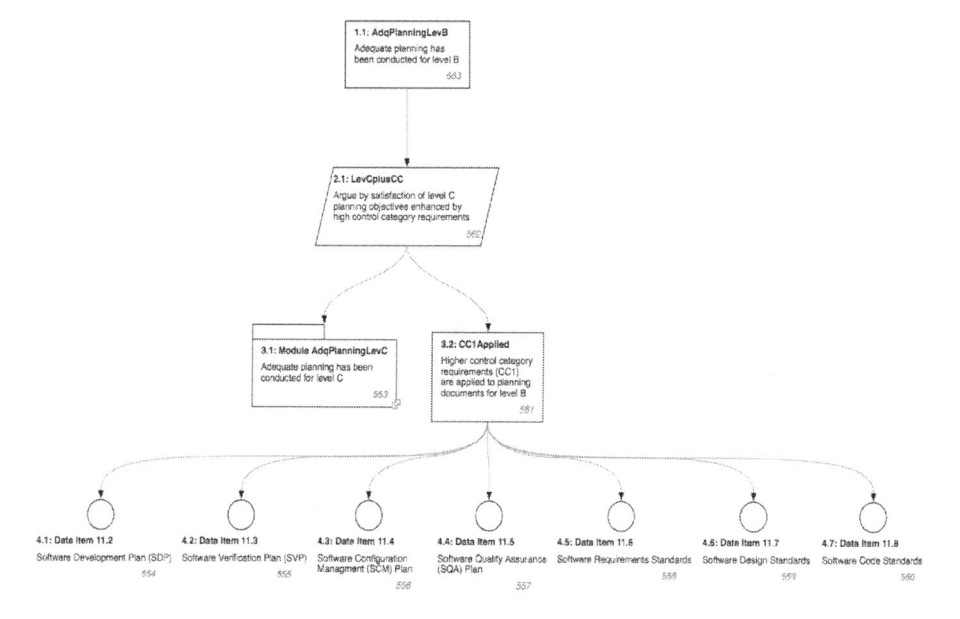

Fig. 6. Level B adequate planning argument in GSN

The objectives for planning at Level B are the same as the objectives for Level C. The only difference lies in the raising of the control category that applies to the seven planning-related data items, which are shown here as evidence items.

3.3.5 Level A verification of verification process results argument

A final excerpt from the e78-1.6 assurance case is given in Figure 7. This structure constitutes the verification of verification process results module of the confidence argument for Level A.

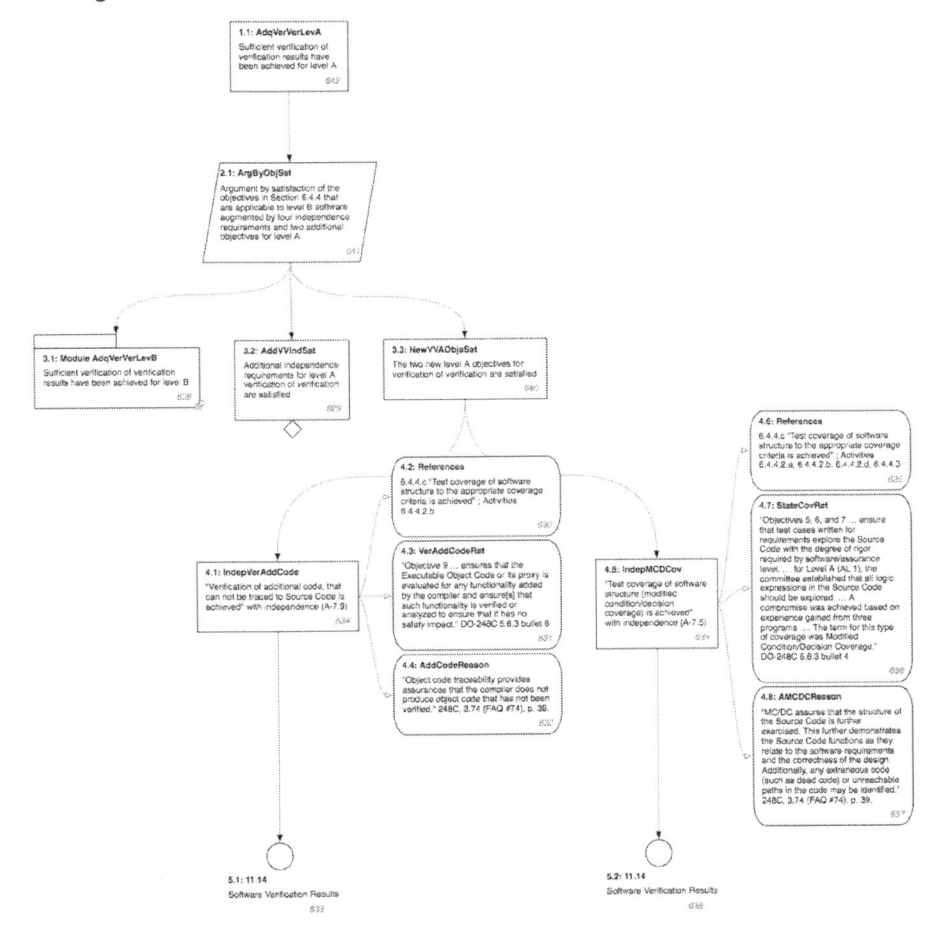

Fig. 7. Level A verification of verification process results argument in GSN

Verification of verification process results for Level A differs from Level B only in having additional requirements for independence (which are not elaborated in the figure, but are in the full assurance case), and two new objectives: verifying untraceable code (**IndepVerAddCode**) and achieving modified condition / decision coverage (**IndepMCDCov**), which must be done with independence.

4 Next steps and concluding remarks

The e78-1.6 assurance case discussed in this paper is not the final product of the Explicate '78 research. The case needs to be subjected to careful scrutiny by aviation industry and regulator experts, as well as assurance case and GSN experts. For the former, the existing textual representations most likely will need to be expanded to include the entire case. For the latter, the somewhat loose use of GSN elements that characterize the current case will likely need to be tightened.

The current case, however, seems to be sufficiently stable and complete to permit two concurrent activities to be undertaken during the heightened scrutiny period:

Beginning to evaluate the sufficiency of the case, not just as an accurate reflection of what DO–178C requires, but also as to whether what it requires is strong enough at each software level to provide justified assurance that software that complies with the document will perform 'its intended function with a level of confidence in safety that complies with airworthiness requirements.

Extending the existing case to include the guidance from one or more of the supplement documents.

If all goes well, good progress on all of these activities will be made before this paper is published. The goal is to complete the research before the end of 2015.

At least four benefits may arise from successful completion of this research, two of which are specific to DO–178C, and two of which are more general. First, the existence of an explicit assurance case for the DO–178C guidance should facilitate intelligent conversations about the relative efficacy of DO–178C and proposed alternative approaches for demonstrating compliance with airworthiness regulations. The likelihood of this benefit truly happening increases with the number of people within industry and the regulatory authorities who accept the Explicate '78 assurance case as an accurate reflection of the guidance.

Second, effectively analysing the adequacy of the assurance case should provide a solid foundation for future modifications to the guidance. When the time comes to create DO–178D, perhaps the Explicate '78 results will help provide the committee with a more structured and systematic basis for making changes than an unordered list of issues.

Third, more generally the existence of an assurance case representation for one guidance document may motivate the creation of such representations for other guidance documents. This, in turn, may result in clearer understanding of and more systematic updates to such documents.

Fourth, and most generally of all, perhaps the Explicate '78 work may help serve as a catalyst for prompting improved cooperation and mutual understanding between supporters of prescriptive standards and supporters of goal-based standards. One might even go so far as to hope for a lasting peace.

Acknowledgments This work is partially funded by the Reimbursable Interagency Agreement DTFACT-10-X-00008, Modification 0004, Space Act IAI-1073, between the Federal Aviation Administration and the National Aeronautics and Space Administration, Langley Research Center, for Design, Verification, and Validation of Advanced Digital Airborne Systems Technology: Annex 2, Assurance Case Applicability to Digital Systems. This paper, however, did not undergo any official review by FAA personnel.

References

Ankrum T, Kromholz A (2005) Structured Assurance Cases: Three Common Standards. Proceedings of the Ninth IEEE International Symposium on High-Assurance Systems Engineering (HASE'05). Heidelberg, Germany

Ayoub, A, Chang, J, Sokolsky, O, & Lee, I (2013) Assessing the Overall Sufficiency of Safety Arguments. Assuring the Safety of Systems: Proceedings of the Twenty-first Safety Critical Systems Symposium. C. Dale & T. Anderson (Eds.). February 5–7. Bristol, UK. Springer

Bloomfield R, Bishop P (2010) Safety and Assurance Cases: Past, Present and Possible Future. Making Systems Safer. C. Dale and T. Anderson (eds). Springer-Verlag

Denney, E, Pai, G, Habli, I, Kelly, T, & Knight, J (2013). 1st International Workshop on Assurance Cases for Software-intensive Systems (ASSURE 2013). Proceedings of the 2013 International Conference on Software Engineering. May 18–26. San Francisco, California.

European Aviation Safety Agency (2013) AMC 20-115C Software Considerations for Certification of Airborne Systems and Equipment. ED Decision 2013/026/R http://easa.europa.eu/system/files/dfu/Annex%20II%20-%20AMC%2020-115C.pdf (last accessed December 2, 2014)

Federal Aviation Administration (2013a) Standard Airworthiness Certification: Regulations – Title 14 Code of Federal Regulations. http://www.faa.gov/aircraft/air_cert/airworthiness_certification/std_awcert/std_awcert_regs/regs/ (last accessed December 5, 2014)

Federal Aviation Administration (2013b) Advisory Circular 20-115C Airborne Software Assurance. http://www.faa.gov/documentLibrary/media/Advisory_Circular/AC_20-115C.pdf (last accessed December 2, 2014)

Galloway A, Paige R, Tudor, N, Weaver R, McDermid, J. (2005) Proof vs. Testing in the Context of Safety Standards. The 24th Digital Avionics Systems Conference (DASC), Washington D.C.

Goodenough J, Weinstock C, Klein A (2012) Toward a Theory of Assurance Case Confidence. CMU-SEI-2002-TR-002, September

Graydon, P (2014) Towards a Clearer Understanding of Context and Its Role in Assurance Argument Confidence. Computer Safety, Reliability, and Security, 139-154.

GSN Committee (2011) Draft GSN Standard Version 1.0. http://www.goalstructuringnotation.info/ (last accessed December 2, 2014)

Haddon-Cave C (2009) The Nimrod Review. London: The Stationary Office http://www.official-documents.gov.uk/document/hc0809/hc10/1025/1025.pdf (last accessed December 1, 2014).

Hawkins R, Habli I, Kelly T, McDermid J (2013) Assurance cases and prescriptive software safety certification: A comparative study. Safety Science. Vol 59

Hawkins R, Kelly T (2009) A Systematic Approach for Developing Software Safety Arguments. Proceedings of the 27th International System Safety Conference. Huntsville, Alabama

Hawkins R, Kelly T, Knight J, Graydon P (2011) A New Approach to Creating Clear Safety Arguments. Advances in Systems Safety. C. Dale and T. Anderson (eds). Springer-Verlag

Holloway CM (2013) Making the Implicit Explicit: Towards an Assurance Case for DO–178C. Proceedings of the 31st International System Safety Conference. August 12-16. Boston, Massachusetts (ref. z)

Holloway CM (2012) Towards Understanding the DO–178C / ED–12C Assurance Case. 7th IET International Conference on System Safety, Incorporating the Cyber Security Conference. Edinburgh

Holloway CM (2008) Safety Case Notations: Alternatives for the Non-Graphically Inclined? Proceedings of the 3rd IET International System Safety Conference. Birmingham, UK

Knight J (2012) Fundamentals of Dependable Computing for Software Engineers. Boca Raton, Florida: CRC Press

Matsuno, Y (2014) A Design and Implementation of an Assurance Case Language. Dependable Systems and Networks (DSN). Atlanta, Georgia

RTCA (1992) Software Considerations in Airborne Systems and Equipment Certification. DO–178B.

RTCA (2011a) Software Considerations in Airborne Systems and Equipment Certification. DO–178C.

RTCA (2011b) Supporting Information for DO–178C and DO–278A. DO–248C

RTCA (2011c) Software Integrity Assurance Considerations for Communication, Navigation, Surveillance, and Air Traffic Management (CNS/ATM) Systems. DO–278A

RTCA (2011d) Software Tool Qualification Considerations. DO–330

RTCA (2011e) Model-Based Development and Verification Supplement to DO–178C and DO–278A. DO–331

RTCA (2011f) Object-Oriented Technology and Related Techniques Supplement to DO–178C and DO–278A. DO–332

RTCA (2011g) Formal Methods Supplement to DO–178C and DO–278A. DO–333

Rushby, J (2013) Logic and epistemology in safety cases. Computer Safety, Reliability, and Security, 32nd SAFECOMP. Toulouse, France

Rushby J (2011) New Challenges in Certification of Aircraft Software. Proceedings of the 11th International Conference on Embedded Software (EMSOFT). Taipei, Taiwan

SAE International (1996) Guidelines and Methods for Conducting the Safety Assessment Process on Civil Airborne Systems and Equipment. SAE ARP 4761

SAE International (2010) Guidelines for Development of Civil Aircraft and Systems. SAE ARP 4754a

Toulmin S (2003) The Uses of Argument, Updated Edition. Cambridge University Press

UK Ministry of Defence (2007) Defence Standard 00-56 Issue 4: Safety Management Requirements for Defence Systems

Yuan T, Kelly T (2011) Argument Schemes in Computer System Safety Engineering. Informal Logic 31 (2)

Using a Goal-Based Approach to Improve the IEC 61508-3 Software Safety Standard

Thor Myklebust
Information and Communication Technology SINTEF Trondheim, Norway[1]

Tor Stålhane
Norwegian University of Science and Technology Trondheim, Norway

Børge Haugset, Geir Kjetil Hanssen
Information and Communication Technology SINTEF Trondheim, Norway

Abstract *In this paper we argue that the methods and techniques specified in the annexes in IEC 61508-3 are just sound software engineering principles. Problems when developing safety critical software are not caused by lack of adherence to the standard per se but by ignorance of sound engineering principles related to the specified techniques. Further we argue that IEC 61508-3 should be more flexible regarding the safety lifecycle requirements by mentioning the use of modern software development practices together with the V-model.*

1 Introduction

In our opinion, the most important things needed when making safety critical software is general, sound engineering competence, combined with competence in software development and the application domain, good communication within the development team and mindfulness of safety. It is our opinion that standards, such as IEC 61508-3 (IEC 2010), are useful since they list more or less the best techniques and methods available at the time when the standards were written. This is their strong point but also their weak point. The weakness becomes apparent when we want to apply new tools, techniques or methods. Software engineering research has gradually acknowledged the importance of context as a factor in determining success of methods, and that it is impossible to determine which method is best - it all depends on context (Dybå 2013).

Different project development contexts may give rise to entirely different best practices. At that point in time, the standard moves from being helpful – what are the recommended current "best" practices – to becoming a hindrance. For new

---

[1] thor.myklebust@sintef.no

© SINTEF/NTNU 2015. Published by the Safety-Critical Systems Club. All Rights Reserved

methods not listed in the standard, the assessor has to be consulted. This complicates the process and is highly dependent on the assessor subjective opinion. There are two ways to get past this problem:

- We can work to change the standard and thus include the new techniques or methods into the standard. This is our current approach, as one of the authors takes part in the maintenance work of IEC 61508-3 that started in autumn 2014, but it is just a stop-gap measure. We will face the same problem again when something new comes along.
- Move towards a goal-oriented standard (IMO 2014). This will allow us to use whatever tools, techniques and methods that are appropriate as long as we meet the standard's designated goals

Appealing as the second alternative is, it creates one new problem. A typical statement in a goal-oriented standard could be "The methods for testing and the volume of tests must be sufficient for the defined SIL". The question is who shall decide what is sufficient – the regulators, developers, the customer or the assessor?

2 Background

Below we present information related to the main topics of this paper, the software safety standard IEC 61508-3, information related to goal-based standards and agile development of software.

2.1 IEC 61508-3

The lifecycle requirements are presented in chapter 7 in IEC 61508-3:2010. The objectives are to structure the development of the software into defined phases and activities. The requirements are based on the waterfall process methodology although "any software lifecycle model may be used provided all the objectives and requirements of this clause are met". The current standard has not succeeded in presenting the requirements as model independent since the requirements are presented according to the waterfall model, including the V-model. This makes it difficult for the manufacturers to use other models although we have indicated that with a little flexibility there are no large obstacles for using agile development models like Scrum for safety critical software (Stålhane et.al. 2012).

Chapter 7.1.2.7 in the standard requires that appropriate techniques and measures shall be used. Techniques or measures are highly recommended for the relevant safety integrity level. If a technique or measure is not used, then the rationale behind not using it should be detailed with having in mind that the large number of factors that affect software systematic capability. It is not possible to give an algorithm for combining the techniques and measures that will be correct for any given application. This should be performed during the safety planning and agreed with the assessor. As a result the standard is in practice a prescriptive standard.

2.2 Goal based standards

In the last two decades there has been an increasing tendency towards a goal-based approach to regulation and standards (requirements for the manufacturers, what they have to do – that includes alternative ways of achieving compliance) compared to the earlier prescriptive regulations and standards (requirements that have to be met if a user wishes to claim compliance with the standard). The reasons behind a goal-based approach are rapid technology changes, new development processes and the legal viewpoint. Too restrictive standards may be viewed as a barrier to trade.

There has been some research on goal-based standards together with the change of some of the safety standards towards a goal-based approach. Already issued goal-based standards are e.g. IMO (International Maritime Organization) standards (IMO 2014) and the defence standard 00-56 (MOD 2004). In the research papers (Grouped References), several aspects related to the use of Goal-based standards have been discussed, including: level of argument and evidence, goal-based safety cases and challenges related to COTS equipment (Commercial Off The Shelf). Based on this survey there seems to be little research on the use of new software development methods and goal-based standards. Further, the "Techniques & Measures" in Annex A and B in IEC 61508-3 have not been studied by the research community related to goal-based standards.

2.3 Agile development

Agile software development is a way of organizing the development process, emphasizing direct and frequent communication, frequent deliveries of working software increments, short iterations, active customer engagement throughout the whole development life cycle and change responsiveness rather than change avoidance. This can be seen as a contrast to waterfall-like models, which emphasize thorough and detailed planning, and design upfront and consecutive plan conformance. Several agile methods are in use whereof *extreme programming (XP)* (Beck 2004) and *Scrum* (Schwaber 2001) are the most commonly used for development of non safety critical systems. Figure 1 explains the basic concepts of an agile development model.

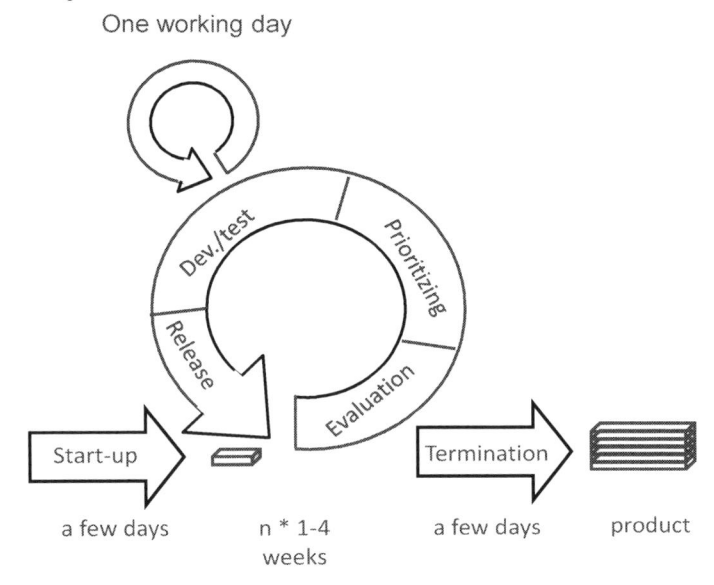

Fig.1: The basic agile software development model

The main constructs of this model are (based on Scrum):

- Initial planning is short and results in a prioritized list of requirements for the system called *the product backlog*. Developers also develop *estimates* per item.
- Development is organized as a series for *sprints* (iterations) that lasts a few weeks. Each sprint should have the same length throughout the project.
- Each sprint starts with a *sprint planning meeting* where the top items from the prioritized product backlog is moved over to the *sprint backlog* – adding up to the amount of resources available for the period. These requirements will be implemented in the following sprint.
- Each working day starts with a *scrum*, which is a short meeting where each member of the development team explains what she/he did the previous work day, any impediments or problems that need to be solved, and planned work for the work day.
- Each sprint *releases* an *increment* which is a running or demonstrable part of the final system.
- The increment is *demonstrated* for the customer(s) and other stakeholders, who decide which backlog items that have been resolved and which that need further work. Based on the results from the demonstration the next sprint is

planned. The product backlog is revised by the customer and is potentially changed / reprioritized. This initiates the sprint-planning meeting for the next sprint.
- When all product backlog items are resolved and / or all available resources are spent the final product are released. Final tests can be run to ensure completeness.

The adaptation of Scrum to development of safety-critical systems, which we call "SafeScrum", is motivated by the need to make it possible to use methods that are flexible with respect to planning, documentation and specification while still being acceptable to IEC 61508, as well as making Scrum a practically useful approach for developing safety-critical systems. The rest of this section explains the components and concepts of this combined approach.

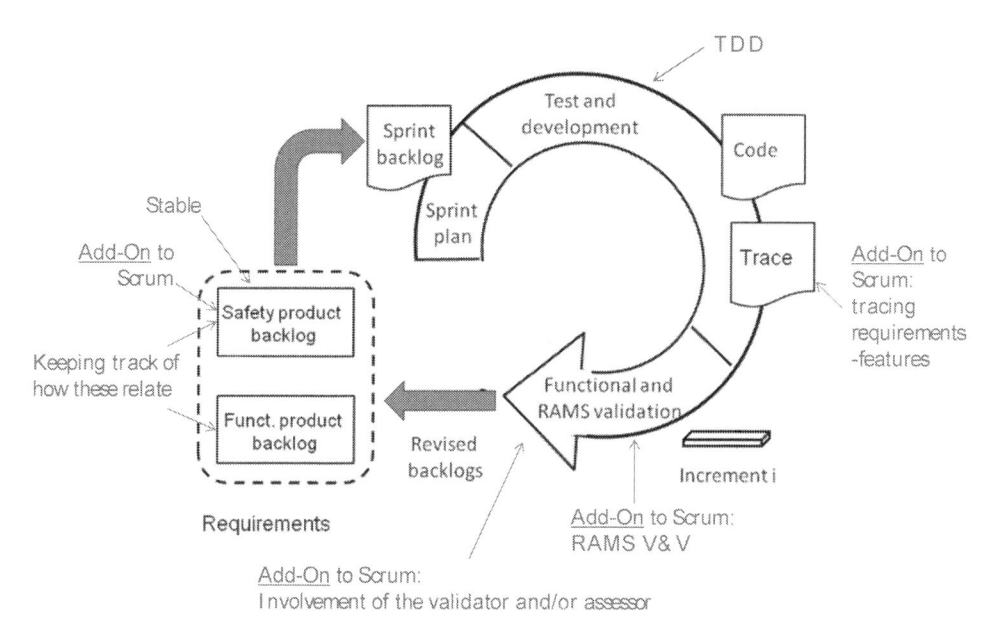

Fig.2: The SafeScrum model. TDD: Test Driven Development

The safety requirements and other requirements are documented as product backlogs. A product backlog lists all functional and safety related system requirements, prioritized by importance. The safety requirements are quite stable (relevant regulations and safety standards are normally stable during the project), while the functional requirements can change considerably over time. Development with a high probability of changes to requirements will favor an agile approach.

All risk and safety analyses on the system level are done outside the SafeScrum process, including the analysis needed to decide the SIL level. Software is considered during the initial risk analysis and all later analysis. Just as for testing, safety analysis also improves when it is done iteratively and for small increments – see (Morsicano).

Due to the focus on safety requirements, we use two related product backlogs, one functional product backlog, which is typical for Scrum projects, and one safety product backlog, to handle safety requirements. We will keep track of how each item in the functional product backlog relates to the items in the safety product backlog, i.e. which safety requirements that are affected by which functional requirements. These two backlogs do not necessarily need to be separated by more than different tags within the same requirement tracking tool.

The core of the Scrum process is the repeated iterations. Each iteration is a mini waterfall project or a mini V-model, and consists of planning, development, testing, and verification. For the development of safety critical systems, traceability between system/code and backlog items, both functional requirements and safety requirements, is needed. The documentation and maintenance of trace information is introduced as a separate activity in each sprint – see fig. 3. In order to be performed in an efficient manner, traceability requires, in practice, the use of a supporting tool.

The iteration starts with the selection of the top prioritized items from the product backlog. In the case of SafeScrum, items in the functional product backlog may refer to items in the safety product backlog. The staffing of the development team and the duration of the sprint (14 days is common), together with the estimates of each item decides which items that can be selected for development. The selected items constitute the sprint backlog, which ideally should not be changed during the sprint. The development phase of the sprint is based on developers selecting items from the sprint backlog, and producing code to address the items.

A sprint should always produce an increment, which is a piece of the final system, for example executable code. The sprint ends by demonstrating and validating the outcome to assess whether it meets the items in the sprint backlog. Some items may be found to be completed and can be checked out while others may need further refinement in a later sprint and goes back into the backlog. To make Scrum conform to IEC 61508, we propose that the final validation in each iteration is done both as a validation of the functional requirements and as a RAMS validation, to address specific safety issues. If appropriate, the independent safety validator may take part in this validation for each sprint. If we discover deviations from the relevant standards or confusions, the assessor should be involved as quickly as possible. Running such an iterative and incremental approach means that the development project can be continuously re-planned based on the most recent experience with the growing product.

As the final step, when all the sprints are completed, a final RAMS validation will be done. Given that most of the developed system has been incrementally validated during the sprints, we expect the final RAMS validation to be less extensive than when using other development paradigms. This will also help us to reduce the time and cost needed for certification.

The key benefits of this combination of a safety-oriented approach and a process model for agile software development are that the process enables:

- Continuous communication between customers, the development team and the test team(s).
- Re-planning based on the most recent understanding of the requirements and the system under development.
- Mapping of functional and safety requirements.
- Code-requirements traceability.
- Coordination of work and responsibilities between the three key roles; the development team, the customer and the assessor.
- Test-driven development of safety critical systems.

All of these points will help us to get a more visible process and thus better control over the development process.

3 Development of a hypothesis

Approach

The four authors have together participated in several development projects where the IEC 61508 standard has been used, both as developers and assessor. Our current analysis of IEC 61508-3 is based on a thorough analysis of the standard, performed as follows:

- All standard requirements in annexes A and B of IEC 61508-3, pertaining to software that were categorized as HR (highly recommended) for SIL 3 were inserted into a table.

Since we are discussing common software development practices, all requirements that were only related to the development of safety critical systems are removed as indicated in the column marked with "Safety".

For each of the remaining requirements, we inserted our interpretation of the intent of this requirement – see example in table 1 below. Thereafter, we organized them into the traditional categories of software development activities – analysis and design, reuse, coding, validation and verification (V&V) and finally maintenance.

- The list of intents was checked to see if there were intents that were not part of sound engineering practices. Our conclusion is that there was none.

Table 1: Example of requirements walk-through

A9 - Software Verification (IEC 61508-3)			
Requirement	SIL 3	Software development interpretation	Safety
Formal proof	No	-	Yes
Animation of specification and design	No	-	Yes
Static analysis Boundary value analysis • Control and data flow analysis • Design review • Formal inspection	Yes	*Rigour may range from language subset enforcement to mathematical formal analysis. Formal inspection is seldom used*	No
Dynamic analysis and testing • Boundary value analysis • Structure based testing	Yes	*White box testing. The requirement is supported by the use of tools like Junit (http://junit.org/)*	No
Forward and backward traceability between the software design specification and the software verification	Yes	-	Yes
Offline numerical analysis	Yes	-	Yes

Results

As stated in the start of this section, all IEC 61508–3 requirements are mapped onto standard software development methods and techniques – analysis and design, reuse, coding, validation and verification (V&V) and finally maintenance:

Analysis and design
- Design standard. Will be enforced by a computer-aided design tool.
- Semi-formal methods – e.g. UML (Unified Modelling Language). Sequence diagrams and finite state machines are part of UML.
- Structured method, structured programming. A modular approach, encapsulation, one entry-point – one exit-point and a fully defined interface are all part of structured programming
- Design review. Most projects do a design review. The level of details and effort will, however, vary.

Reuse
- Use of trusted software modules and components. Common practice

Coding
- Strongly typed programming language. Most programming languages are now strongly typed
- Language subset. Most companies use just a language subset.
- Coding standards. Most projects have one. Scrum requires one. Coding standards are available both in books, reports and can be bought.

Validation (testing) and verification – V&V
- Static analysis. Rigour may range from language subset enforcement to mathematical formal analysis. The analysis shall include boundary value analysis, control and data flow analysis and dynamic analysis and testing. This can be done using tools such as QA-C and QA-CPP (www.programmingresearch.com/).
- White box testing and structure based testing. The requirement is supported by the use of tools like Junit.
- Data recording and analysis. Fault records, test logs and software baseline info. Used in all or most testing activities.
- Functional and black box testing. Always used
  - Equivalence class and partitioning testing are important part of black box testing but often not done formally.
- Interface testing. Is part of black box testing
- Test management and automatic tools, requires that coverage targets are defined and met. Coverage targets is an important part of white box testing.

Maintenance
- Impact analysis. Is mostly used by the manufacturers.
- Re-verify changed and affected modules. Is always used.
- Revalidate complete system. Is always used.
- Software configuration management. Common practice.
- Data recording and analysis. Just common sense / diligence. Fault records, test logs and software baseline info. Used in all or most testing activities. This also includes checks for:
  - Completeness of modification
  - Correctness of modification

Conclusion

We have shown that all IEC 61508 – part 3 requirements for software development in annexes A and B are just sound and established software development practice. We do not, however, know which methods and tools that will be available in the (near) future. In addition, it is important to be aware that the standard requirement e.g. boundary value analysis, says nothing on how much, to what level and how extensive it should be.

Thus, it would have been more sensible and practical to replace a large part of annexes A and B with a statement to the effect that the developers should use well tested methods for software development and add a requirement on the developers' knowledge and experience. A good example of how this can be done is tables B2 and B3 in EN 50128 (EN 50128:2011). In these tables the standard defines the responsibilities and key competencies of all roles involved. Some of the key competencies for a developer (implementor) is, for instance, competence in the implementation language and support tools and understanding the relevant parts of EN 50128.

Hypotheses

Based on the discussions above, we offer the following hypotheses:

H1: When the development of safety critical software fails, either by not delivering a product or delivering a product that did not obtain a certificate, this is due to lack of adherence to the IEC 61508 standard and not due to lack of adherence to sound software engineering practices.

H2: Strict adherence to a standard may prevent developers from choosing methods and techniques that are optimal for the situation at hand.

To test the hypotheses, we need documents from failed projects that can be used to gain understanding of how, why and in what respect the project failed. Such documents are for instance post mortem reports (Birk 2002) and other types of project assessment reports. In addition, we should interview relevant project personnel in cases where the project was terminated less than a year ago. Given access to this information, the two hypotheses can be tested as follows:

H1: We will reject H1 if data in more than 90% of the projects reviewed indicate that the project's problems stem from lack of sound engineering practices.

H2: We will reject H2 if data from more than 90% of the projects reviewed indicate that the project has chosen methods that are not optimal for the project due to strict adherence to the standard.

Threats to validity

The hypotheses will be tested based on our interpretation of available documents and our interpretation of interviews, where the available information will be based on project participants' sometimes unreliable memory.

4 Software lifecycle models

The second edition of IEC 61508-3 (IEC 61508-3:2010) has moved towards a more goal-based approach than earlier editions of this standard (IEC 61508-3:1998). The main change was that a new chapter 7.1.2.2 was included, see Table 2 below. According to IEC 61508-3 (IEC 61508-3:2010), any lifecycle may be used provided that all the objectives and requirements are met. Although this is stated in the standard, it presents the requirements as if the v-model is the only model to be used. In the Table 2 below we have presented the current requirements together with suggested future requirements to make the standard Goal-based.

Table 2: Goal-based lifecycle requirements

Cl.	Requirements in IEC 61508-3	Goal-based requirements	Comments
7 7.1	**Software safety lifecycle requirements** General	No change in the Title of the chapter and subchapter	-
7.1.2.1	"A safety lifecycle for the development of software shall be selected and specified during safety planning in accordance with clause 6 of IEC 61508-1.	No change	Our previous paper (Stålhane et.al. 2012) present as an example of an agile safety lifecycle. A company introducing a new method not mentioned in IEC 61508-3 should include their adaptation in their quality/safety system. In addition they should justify the use of the new method as we also show (IEC 61508-3:2010)
7.1.2.2	"Any software lifecycle model may be used provided all the objectives and requirements of this clause are met."	No change	-

Cl.	Requirements in IEC 61508-3	Goal-based requirements	Comments
7.1.2.3	"Each phase of the software safety lifecycle shall be divided into elementary activities with the scope, inputs and outputs specified for each phase.

NOTE See Figures 3, 4 and Table 1." | No change in the requirement, but add two notes:

"NOTE 1: A software lifecycle typically includes a requirements phase, development phase, test phase, integration phase, installation phase and a modification phase".

Note 2: A software lifecycle in a typical Sprint includes an evaluation phase, prioritizing phase, development and test phase and release phase. | None of the references mentioned in the column "Requirements subclause" in Table 1 is to ch.7.1 for lifecycles. The evaluation of Table 1 is therefore considered outside the scope of this paper. |
| 7.1.2.4 | "Provided that the software safety lifecycle satisfies the requirements of table 1, it is acceptable to tailor the V-model (see figure 6) to take account of the safety integrity and the complexity of the project. | Provided that the software safety lifecycle satisfies the requirements in chapter 7, it is acceptable to tailor the model chosen (e.g. V-model or Scrum) to take account of the safety integrity and the complexity of the project. | Two notes mentioned in 7.1.2.4 are not included in column 2 |

Cl.	Requirements in IEC 61508-3	Goal-based requirements	Comments
7.1.2.5	"Any customization of the software safety lifecycle shall be justified on the basis of functional safety."	No change	See (Stålhane et.al. 2012) for an example of a sufficient justification
7.1.2.6	"Quality and safety assurance procedures shall be integrated into safety lifecycle activities."	No change	-
7.1.2.7	"For each lifecycle phase, appropriate techniques and measures shall be used. Annexes A and B provide a guide to the selection of techniques and measures, and references to IEC 61508-6 and IEC 61508-7. IEC 61508-6 and IEC 61508-7 give recommendations on specific techniques to achieve the properties required for systematic safety integrity. Selecting techniques from these recommendations does not by itself guarantee that the required safety integrity will be achieved.	No change	One note mentioned in 7.1.2.7 is not included in column 2
7.1.2.8	"The results of the activities in the software safety lifecycle shall be documented (see clause 5).	No change	One note mentioned in 7.1.2.8 is not included in column 2

Cl.	Requirements in IEC 61508-3	Goal-based requirements	Comments
7.1.2.9	"If at any phase of the software safety lifecycle, a modification is required pertaining to an earlier lifecycle phase, then an impact analysis shall determine (1) which software modules are impacted, and (2) which earlier safety lifecycle activities shall be repeated.	No change, but add one note: Note 2: If the earlier lifecycle is one of the phases in the sprint, the requirements included should be included in a later sprint For further information on Impact analysis and Agile methods, see (23) – (25).	One note mentioned in 7.1.2.9 is not included in column 2

6 Conclusion

Our conclusion is that IEC 61508-3's handling of software development in its annexes is just a codification of current sound engineering practices for software development. The reason for this is obvious: The standards have been created based on what the committee participants considered important practices and artefacts. In addition, they were created without leaning too much on actual research into each aspect. Even though IEC 61508 has been updated once, it has still been focused on the waterfall process. It could also in principle include items not actually needed as well as a lack of other important aspects or 'other ways of solving the equation'. It described the world as these groups of people saw it at the current time, given their background and knowledge. We believe time is ripe for change from a method-centric to goal-based assessment focus. We believe this could work because these practices together make the stakeholders and developers look more closely into what they are making. They scrutinize all matters, double-check the code and test it thoroughly. However, that does not mean that that particular set of practices is the only way of developing software.

We have analysed the lifecycle requirements in IEC 61508-3 ch.7 and suggested how the requirements could be modified to ensure that the next edition of IEC 61508-3 becomes a goal-based standard. Only ch. 7.1.2.4 has to be changed.

References

Beck, K and Andres, C (2004). Extreme programming explained: embrace change, 2nd Edition. 2004, Boston: Addison-Wesley Professional

Birk, A., Dingsøyr, T., and Stålhane, T. (2002), *Postmortem: Never Leave a Project without It.* IEEE Software, 2002. 19(3): p. 43 - 45.

Dybå, T (2013), "Contextualizing empirical evidence", IEEE Software, Vol.30, no 1, pp81-83, Jan 2013

EN 50128:2011 Railway applications – Communication, signalling and processing systems – Software for railway control and protection systems. Edition 2.

IEC (1998) IEC 61508-3:1998. Functional safety of electrical/electronic/programmable electronic safety-related systems – Part 3: Software requirements. Ed.1

IEC (2010) IEC 61508-3:2010. Functional safety of electrical/electronic/programmable electronic safety-related systems – Part 3: Software requirements. Ed.2

IMO (2014), Goal-based construction standards for new ships, www.imo.org/OurWork/Safety/SafetyTopics/Pages/Goal-BasedStandards.aspx, accessed December 2014

MOD (2004), Issue 3 of UK Defence Standard (DS) 00-56, published at the end of 2004

Morsicano R. and Shoemaker, B. Tutorial: Agile Methods in Regulated and Safety-Critical Environments. ShoeBar Associates

Myklebust, T., Stålhane, T, Hanssen, G.K. and Haugset, B. (2014), Change Impact Analysis as required by safety standards, what to do? PSAM 12 Hawaii 2014

Schwaber, K. and Beedle, M (2001). Agile Software Development with Scrum. 2001, New Jersey: Prentice Hall

Stålhane, T., Hanssen, G., and Myklebust, T.: The Application of SafeScrum to IEC 61508 Certifiable Software – part 1 and 2. Safety Sytems. The SCSC Club Newsletter, vol 23, no 1 and 2

Stålhane, T., Hanssen, G.K., Myklebust, T. and Haugset, B. (2014), Agile Change Impact Analysis of Safety Critical Software. SafeComp, SASSUR 2014

Stålhane, T., Katta, V. and Myklebust, T. (2014). Change Impact Analysis in Agile Development. EHPG Røros 2014

Stålhane T, Myklebust T, Hanssen G (2012). The application of Safe Scrum to IEC 61508 certifiable software. PSAM11/ESREL 2012. Helsinki June 2012.

Grouped References

[1] McDermid, J. and Rae, A. (2012): Goal-Based Safety Standards: Promises and Pitfalls. In proceedings of Twentieth Safety-Critical Systems Symposium. 2012. Bristol, UK: Springer.

[2] Stensrud E., Skramstad T., Li Jigyue and Xie Jing (2011). Towars Goal-based Software Safety Certification Based on Prescriptive Standards. 2011 First International Workshop on Software Certification.

[3] Becht H. (2012) Moving Towards Goal-Based Safety Management. ASSC2011
[4] McDermid J. and Rae A (2012). Goal-Based Safety Standards: Promises and Pitfalls. Achieving Systems Safety 2012, pp 257-270
[5] Coglianese C, Nash J, Olmestad T (2002) Performance-based regulation: prospects and limitations in health, safety and environmental protection. KSG Working paper series No RWP02-050. http/ssrn.com/abstract=392400. Accessed 2014-08-05.
[6] Kelly T, McDermid J and Weaver R (2005). Goal-based safety standards: Opportunities and challenges. In Proc 23rd Int system safety eng. Conf, system safety society. San Diego 2005.
[7] Hoppe H. Goal-based standards – A new approach to the international regulation of ship construction
[8] Penny, J., Eaton, A., Bishop, P., Bloomfield, R., "The Practicalities of Goal-Based Safety Regulation", Proc. Ninth Safety-critical Systems Symposium (SSS 01), Bristol, UK, 6-8 Feb, pp. 35-48, New York: Springer, ISBN: 1-85233-411-8, 2001
[9] Melkild E. M. B (2013). Goal and evidence based dependability assessment. NTNU Thesis 2013
[10] Menon C, Hawkins R, and McDermid J (2009). Defence Standard 00-56 Issue 4: Towards Evidence-Based Safety Standards. SSS-2009.

Risks People Take and Games People Play

Peter Bernard Ladkin

University of Bielefeld and Causalis Limited

Bielefeld, Germany and London, UK

Abstract *In July 2014, a commercial transport aircraft, Malaysia Airlines Flight 17, in cruise flight over Ukraine, had its flight abruptly terminated through "impacts from a large number of high-energy objects from outside the aircraft",* The *suspicion is that it was shot down. Three other commercial aircraft on international flights were in the same control sector at the time; other airlines had chosen to avoid the area. I argue that the kind of risk analysis one must perform to assess such possible security threats cannot be of the IEC 61508 type. I propose Meta-Game Theoretic Analysis, MGTA.*

1 What is Risk Assessment? An International Standard or Two

Safety assessment of critical systems in commercial aviation has been based for a long time on risk assessment. Since the late 1990's, the international standard for functional safety of electrotechnical systems IEC 61508 has also propagated an approach based on assessing risk (IEC 2010). Indeed, there is a general guide for electrotechnical standards which incorporate safety aspects, prepared by the Advisory Committee on Safety of the IEC, the international electrotechnical standardisation body, which requires that all such standards incorporate a risk assessment (ISO/IEC 2014).

A risk assessment according to the 2014 edition of Guide 51 (op. cit.) proceeds as follows:

1. You identify hazards;

   Loop:

2. You estimate risk;
3. You evaluate risk;

© Peter Bernard Ladkin 2015.
Published by the Safety-Critical Systems Club. All Rights Reserved

4. You reduce risk where intolerable;

   until <residual risk is tolerable>

5. You validate and document your reasoning along with the evidence.

In evaluating whether residual risk is tolerable, a nod is given to ALARP, and to social conventions concerning tolerability as well as other factors.

The use of technical terms here is as follows. A *risk analysis* comprises Steps 1 and 2, and is said to be a *systematic use of available information to identify hazards and to estimate the risk*. A *risk assessment* is a risk analysis followed by a risk evaluation and comprises Steps 1-4 above.

It's worth saying a couple more words about the underlying technical vocabulary, because it coheres with that of IEC 61508, which is not at time of writing incorporated into the International Electrotechnical Vocabulary (IEC various) and it varies from other, more common and maybe more intuitive, vocabulary.

Harm is what you think it is. It used to be restricted to persons, but in the last decade or so has expanded to include damage to infrastructure and environment, other words almost any kind of loss. A *hazard* is a potential source of harm; a *hazardous event* is an event that can cause harm. What is meant here is that a hazard is a state, or an event, or a combination, from which harm may result, and a hazardous event is something that happens which may, but must not, result in harm, and the harm, if any, resulting from a hazardous event is variable. So a hazard can be a sharp bend in the road; or a sharp bend in the road without a speed restriction; or a sharp bend in the road without a speed restriction and a car coming towards it faster than it can negotiate the corner. A hazardous event can be a car coming towards the sharp bend faster than it can negotiate the corner (but presumably not if this is already considered part of the hazard); or it can be the car coming off the corner and hitting the wall. That may not result in harm if everyone is belted in and the airbags deploy; equally it will result in harm if neither is the case. And the harm that results is dependent on the speed of collision as well as other factors.

It is important to note that there is a choice of what to construe as a hazard. Such a choice is amongst other factors practically bound up with the possibilities for prophylaxis. A hazard identified earlier in a possible accident progression, and then avoided or mitigated, may be easier to document and handle. But such early intervention may exclude certain system behaviors that would have been OK, and one would thereby have taken unnecessary action. Leaving the identification of a hazard to later in a possible accident sequence, when it becomes clearer that something bad is about to happen, may avoid unnecessary earlier intervention, but may also require a more resource-intensive reaction to avoid or mitigate an accident.

There is much in this vocabulary to quibble with; my preferred vocabulary is published elsewhere (Ladkin 2008). But most of the necessary concepts are here somehow.

It is not defined what a risk estimation is, but *risk* is a *combination of the probability of occurrence of harm and the severity of that harm*; it is not said how they are to be combined. One option is that of de Moivre: "*The Risk of losing any sum is the reverse of Expectation; and the true measure of it is, the product of the Sum multiplied by the Probability of the Loss*" (de Moivre 1711), in modern terms the expected value of loss. Multiplication of risks associated with individual hazards, followed by a sum over all individual hazards, is a common way of deriving that expectation. A problem is that the enumerated hazards might not be probabilistically independent, but we shall let that be.

So a risk estimation is an estimate of risk. According to the de Moivre model, that would be an estimate of the expected value of harm. And if you have another combinator in mind, an estimate of the value of that combinator.

2 The Central Role of Probability

Notice the dependence of all this on notions of probability. You will need some theory about probability to fill all this out. The notion of probability has itself a wide variety of interpretations. Good explanations of the varying conceptions may be found in (Hacking 2001).We shall consider three.

There is the Laplacian interpretation, in which a probability is physically inherent in objects. A fair die, because of its careful construction, has an inherent probability of one-sixth of landing with any given face showing. The word "has" is possessive and here exactly right: the probability is a property of the die. A biased die has different probabilities for some faces; say a slightly-less-than-one-sixth probability of landing with 6 showing and a slightly-more-than-one-sixth chance of landing with 1 showing.

Then there is the frequentist interpretation, associated with Jerzy Neyman. Probability is associated with events, and is a statement of how often a specific type of event occurs. How frequently your bicycle tire punctures, say. If you go out on a ride and estimate the probability of a puncture as one in four, or one-quarter, depending on how you present probabilities, you are saying according to this interpretation that when you do a lot of these specified kinds of rides, ceteris paribus you'd experience a puncture on about a quarter of them.

The third kind of interpretation is the Bayesian, or subjectivist, interpretation, associated with de Finetti, Savage, and in Britain especially D.V. Lindley, after the Reverend Thomas Bayes and his theorem. This says that a probability is a statement of a degree of rational belief. Here, the word "rational" is normative: one is expected to form a belief on account of reasons and evidence, and update that estimate as evidence becomes available. You've seen one black swan and one white swan in your life. You know (somehow; by authority, or by painstaking genetic analysis) that a swan must be white or black and not both and not vaguely neither, so it is certain that any given swan is white or is black. You rationally assign the probability of a swan being white as one-half; identical with the probability of a swan being black, based on your experience to date. Then you see a lot more white swans, and no more black swans. For each swan you see, you update your estimates of the probabilities of whiteness or blackness according to Bayes's rule (a process called Bayesian updating), subject to the a priori constraint that it is certain that any given swan is white or is black and not both.

For events which are repeatable and frequent, there are theorems of probability theory which entail that all these conceptions come up with more or less the same values of probability for classes of such events. However, people building safety-critical systems are concerned with harmful events, or more precisely harm-loaded events (those events in which it is happenstance, or independent of the event itself, whether harm is caused or not, such as the car hitting the wall at speed). And such events are neither desirably frequent nor desirably repeatable.

For civil transport aircraft, one speaks not of hazardous events and their consequences, like the IEC, but rather of events resulting in specific effects. Extremely improbable effects are those unlikely to arise in the life of the fleet (all aircraft of a given type); extremely remote effects maybe once or so; remote effects maybe once per aircraft life (and many times in the life of the fleet). The certification regulations require is that a single failure that results in a catastrophic effect must be extremely improbable. Certification requires the constructor to show that this is so. Other effect severities are hazardous, major and minor (not that this notion of "hazardous" is different from that in IEC 61508). A classic introduction to these conceptions is (Lloyd and Tye, 1982).

A Laplacian interpretation applying to, say, the wing of a modern airliner may be plausible, as follows. The structures are designed to have it break under a specific load distribution at just over "ultimate load", which is defined to be 1.5 times "limit load", which itself is a number fixed at design time and which is purported to represent the highest loads to which the structure could be subjected during anticipated operations. And wings do so break at or above "ultimate load", during the required destructive test. They are engineered to withstand the required load, but no more, and this seems to be well achieved. Then the wing (rather, its intact successors) goes on to fly in uncontrolled but moderately well understood aerial environments, which can be argued to have probabilistic aspects.

So the wing is like the die; the engineering structure is well understood, as are the general characteristics of a throw, respectively of the weather, but the precise characteristics – the actual motion of the hand during the throw; respectively the precise behavior of the atmosphere during the flight – remain not sufficiently determined to render a deterministic calculation plausible. But notice here the justification in terms of what is known. The Bayesian approach takes the phenomenon of known information more rigorously, and arguably leads to a better intellectual fit.

A frequentist interpretation of wings breaking seems nowadays inapplicable, even implausible – wings just don't break in commercial service (any more), just like the regulation requires them not to. So the frequency is zero. (There are exceptions to this, but not in commercial transport.) But suppose one were to break, sometime. Then what's the frequency? One in ... what? Were the ceteris paribus conditions satisfied on that one occasion? Or were there particular conditions? How do you decide whether causal conditions have a probabilistic nature or a particular, exceptional nature?

It seems we are best advised to be Laplacians or Bayesians. But the Laplacian construal is nowadays "denigrated", so I guess we would have to be Bayesians.

But say you are inspecting a newly-built homebuilt aircraft for airworthiness. You can't see any of the composite lay-up of the wings – it's all hidden. So you interview the owner and form a rational belief about hisher construction capabilities and the care taken. It all looks good; you declare the aircraft airworthy. The owner goes up on a test flight and promptly a wing breaks off. You calmly update your estimate as the Reverend Bayes said you should...

Surely, given that the design is in order, the chance of the wing breaking as it did depends, not on your beliefs, but on how the wing was built, objectively? The owner didn't take as much care as heshe said during building; heshe screwed up badly in one place and didn't realise it. It seems we're back to Laplace: it's the airplane that has been built well or badly and the – what shall I call it? - propensity to break, the greater or lesser chance of breakage, is inherent in the structure. It seems that the construction and its thereby inherent propensities to fail matter rather more concretely than an assessor's beliefs.

3 The Way It Is Done in Aeroplane Certification

The acceptable means of showing compliance with aviation regulations are codified and formulated explicitly by the main airworthiness certification agencies, the US FAA and the European EASA. FAA rules are in 14 CFR Part 25 (United States Government, various dates). EASA rules may be found in the EASA Certification Specification CS-25 (European Aviation Certification Authority, various). They specify what is called in other contexts a risk matrix, a discretisation of effects against occurrence likelihood:

- Catastrophic effects must be extremely improbable
- (EASA) Hazardous effects must be extremely remote and major effects remote; or
  (FAA) major effects must be remote/improbable
- Minor effects may be probable, or even frequent.

Nowadays, a specific numerical probability per flight hour is associated with the qualitative probabilities.

But in fact what mostly happens is something rather different. Going back to the wing, recall that it must withstand ultimate load, defined to be limit load times 1.5, where limit load is an estimate of the highest loads to be plausibly experienced in service. A wing is built, and loaded until it breaks. And that should occur at equal to or higher than ultimate load. It is assumed (and checked and controlled!) that manufacturing-process quality, along with timely (checked and controlled!) in-service replacement of life-limited parts, ensures that all wings are interchangeable in terms of load withstood. That has everything to do with engineering and control and nothing at all to do with probability. You believe that it won't break because you built it that way and have enough experience to know that that suffices. And you test that understanding precisely once. (Actually, it turns out on a recent certification it was acceptable to have the wing break at slightly below ultimate load, then perform a redesign and show by means of extensive computer simulations that the strength of the wing had thereby been increased to withstand ultimate load, without destructively testing the redesign.) All this is taken to show that the possibility of a wing failing to fulfil its function in flight is extremely improbable; that is, it won't happen during the fleet lifetime, as far as anyone can tell. Note that there is no intellectual connection here with probabilistic criteria per se. Engineering design, simulation and deterministic test is deemed satisfactory to fulfil a criterion, itself expressed but not enforced in terms of likelihood.

Perceptive readers will note I have glossed over some of the subtleties in airworthiness certification, but I believe the story as I have told it suffices for my purpose here. In short, the notion of probability or likelihood is problematic when referring to very rare events. When possible in aerospace, we far prefer to have designs which we can plausibly argue on the basis of design and construction will withstand all occurrences of adverse events in their lifetimes.

Except of course when some other people have designed an object which is intended to cause your structure to fail, and is built according to similar principles as above to execute that function. Which we now consider.

4 Risk of a Different Variety: Security Risk

On 17 July 2014, a Boeing 777 operating as Malaysian Airlines Flight 17 between Amsterdam and Kuala Lumpur was destroyed in and over Eastern Ukraine. Witness reports and the fact that the wreckage was strewn over a very large area point unequivocally to in-flight disintegration. *"Damage observed on the forward fuselage and cockpit section of the aircraft appears to indicate that there were impacts from a large number of high-energy objects from outside the aircraft"* (Dutch Safety Board 2014). An admirably careful statement. Put another way, pieces of wreckage photographed by reliable observers show damage such as caused by shrapnel from the detonation of an explosive projectile with a proximity fuse. The Report also says there were no indications of any problems or malfunctions before the abrupt end of recording on the data recorders (op. cit. Section 3, Summary of Findings). In other words, it is almost certain that somebody shot the flight down. There was and is an armed insurrection occurring in the area, with fighting between sovereign Ukrainian forces and heavily-armed "rebels" who appeared to be led by Russian citizens.

Photo by Alan Wilson Licensed under Creative Commons

Fig 1. The Incident Aircraft, Boeing 777-200 9M-MRD

Ukraine is sovereign over the airspace in which MH 17 was flying. Many airlines had been using the airway, L980, and adjacent airways. Indeed, when destroyed, MH 17 was at Flight Level (FL) 330, a nominal 33,000 ft above mean sea level in a "standard atmosphere", in Section 4 of the CTA (Control Area) Dnipropetrovsk (known to aviators as Dnipro Control). In the same sector at that time were a same-direction Boeing 777 at FL 330 about 100km southwest on airway M70 heading towards waypoint PW, a same-direction Boeing 777 at FL 350 about 30km northwest, and an opposite-direction A330 at FL 400, 50km east-north-east on airway A102. (op. cit., Figure 2, p12). (Note: A report in the weekly journal Aviation Week and Space Technology from a week or two after the accident had MH 17 14 nautical miles or so in trail of a Singapore Airlines aircraft at FL 350, and about 8 nautical miles abeam of an opposite-direction Air India aircraft on another airway. (Schofield et al., 2014). The divergences between the two reports show again how difficult it is to establish facts about such events, even though the relevant information is ostensibly readily available from multiple sources.

At the time, there was a Temporary Restricted Area from the surface to FL 260, valid from July 1 through July 28. The existence of this area was distributed by NOTAM (Notice to Airmen, the international standard informational service). On 14 July, a further TRA existed from FL 260 up to FL 320, valid until 14 August, covering the eastern part of the area covered by the first TRA. All the flights passing through Sector 4 of Dnipro Control were conforming with both NOTAMs, as indeed to be expected with commercial flights under positive control.

Some airlines had previously performed a "risk analysis" and had been avoiding overflying the area, such as, it was reported, Qantas and BA. Other airlines avoided the area afterwards.

Photo by neuwieser Licensed under Creative Commons

Fig.2 A Sister Aircraft in Flight

MH 17 had filed a flight plan with requested FL 350 in the area, but when in contact with Dnipro Control at FL 330 was unable to transition to FL 350 and continued on FL 330 (op. cit.).

5 Security Risk Analysis: What's With Probability?

What kind of risk analysis can it have been which had been performed by those airlines avoiding the area? Could it have been one as described above? Let us try:

- Identify the hazards:
  - Getting shot down by a ground-based missile
  - Getting shot down by another aircraft
  - Getting shot down by ground-based artillery or flak
- Severity of all these events is the same: catastrophic, everyone on board dead, hull loss, damage on the ground, perhaps harm to people on the ground
- Estimate the risk: as defined, "combine" probability of each hazard with severity. So what is the probability of each hazard?
- What is the probability of getting shot down by a ground-based missile? Zero if there aren't any in the area with the capability of reaching a target at FL 330. Someone explained to a journal that the commercial-aviation industry relied on sovereign militaries to control their assets -- does that mean zero probability if the only such missiles in the area are maintained by sovereign militaries? Well, not quite. Siberian Airlines (Sibir) Flight 1812 was shot down from FL 360 on 4 October 2001 over the Black Sea on its way to Novosibirsk from Tel Aviv (Aviation Safety Network, no date). The aggressive object was a missile operated by the Ukrainian military during military exercises, which locked on to the airliner rather than its intended target. OK; so the chance is not zero. What is, then, the prob-

ability? One in ... what? Can one possibly tell? What are the ceteris paribus conditions that say "a Flight 1812-type incident could occur here"?
- What is the probability of getting shot down by another aircraft? Ukrainian military aggressor aircraft, specifically Su-25 Frogfoots, use the airspace. But Frogfoots have an effective service ceiling some 10,000 ft lower and as far as we know can't "shoot up" (see, for example, (Sweetman, 2014), or details in (Locklin, 2014)). Besides, why would such an aircraft try such a thing? There are no "rebels" up there at FL 330. A Russian or Ukrainian fighter aircraft could be up there; indeed there were previous unconfirmed reports of unauthorised Ukrainian-airspace intrusions by Russian military aircraft. But what would aircraft under strict sovereign-state control possibly be doing up there shooting at traffic at FL 330? As far as anyone has seen or said so far, there were no such aircraft up there at FL 330 in Ukrainian airspace anywhere in the neighbourhood.

The precursor state to Russia, the Soviet Union, had shot down civilian airliners. The first was a Korean airliner violating Russian airspace, which refused an interception using internationally-recognised manoeuvres and was consequently shot at by an interceptor, on 20 April 1978 (Aviation Safety Network, no date). The fire killed two people. The aircraft was not destroyed, but landed relatively safely off-airport on a frozen lake. The second was also a Korean airliner, a Boeing 747 which had also violated Russian airspace, crucially in the neighborhood of and around the time of an important missile test. The aircraft was shot down by an interceptor who had mistaken it for a US military intruder, a reconnaissance aircraft also built by Boeing, of Boeing 707 size, and believed it was manoeuvring to avoid interception. That was on 1 September, 1983 (Aviation Safety Network, no date).

Photo by
Rob Schleiffert
Licensed under
Creative Commons

Fig 3. An Su-25 Frogfoot Aircraft

However, ceteris paribus conditions are nowhere near satisfied. Neither of the shot-down airliners was on or indeed near internationally-recognised civil airways for which it had a clearance. Both were formally intercepted using internationally accepted protocols. One airliner refused the interception; the other airliner was honestly judged to be actively avoiding one on a dark and somewhat cloudy night. Both incidents occurred during the "Cold War", during which the Soviet Union was on one side and South Korea, considered by the Soviets as something of a protégé of the United States, definitively on the other. The Soviet Union believed itself, with reason, to be at times actively intruded upon, sometimes by civilian assets performing military tasks under subterfuge. (And indeed vice versa.)

In stark contrast with these circumstances, MH 17 was following a recognised airway at a cleared Flight Level on a filed flight plan and was not violating, or about to violate, anyone's sovereign airspace without clearance. Neither is it plausible to imagine it was trying to perform military tasks by subterfuge. Neither was Malaysia on one side of a "Cold War" with Russia on the other.

At time of writing, Russia has in fact claimed that MH 17 was shot down by a Ukrainian Frogfoot. Russia has published radar data they claim is proof, which has been assessed by reliable third parties who are less than convinced by it. The United States claims to have proof that MH 17 was shot down by a surface-to-air missile launched from Eastern Ukraine. The United States is known to have assets which could establish this beyond reasonable doubt, but at time of writing this information has not been published and independently verified.

Photo by: Ajvol
in the public domain

Fig.4. A Buk-M1-2 Launcher

Photo by Vitali V. Kuzmin in the public domain

Fig.5. A Complete Buk-M1-2 System, Comprising Multiple Vehicles

On 20th October, 2014, I discovered through my local newspaper that the head of the German Federal Intelligence Service, BND (Bundesnachrichtendienst) told its parliamentary oversight committee on October 8 that MH 17 had been shot down by separatists using a Buk system which they had obtained through plundering a Ukrainian military base. It is said that convincing evidence was presented (Gude and Schmid 2014).

- What is the probability of getting shot down by ground-based artillery? Nobody thinks that anyone has any artillery assets in the area that can reach up to FL 330. Even if there were, people estimate chances of getting a ballistic hit at close to zero. Ballistic projectiles are intended for buildings and very slow-moving objects such as battleships, not for high-performance aircraft.

- What is the probability of getting shot down by flak? Up there at FL 330, almost zero. Besides, as far as anybody knows there are no flak delivery assets in the area.

So where is here the probability value? As far as I can see, and I am suggesting as far as the reader can see also, there isn't one. A Guide 51-type or IEC 61508-type risk analysis is not what is being performed when analysing such risks.

6 What's Really Going On

So what reasoning is being used here? I have just performed something like the following:

1. It is observed that hostile military engagements are taking place in the area.
2. The area in which those engagements are taking place, or to which they could plausibly spread, is circumscribed.
3. A hoped-complete list of hazardous events occurring through hostile military acts to commercial aviation flying in open civil airspace is enumerated.
4. Scenarios leading to those hazardous events are constructed.
5. The plausibility of each scenario is assessed.
6. Plausibilities are ranked. First, plausible-implausible. Then, more plausible-less plausible.
7. A discrete decision is made based on those plausibilities: use the airspace/don't use the airspace.

Up to Step 3, that is what the IEC documents on engineering risk say to do under hazard identification. But then the method diverges. Scenarios are not necessarily considered in IEC methodology. Some may consider Fault Tree Analysis followed by Event Tree Analysis to be a form of scenario construction, but I suggest that the current type of scenario construction is significantly more detailed than what occurs in a typical FTA/ETA. One could, perhaps, consider a qualitative fault tree as some kind of enumeration of scenarios, or at least as a structure which yields such an enumeration. But the scenarios considered here are not possible causes hierarchically ordered in subsystems, as in an FTA. Neither are they abstract possible futures as in an ETA. They are temporal scenarios with actors performing actions according to motivations and reasons and other human characteristics. Then, some decision is made on the basis of that analysis: do or don't.

What is most important about that decision is that it is the Right One: don't fly there if somebody's maybe going to get shot at in any place where you are going to be.

In a probability-based analysis, one could make all the rational decisions based on probabilities and still get stung on your first outing. Your analysis is valid according to the IEC conception. You took a risk and then you lost the bet. So go ahead, do it again! Your analysis is still valid. Toss the die!

Contrast this with commercial transport aircraft certification. The rules say: your airplane will do this-and-this. And furthermore the evidence will be documented. The evidence deemed acceptable may be probabilistic and is retained and available. So rational decisions were made based on evidence couched in terms of probabilities, as in the IEC approach. Say you go out and get stung on your first outing. The judgement is different: your airplane is not airworthy; make it airworthy and you can go fly it again (this is accomplished by means of instruments called Airworthiness Directives, which are remedies mandated for all operators of the aircraft type to restore and/or maintain airworthiness of their aircraft. If you don't fulfil an AD, your aircraft is not airworthy and may not be flown.) This outcome is different from the IEC outcome of a critical failure. You can't just go ahead and do it again; you must remedy.

The current airspace-use situation we are considering is comparable with the aircraft airworthiness procedures in that immediate remedy is required: the airspace is closed to civil traffic, and even if it weren't it would be doubtful if anyone would be using it. But it diverges in that it is called a risk analysis; aircraft certification is not called "risk analysis" by anyone, and the process is not treated as if it were. Testing a wing to destruction is not analysing risks; it is assuring ourselves that the engineering is sound and a wing will not break in service because it is functionally identical (through process and quality control) to the successful-test object. And, conversely, a decision to use airspace is not called "traversal-worthiness certification" and neither will it be.

In truth, the probabilistic risk of getting shot down over Eastern Ukraine was low, even under a reasoned belief that there were high-altitude surface-to-air missiles (SAMs) in the hands of unreliable combatants. Troop and equipment movements had been seen at the weekend, 12-13 July (but it is not known at time of writing what the contemporary analysis had concluded), and a Ukrainian military transport had been shot down at FL 210 on Monday 14 July. Hundreds of airplanes had flown the routes over Eastern Ukraine in the meantime; four were flying it at the time of shootdown. And only one of those aircraft was shot down. An $O(10^{-2})$ risk is high compared with other estimates of risks in aviation, but in objective terms one might question whether such an event is likely.

It is also plausible to think that right after MH 17 was shot down, if it was shot down by a SAM then the chances another aircraft would be shot down in the region had plummeted to near zero.

That conclusion is also based on scenario analysis. Such assets were widely assumed to belong to the Russian military. It is true that "rebels" had boasted of capturing some Ukrainian Buk SAMs in June but this had remained unverified and it would have been unlikely they could operate them effectively without having some sort of rudimentary training which would not have been available. So if Buk SAMs were available to rebels, it is likely they would have been Russian assets and thus recommandeered immediately after the shootdown for many reasons; and it is not regarded as plausible that Russian military assets under direct Russian control, as re-commandeered devices would have been, would be used to shoot down civil aircraft. (But contrast this reasoning with the reported claim by the Head of the BND in camera, noted above.)

It is not regarded as plausible, but it could happen. Some odd soldier of almost any army could get drunk or suicidal or both, and fantasise about going out in a blaze of notoriety, like the 9/11 terrorists. And succeed, as two out of four cohorts of the 9/11 perpetrators did. This possibility appears not to be sufficient reason for any of the world's airlines to avoid Russian airspace. Neither did the shootdown of Siberian 1812 in 2001 cause Russian or any other airlines to avoid Ukrainian airspace; a repeat was not regarded as plausible.

Why not? I believe it has to do with people, motivational and goal analysis, and assessments of capabilities at fulfilling goals. Put crudely, the Soviet Union had screwed up badly with KAL 007 in 1983; that was never going to happen again. Ukraine had screwed up badly with Siberian 1812 in 2001; that was never going to happen again. Controls were already in place and must be followed more precisely. Whereas two of four cohorts of 9/11 aggressors had achieved what appeared to be explicit goals. Few controls were in place and it was unknown whether others with similar goals were still out there. World civil air traffic stopped, and restarted slowly with considerably more assessment and control, including previously unthinkable measures such as giving the USAF rules of engagement to shoot down civilian transport aircraft.

When we are in the realm of personal and organisational goals, motivations, means and so forth, we are no longer in the realm of probabilistic assessment. Probabilistic assessment is based ultimately on a notion of a random variable and while goal, motivational and strategic analysis may rely somewhat on uncertainties, as in "taking a chance", it hardly relies on any notion of randomness. "Purposeful" behavior is indeed the contrary of "random" behavior.

Consider another example, from a different realm. The chances that your WWW server suffers a surfeit of incompletely-formed TCP handshaking packets are low; but they are very high to almost certain if your server is the target of a DDoS attack. The difference between the two situations is not probabilistic, or generally in any way related to chance, but rather concerned with some specific agent's purpose and means at that point in time. Analysis is concerned, not with bursty behavior on communications networks, but with whether there is an agent who had reason and means to elicit the behavior and why. Far from being a probabilistic random variable, it is more like an almost-Boolean environmental variable: are you currently subject to DDoS attack, or not?

Furthermore, there are no uniform assumptions one may make about chances in the background. If your civil aircraft has been subject to rocket attack in Eastern Ukraine, I have just argued that it is very unlikely you or anyone will be subject to further attack. Whereas if you have just survived a DDoS attack, the chances rise that you will be subject to another one soon. Or, to compare like with like, a ManPAD attack on a civilian cargo aircraft deploying around Bagram Air Base in Iraq might be seen to increase the chances that another such aircraft will be so attacked soon. The differences are not to be found in any quasi-objective analysis of inanimate situations; they are to be found in the goals, motivations and means of some of the players. (Or may be all of them. A second ManPAD attack may be canonically thwarted by grounding and guarding all aircraft, as happened for similar reasons immediately after 9/11. The goals, means and motivations of all the participants should likely be considered.)

7 Meta Game Theoretic Analysis (MGTA)

We are in the realm of game theory. Indeed, the classical game theory of non-cooperative games (so-called "game theory" is usually the study of non-cooperative games, contrasted with cooperative game theory, or coordination games, studied by the philosopher David Lewis (Lewis, 1969) as well as the political theorist Thomas Schelling (Schelling, 1960)). But this is not pure game theory, as studied by economists. It is more like a meta-theory of games. First there are methods to choose a game from amongst a variety of possibilities (the "meta" part), then follow methods to choose actions within the chosen game (game theory proper).

There are situation variables, which in some sense set the game being played. Am I currently subject to a DDoS attack? If not, I am administering a server in an unreliable bursty environment and there are lots of things I can choose to try to ameliorate the situation compatible with my goals. If so, then nothing I do for a while will change the environment and I have only two actions available to me: let my server be overwhelmed and clean up whatever mess results; or disconnect my server from its channels (most likely is that there are only a few channels to which I am connected). Are there currently high-altitude SAMs or high-altitude aggressor aircraft available to unreliable players engaged in hostilities in Eastern Ukraine? If yes, my airliner might be shot at/down and I have to think of what I do. If the answer is no, the high-altitude airspace is just like any other airspace anywhere else in the world; free from worry (if I have reliable collision avoidance!). In that case I am in the null or trivial game: payoff is the same whatever I do and whatever the "opposition" does and is exactly the unit value.

In this first step, chances may reappear, as do the dilemmas associated with the interpretation of probabilities which I considered earlier. What is important for my decision making is what I know or have reason to believe. How likely do I think it is that unreliable players in Ukrainian hostilities have SAMs? Say I think there is an 60% chance. It then follows that there is a 40% chance I am playing the null game and a 60% chance I am playing another, more complicated game. Or so I reckon. Whereas the reality is either that unreliable players have SAMs, in which case I am truly in the complicated game and would do best by deciding my actions according to that; or that unreliable players do not have SAMs, in which case I am truly in the null game and can return to my Sudoku without further ado. Thus this reckoning of chance is an assessment of my uncertainty. We are unequivocally in the realm of Bayesian probability.

Can I collapse this twofold structure into a single structure, say Decision Theory? I don't believe so, for the reasons in the last paragraph. Best is to know what game you are actually, objectively in, and to choose your actions according to that game. In the case of a DDoS attack, I know and can choose. In the case of unreliable players with or without SAMs, I don't know, but it were best if I did. If the reality is the null game, I am optimally well off by deciding this correctly. And as MH 17 shows (if the most plausible scenario at time of writing is correct), I am not necessarily well off by deciding this incorrectly.

Can I, then, assume one game or the other? The perils of assuming the null game (no SAMs) are by now obvious. What of the other choice, that in the absence of knowledge I assume the "worst" game from amongst the possibilities? People have done the work for us – if we are to assume the worst and avoid all areas of hostilities in which the weaponry is not publicly known, I would have to get to anywhere east of Kiev more or less by flying around the Cape of Good Hope (see, for example, the graphic (Times 2014)). If an individual airline were to choose to follow this option, I would lose custom and fold quickly. If the world's airlines were collectively to choose to follow this option, then international business would instantly suffer a step change for the worse: all personal dealing suddenly becomes far more expensive in both time and money and costs of international business suddenly rise. Except for suppliers of aviation fuel, who are laughing all the way to the bank. Neither of those seem particularly attractive, let alone ideal, options.

If I am in the situation, though, in which I have two choices of game and one of those happens to be the null game, then there is a way I can "play both". The null game says always: do nothing, and ye shall neither suffer nor gain. So I can play the other, non-null game but weight my payoffs by my assessed chances that I am in that game. Then I can consider that I am somehow in both games simultaneously: I am in the one game to 60%, by obtaining a 60% payoff for my actions, and also in the null game to 40%, by obtaining 40% unit payoff.

So I can solve – let us call it - the "Ukraine" problem simply by playing the "SAMs-yes" game to the weighted value of my belief that SAMs-yes.

But consider the following situation. On the ground, there are two combatants and one SAM base whose operators are effectively commanded by whichever combatant has control at any one time (the operators behave neutrally in order to, they hope, "save their skins"); and control changes hands regularly and, let us suppose, evenly (50% each). Suppose the one combatant dislikes airlines whose names start with A, C, E, G, I... and the other combatant dislikes airlines whose names start with B, D, F, H, J... As an airline, I am either in the null game or the firing line. I am pretty much forced to choose my game; if I choose a weighted average then half my planes are shot down and I lose custom and fold (not to speak of the distress caused my shot-down passengers and their relatives, for which I and my insurers are also liable). In this case it seems I cannot avoid choosing my game explicitly.

So in general we should expect that an explicit choice of game should be made. The weighted-belief approach works for the specific case in which one game is the null game, but not in general.

It follows that, in general, there are two separate, non-conjoinable steps to the analysis.

1. Choose your game;
2. Choose your actions in the selected game.

Airlines performing risk assessment on the "Ukraine" problem may have been able to use the risk analysis afforded when there are just two games, one of which is the null game, but in general this is not possible. Generally: choose your game; then choose your action.

It follows that security-risk analysis is fundamentally different from IEC-type safety-risk analysis. Game definition and choice, then action choice; respectively situation probability assessment.

8 MH 17 and Consequences in Light of MGTA

This construal of security-risk analysis already yields some results which run counter to the "prevailing wisdom". Many aviation professionals have been proposing that ICAO Do Something about the analysis of airspace usage risks apropos MH 17 (Schofield et al., op. cit.). First, there is arguably a misconstrual of ICAO's structure; second, there is a false expectation of effect, according to our analysis above.

First, to ICAO's responsibilities and capabilities. ICAO cannot Do Anything in the sense intended. ICAO is not a sovereign entity, it is a talking shop for sovereign entities in which things only happen if they are agreed amongst all sovereign members. Almost every nation belongs to ICAO. It advertises and propagates matters on which there is universal consensus. ICAO cannot issue recommendations not to fly over Ukraine (and recommendations are all it can issue) unless almost every member nation besides Ukraine decides it is not a good idea to fly over Ukraine (of course if Ukraine itself decides so, it may enforce the measure without any consultation). And if almost every member has so decided, then all airlines have already been informed of that advice by their sovereign and are following it for reasons of due diligence, not to speak of their insurance contracts. So such an ICAO recommendation would be a no-op; it would already be a done deal.

If people really yearn for an ICAO determination, this is no argument against that; let them have one by all means. It is only an argument that an ICAO determination would change nothing "on the ground" (that is, in the air).

More problematic is that such a determination would be insidious, in that it determines the game to be played. No matter what agreement might be reached in ICAO on a way to assign usage risk, there is a new game to be played based on the determined risk, and that new game is riskier for airlines.

For along with a determination of risk will come inevitably an assignment of responsibility. If there is a risk, say of 5%, of being shot down in Sovereign Airspace Q, then Q's sovereign, and/or the canonical risk assessors, and/or the airline which proceeded across Q's airspace according to the canonical risk assessment, will be held liable to some mathematical formula. This will happen because they are regarded as the pertinent actors in the regrettable decision to fly across – and thereby get shot down.

Let us suppose there is some "standard" assignment of airspace-usage risk. Say, as determined by ICAO for those entities who wish for this, but for the purposes of this argument determined by any means. Along with this assignment will come the liabilities associated with this assignment, as above.

Suppose further that Sovereigns A and B are at loggerheads. Sovereign B knows how Sovereign A calculates airspace usage risk; namely, according to the "standard". Sovereign B unattributably infiltrates A and brings down a commercial airplane, in order that A will be actively internationally criticised for screwing up the risk analysis, thus causing airlines to avoid A's airspace and diminishing that source of revenue for A, as well as possibly causing Sovereign A to pay compensation for the shootdown on the basis of the assignment of responsibility recounted above, which can nowadays run into ten-digit dollar sums.

That's a great win for Sovereign B at expense of Sovereign A. It would be appropriate to consider Sovereign A a victim (an undeserving loser) of that game. And the game can only be played if there is a "standard" risk assessment of airspace usage, as wished by those who want ICAO to establish a standard.

So, beware of what you wish for!

9 Some Other Examples

I have applied MGTA to the phenomenon of ATM phantom withdrawals from demand-deposit bank accounts, and how they are handled by customers and banks. I have also applied it to the phenomenon of "security theatre" with respect to implanted/implantable digital medical devices, in which security "researchers" graphically demonstrate "vulnerabilities" with devices, such as hackers on the street reprogramming a heart defibrillator or an implanted insulin pump remotely. There is no space here, though, to recount those studies.

Acknowledgements Thank you to Mike Parsons, for inviting me to give a Keynote talk at the 2015 SCSC Safety-Critical Systems Symposium and expressing the wish for me to say something non-trivial about MH 17. I hope I succeeded. I thank Harold Thimbleby and Ken Hoyme for initial reviews of this work (including the phantom-withdrawal and medical-implant examples). Ken chairs the AAMI committee on medical device security, and gave me extensive references to what is being done. He also provided incisive comments way beyond the usual scope of review; the paper has been thereby improved, and I am particularly grateful.

References

Aviation Safety Network (no date) Sibir 2001 incident, Entry http://aviation-safety.net/database/record.php?id=20011004-0 , accessed 2014-08-08.
Aviation Safety Network (no date) KAL 1978 incident, Entry http://aviation-safety.net/database/record.php?id=19780420-1 , accessed 2014-11-18.
Aviation Safety Network (no date) KAL 1983 incident, Entry http://aviation-safety.net/database/record.php?id=19830901-0 , accessed 2014-11-18.
de Moivre, A. (1711) de Mensura Sortis, Phil. Trans. Roy. Soc, 1711. Reprinted with commentary in (de Moivre and Hald, 1984).
de Moivre, A. and Hald, A. (1984) A. de Moivre: 'De Mensura Sortis' or 'On the Measurement of Chance', A. Hald, Abraham de Moivre, Bruce McClintock, International Statistical Review / Revue Internationale de Statistique 52(3):229-262, Dec.1984.
Dutch Safety Board (2014) Preliminary Report, Crash involving Malaysia Airlines Boeing 777-200 flight MH17, Hrabove, Ukraine, 17 July 2014, Dutch Safety Board, The Hague, September 2014, available from http://www.onderzoeksraad.nl/uploads/phase-docs/701/b3923acad0ceprem-rapport-mh-17-en-interactief.pdf
European Aviation Certification Authority (various) EASA Certification Specification CS-25, the initial version of which is available at
http://easa.europa.eu/system/files/dfu/decision_ED_2003_02_RM.pdf
Gude, H., and Schmid, F. (2014) Deadly Ukraine Crash: German Intelligence Claims Pro-Russian Separatists Downed MH17, der Spiegel on-line, October 19, 2014,

http://www.spiegel.de/international/europe/german-intelligence-blames-pro-russian-separatists-for-mh17-downing-a-997972.html

Hacking, I. (2001) An Introduction to Probability and Inductive Logic, Cambridge University Press.)

IEC (2010) International Electrotechnical Commission IEC 61508:2010 Functional Safety of electrical/electronic/programmable electronic safety-related systems. Seven parts, available from http://www.iec.ch/functionalsafety/standards/page2.htm

IEC (various) International Electrotechnical Commission IEC 60050, International Electrotechnical Vocabulary, many dates. Available on-line with some delay at http://www.electropedia.org

ISO/IEC (2014) International Organisation for Standardization/International Electrotechnical Commision ISO/IEC Guide 51:2014. Safety aspects – Guidelines for their inclusion in standards, ISO & IEC, 2014. Available from http://www.iso.org/iso/catalogue_detail.htm?csnumber=53940).

Ladkin, P. B. (2008) Definitions for Safety Engineering, Causalis Limited. Available at http://www.causalis.com/90-publications/DefinitionsForSafetyEngineering.pdf

Lewis, D. (1969) Convention: A Philosophical Study, Harvard University Press.

Lloyd, E. and Tye, W.(1982) Systematic Safety, CAA Publications, London.

Locklin, S. (2014) Can the Su-25 intercept and shoot down a 777?, Locklin on Science blog, 2014-07-21. Available at http://scottlocklin.wordpress.com/2014/07/21/can-the-su-25-intercept-and-shoot-down-a-777/

Schelling, T. (1960) The Strategy of Conflict, Harvard University Press.

Schofield et al. (2014) Schofield, A., Flottau, J., Buyck, C., Broderick, S., and Croft, J., Ukraine Shootdown May Spur Risk-Assessment Reform, Aviation Week & Space Technology, 2014-07-28. Available at http://aviationweek.com/commercial-aviation/ukraine-shootdown-may-spur-risk-assessment-reform

Sweetman, W. (2014) How An Su-25 Can Shoot Down a Faster, Higher-flying Aircraft, Ares Blog, Aviation Week and Space Technology, 2014-07-24. Available at http://aviationweek.com/blog/how-su-25-can-shoot-down-faster-higher-flying-aircraft

Times (2014), The Times, 2014-08-08. The article is behind a paywall; the graphic at http://www.thetimes.co.uk/tto/multimedia/archive/00740/inline_937c3442-15d_740840a.jpg is not.

United States Government (various) 14 CFR Part 25, available at for example http://www.ecfr.gov/cgi-bin/text-idx?tpl=/ecfrbrowse/Title14/14cfr25_main_02.tpl

Risk Tolerance

A tale of professional yachtsman and meddling bastards

Les Chambers[1]

Chambers & Associates Pty Ltd

Queensland, Australia

Abstract *This paper is a personal reflection on the capacity of human beings to tolerate risk. My thoughts were triggered and enriched by an Atlantic crossing under sail aboard the yacht Northern Child. Among the wild men and women of ocean yacht racing, I was forced to reassess my engineering attitudes. In eighteen days at sea I reflected on the factors that determine risk tolerance in a particular enterprise and why tolerance varies so markedly across human endeavours. It occurred to me that the main driver of risk tolerance is emotional rather than technical, a state of affairs that engineers are ill-equipped to manage. This led me to research human decision-making processes and the power of metaphor not only in learning and persuasion but also in synthesising safer systems. Too much happened on this voyage to tell here. Instead I present the set of vignettes that informed my personal belief that engineering knowledge must be broadened to include human behavioural modelling and the synthesis of metaphor. And most important of all, our scope of works must go deeper. In particular, the functional safety discipline must be more than a broad-brush administrative framework. Its tropes must be imprinted on the minds of all the people doing the work.*

---

[1] les@chambers.com.au
www.systemsengineeringblog.com/
www.chambers.com.au

© Les Chambers 2015. Published by the Safety-Critical Systems Club. All Rights Reserved

1 Prologue

To know just one eternal truth and to find your way by its light is a beautiful thing.

After many nights of struggle on the ocean, in the wild, I have found a way to wrestle a constellation, and pin the three stars of Orion's belt stationary to a mat of black sky beneath the spreaders.

In the predawn hours of the northern hemisphere this means something to a helmsman: a syntax of navigation older than the Vikings. The truth is: I'm solid on course 270, no need to check the compass.

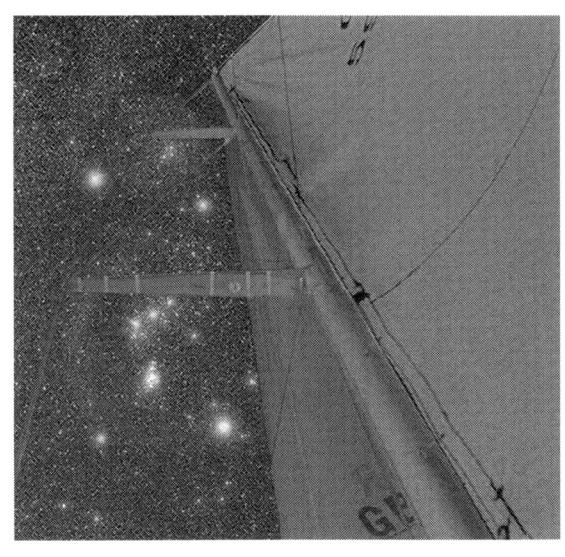

Fig. 1. The three stars of Orion's belt

I feel the transom lift on a swell. They've been rolling in at that angle all night. I rotate the wheel an eighth turn to port and feel the boat slide down the face of the wave. The bow cleaves into the trough with a "wwhoosh". I rotate one eighth to starboard and the stars slide back into place. Behind me the yacht scribes a shallow arc of luminescent wake that soon fades, and we move on before a fair 30-knotter, taking nothing away, and leaving nothing behind.

I smile to myself in the dark. What a turn-up for a systems engineer. Fresh, ten days before, from a life automatic, I can feel myself going native. I'd come aboard expecting a cruise but I'd got a race. I'd expected an autopilot but I got a work assignment: three hours a day on the helm, from a skipper who spat on automation and anyone who might bring it aboard. 'They wouldn't be professional yachtsman,' he said, 'they'd be meddling bastards.'

And slowly in transit on three thousand miles of ocean I can feel the layers of belief peeling back. My hours on the helm have been too much fun; too much to hand over to an algorithm. Could Mark be right? Is the engineering penchant for isolating man from the claws of nature more a hazard than a benefit? Are we breeding out our ability to see off danger with this synthetic existence, where everything we touch is idiot proof? where action does not flow, frictionless, to consequence? where our fears keep us too safe, too passive? Are we creating a society unwilling to use a sharp knife lest it be cut? And most important of all: what will become of us when our machines fail?

In November 2013 I stepped onto the Atlantic aboard the sloop Northern Child and this is what became of me.

2 Emotion

Five days out, on a beam reach with a good 30-knotter from the south, we were travelling well. But after three sleepless nights I was unraveling. 'I can take this for another day or two,' I thought, 'but beyond that, something evil this way comes.' The beginnings of panic. I'd tell myself, 'You're over emotional Les. Breathe. Get a grip.'

Fig. 2. The Northern Child

Emotion. We engineers don't like it. We prefer the life formulaic, in a paradise of determinism. But out on the ocean I was coming to a different point of view: that forswearing emotion is a dangerous form of denial. Left unopposed it destroys us. We parley with it in our critical decision-making and the functional safety specialty is driven by it.

How so? As duty-holders it's our job to reduce risk to a point where it's 'As low as reasonably practicable' (ALARP). Establishing what is 'reasonably practical' involves judgement; weighing the risk against the sacrifice needed to further reduce it. And we define risk as the combination of probability and severity. If the probability is low but the severity is unthinkable we have high-risk. So at the root ball of ALARP we have a subjective, emotive metaphor: severity. What is that?

Some might call ... wrestling the helm of a 51 footer, in a 30 knot wind, on a heaving deck, at 2am on a black night, with the benefit of three hours sleep in three days and a cockpit full of watchmen with nothing better to do than tell you how far you are off course ... subjectively severe. Others might call it paradise.

We'd better look into it.

3 Joining

I found the Northern Child lashed to a pontoon, flat and silent in the giant Marina of Las Palmas, Canary Islands. I was welcomed aboard by Mark the skipper and Kate the first mate. The Northern Child was a sturdy twenty tonne, fifty-one foot Swann, built in Finland thirty years ago, a Rolls-Royce of yachts. I joined early hoping to familiarise myself with the yacht before we set sail for the Bahamas. I had owned a yacht and done some coastal sailing, but never an ocean passage. I looked forward to working with my crewmates in the shakedown cruises.

Time passed and we stayed at the dock. Not a good omen, but I kept my concerns to myself. Instead I concentrated on getting to know my crewmates. There were Uli and Matthias, two German anesthetists, Rolf a Danish IT guy and Johnny a Swedish outward bound instructor. Martin, an affable American, joined us a few hours before we sailed. The remainder of the crew was made up of three Spaniards. The watch schedule was four hours on and four hours off. Kate, Johnny and the Spaniards became the Armada watch. We became the German watch, after the most common language. Mark promoted Rolf to watch leader, a good choice.

Christian, the boat owner, visited and gave us a safety briefing. A whip around the boat that took an hour. We talked about winches, lifejackets and how to jibe a polled out headsail.

It started to rain. The hatches leaked in my cabin and the forward head. Johnny was upset. It was dripping on his bunk. He set about making repairs.

Finally decades of deploying safety critical control systems with months spent in operator training got the better of me. I fronted Mark, 'When's the shakedown cruise?' I asked. 'Oh that's the first three days at sea,' he answered. There were mumblings below decks among the German watch.

Let's review:

A bunch of guys who have never sailed together before are setting out to cross the Atlantic on an unfamiliar boat with leaking hatches and no sail training. What could go wrong? Plenty!

4 Crash Jibe

A yacht at sea presents an endless array of bad outcomes for the unrehearsed. Take jibing the boat: it's a high skill team effort requiring a well trained crew. Expecting it to go well after a ten minute talk, on a level deck in a marina, is the act of a dangerous optimist.

As for what could go wrong: a crash jibe. A crash jibe occurs when, by accident or design, a yacht's stern passes through the eye of the wind. If unconstrained the boom swings violently, and with massive force, from one side of the boat to the other injuring anyone in its path. Arrested by the main sheet, it arrives at its new position with an ominous thunking, crashing sound, sending a shock wave up the mast to the spreaders and beyond. On a good-luck day that's the last you'll hear of it bar some flogging of sail and twanging of rigging. On a bad-luck day the wave energy exceeds the tolerance of the rig, breaking it, usually at the spreaders and, as if from heaven, the rig that was once such a beautiful thing, so whole, so proud and so entire, showers down upon the crew in an ugly disarray of crumpled sail, snaking rigging and the jagged edged spears of a disintegrated mast. Sailors have a euphemism for it: de-rigging.

So the question became: should we be afraid ? and further: what should we do about it?

The German watch went to the pub. We drank. As an ex-unionist I volunteered to take our grievances to the skipper, but the consensus was: no, leave him alone. He was stressed enough already. Christian was fond of telling him, 'Mate, you've got my boat. Don't break it.' Besides, Mark was constrained, Martin had not yet arrived and we needed to stay at the pontoon.

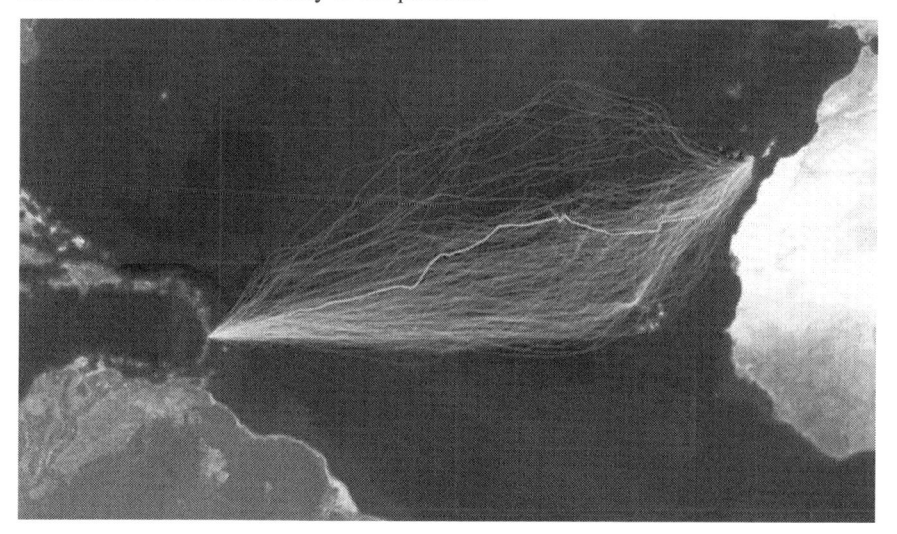

Fig. 3. 2013 Atlantic Rally. Las Palmas to St Lucia

5 First Sight of a Parallel Universe

It was at about this time that I had a sense of falling into a parallel universe with a massive step change in risk tolerance.

There was the twenty-something Australian girl, a lawyer from Sydney, thumbing a ride on Monster Project, a Volvo 70 ocean racer with 17 crew and eight beds. She succeeded. Monster Project came second overall in 11 days. The pre-race weather briefing warned of a storm system forming to the north-west. Most of the fleet avoided it, exiting Las Palmas to the south-west. Monster Project and the other 70 footers headed straight for it (refer Figure 3). 'Yee ha!'

Then, at journey's end: dusk in the bar at Rodney Bay, St Lucia, Bahamas, I met an oracle. He was the first mate of a sister yacht EH01. In a previous life he'd been a scientist, a designer of therapeutic drugs. We had a lot in common. He had lived under same reactive chemical safety disciplines that had ruled my life in my formative professional years. His eyes lit up as he spoke of a passage from Malta to Gibraltar.

6 Approaching Gibraltar

Approaching Gibraltar the Mediterranean narrows, the ocean shallows and the tides oppose the winds. In a bad blow the waves stand up, making them steep in the face. In these narrows, unlike the broad Atlantic, there is nowhere to run, you either surf or suffer and the surfing must be precise: straight down the face of the wave perpendicular to crest. A bad angle meant a skewed path down an almost vertical wall, a boat falling off a wave, mast and sails in the water, waves over the deck and a submarining crew, pissed off and clinging on for dear life.

Lashed in place he'd been at the wheel for many hours suffering several nock-downs[1].

'It's an adrenaline high you just can't get on drugs,' he said. When relieved, he'd gone below and discovered that a body shot through with adrenaline comes down pretty much like a yacht down a wave. It augurs in, to an emotional trough. He was racked with convulsive sobbing, and when he could sob no more, lapsed into, not Shakespeare's: '*Sleep that knits up the ravel'd sleave of care*[2],' but black unconsciousness devoid of dream.

'How did it go?' I asked, 'On balance, did you enjoy the passage?'

'Fantastic!' He replied.

---

[1] A boat has been knocked down when its sails are flat on the water. A knock-down can be due to a sudden extreme gust or a freak wave or a combination of both.

[2] Shakespeare, Macbeth, Act 2, Scene 2

We sat in silence, quiet as a stone, and watched the pink light fall from the sky. There's a warm wind in the Bahamas and forty-five brands of rum; the stuff that leads a man to reflect.

And then he said, 'The reality is Les, if we applied the safety disciplines that you and I experienced in other lives, none of these yachts would ever leave the marina. In fact, no one would dare cross an ocean. No voyages of discovery. No advanced yacht design. No advanced design of any kind. Because you have to take risks and break things to advance anything worthy.'

'I know, not everyone wants this life. They're happy using our inventions, burying their heads in pop culture, bullshit and trivia. But that's not for me. I like it out here on this edge. We don't do stupid things, but we do take calculated risks. It's a matter of risk tolerance. If you want to join us you've got to ramp yours up. Cheers!'

I ordered another rum and reflected on the parallel universes, each with a different risk tolerance. Was variation in ALARP a bad thing? Maybe not. Some environments need high risk tolerances to function, apparently. Try running an aircraft carrier under a commercial aviation regime.

Clearly there are other forces in play. Whereas fear brings risk tolerance down, some other force must drive it up, possibly the scent of a reward; that emotional rush human beings derive from sating a desire.

Bob McKee reflects on it in his book Story (McKee 1999).

> The measure of the value of any human desire is in direct proportion to the risk involved in its pursuit. .... We give the ultimate values to those things that demand the ultimate risks - our freedom, our lives, our souls. ... and to live meaningfully is to be at perpetual risk.

So it seems that the level of ALARP for any given universe represents an equilibrium between the emotional drivers of fear and desire.

The question then becomes: should you fall to earth in a community that tolerates risks that you feel are unreasonable, can the perception of ALARP be changed? can you convince a community to fear more and desire less?

My professors told me, 'To control something you must understand its behaviour. You must have a mental model.' But we're talking humans here, not machines. This is typically where engineers bail out and return to their calculators, waving in the general direction of the humanities building. My point is this: in an environment redolent with safety critical automation, driverless cars and the like, this attitude is not sustainable. Engineers must have knowledge of the fundamental drivers of risk taking. We could start with cognitive scientist, George Lakoff who has models that I find useful.

7 Philosophy in the Flesh

In his book, Philosophy In The Flesh (Lakoff 1999), Lakoff makes three assertions:

1) The mind is inherently embodied. Reason arises from the nature of our brains, bodies and bodily experiences. The architecture of your brain's neural networks determines what concepts you have and hence the kind of reasoning you can do.

> *Ergo it's a stretch to ask a person to fear a hazard if they have not had a bodily experience of its consequences.*

2) Thought is mostly unconscious. It is a rule of thumb among cognitive scientists that ninety-five percent of all thought is unconscious. Human beings are not, for the most part, in conscious control of - or even consciously aware of - their reasoning. Most of their reason is based on various kinds of prototypes, framings, and metaphors.

> *Ergo it is difficult to influence the communal perception of ALARP because it does not involve conscious thought. It's imprinted on the subconscious from past emotional experiences.*

3) We think in metaphors. Metaphors explain 'this' in terms of 'that' where 'this' is something new and 'that' is something familiar. Metaphorical thought makes insight possible and constrains the forms that reason can take. Metaphor is not a minor trope of poetry, but rather a fundamental mechanism of the mind. Metaphor is human thought at peak performance in its ability to make associations and connections. We need it to make sense of complex things, to say a lot with a little and to remember.

A year from now, dear reader, you'll have forgotten these words, but you won't have forgotten that, '*Juliet is the sun*[1],' a compelling metaphor imprinted upon us in 1595.

Good metaphors delight, inform and persuade. When preparing a company for an audit all I need to say is:

'*I am sent with broom before to sweep the dust behind the door*[2].'

8 Calibrating Risk Tolerance

To apply Lakoff, the insightful safety engineer, thrust into a particular universe, on a mission to improve safety, must first set aside his or her personal view of how one should live, and walk the streets in a state of 'no-mind'. That is, a mind

---

[1] Shakespeare, Romeo and Juliet, Act 2, Scene 2
[2] Shakespeare, A Midsummer-Night's Dream, Act 5, Scene 1

cleared of conscious thought, open and receptive to what may come. Accept that you're in their cave, lit by the fires of their egos. Look for compelling stories of triumph and tragedy; their legends. And in doing so find the trail of reasoning that has led them to tolerate various risks.

Mark had sailed the Atlantic eight times, always training the crew on the first three days at sea and never had a problem. Why should he change? Lakoff says that our brains tend to optimise on the basis of what they already have and to add only what is necessary.

So in learning their legends you'll establish an ALARP baseline. Next, to follow Lakoff, assume that no community will move off this baseline without a bodily experience or a compelling metaphor.

9 The Squall

The German watch had the 10pm slot, so as we moved westward, much of our sailing was done at night. At around 9.55pm each evening, two steps up the companionway ladder, head out of the hatch and looking aft, I would stop to take in the surreal. It was most stark on black nights, with a constant blow from the south. Healed over on a beam reach, making a good nine knots through a sloppy sea, a white flare of luminous wash would frame the starboard rails, silhouetting the shapes of my fellow watchmen, huddled like a choir in the rear cockpit. A hooded helmsman stood amongst them bent to the wheel, face in shadow, denatured, hollowed out, simplified, an automaton with a single purpose: to see off the pagan dictate that nothing in nature travels in a straight line.

There was always a sense of power and speed and springing headlong. You could feel the weight of the hull pounding resolute through an ocean. The scene invoked an inverted Leonore, '*Look abroad-the moon shines bright; We and the dead ride fast by night*[1],' yet with an eerie light, not incident from the sky, but rising up from the ocean like the breath of a sea dragon. I had asked Mark why we didn't trail the fishing line at night. 'There are things that come to the surface on dark nights that I don't want to catch,' he said.

The scene jolted me from my reverie with the enormity of where we were and what we were doing; with a foreboding that soon that helmsman would be me.

On one such night, sailing blind, we augured straight into a squall. It fell upon us like an evil spell. The temperature dropped ten degrees in ten seconds. From my perch beneath the spray hood I turned to look forward and watched the mass of water that fell upon us flatten the waves and suck the blue from the sea, leaving everything to glow a ghostly pinging grey. Winds in excess of fifty knots seemed to hit us from all directions at once. The helmsman lost control of the boat. We

---

[1] Leonore. A ballad by Gottfried August Bürger, composed in 1773

crash jibed. None of us had the slightest idea what to do next. Rolf yelled to me, 'Get Mark up here. Now!'

Mark appeared like magic, in a T-shirt and shorts straight out of a dead sleep. He crabbed across the rear deck and stood in the cockpit with one knee on a seat, out of range of the swinging boom. He looked around with a serene expression, thrown into sharp relief against the raging storm. He reached into his pocket and retrieved a pack of cigarettes. Then, in a downpour to rival Niagara Falls, lit a smoke. On the helm he was the calligrapher composing the Haiku[1].

O snail
Climb Mount Fuji,
But slowly, slowly!
    Kobayashi Issa

Within two minutes we were under control and back on course. Handing Rolf the helm he returned below decks to sleep, that '*Balm of hurt minds, great nature's second course*[2].'

The German watch exchanged sheepish glances. Childlike, we had been afraid and cried for our parent who had saved us. In our incompetence we had good reason to be afraid. Mark wasn't afraid.

Mark had his first solo sail at the age of three in a Heron dingy. Of course he'd broached the boat and was pitched overboard to be fished out of the water by a family friend. His father caught his rescuer delivering him to shore and asked, 'What are you doing?' followed by, 'Put him back in the boat!'

Sailing was and is Mark's life. To him the squall was trivia, I doubt if regaining control required much conscious thought.

Fig. 4. 'Morning has broken ...' Mark serenades us on the run into St Lucia

---

[1] A Japanese poetic form. Issa (1763–1827) is considered a Haiku master
[2] Shakespeare's Macbeth, Act 2, Scene 2

In contrast the German watch had sailed into an environment redolent with risks that it did not have the capacity to manage.

> *This scenario describes every single company that has ever introduced software into a safety critical system.*

Like most manifestations of evil, the squall soon passed, the rain ceased and the stars came out. Conversation recommenced in the rear cockpit, but on a more spiritual plane. Who could blame us, we'd just had a life transforming experience.

'What is spirituality anyway?' I asked. The German watch looked to Martin, a staunch Roman Catholic who attended Mass every day and men's fellowship on Saturday. But it was Uli who answered immediately and with great confidence, 'Spirituality is a sense of being part of a system for which you don't make the rules,' he said.

I'm still processing that ...

10 Influencing ALARP

Technology explosion is like the big bang. The light of its meaning continues to reach us millennia after the event. And when the future arrives it's always uneven, known to some universes but not others.

So, as intergalactic traveller, and some-time functional safety consultant, you'll come upon civilisations who have no experience of squalls and no reason to fear them. How do you make them aware? How do you shift that immovable ALARP to a point where they'll take action to reduce risk, when you don't have the legitimate power of a yacht skipper.

The short answer is to face them with truth.

Experience is the best teacher but, if all you have is words, tell stories, wrap your truth in the emotive metaphors of real life. Language has power to transform reality, it teaches, delights and can also persuade. Plato feared its power. He wanted poets kicked out of the ideal republic. '*Allowing the honeyed muse to enter, softens people up making them prey to dangerous and corrupting emotions*[1],' he said. Apple engineers were warned about this. They were told, 'On approach to Steve Jobs you will enter a reality distortion field, and under its sway you WILL believe in the impossible.'

---

[1] Plato, The Republic (Plato 360 BCE)

11 Tom's Eyes

When I first met Tom he was surrounded by a bevy of California girls, in a hot tub, in his backyard, in Walnut Creek, California. It was 1976 and I'd sailed into the San Francisco touchy-feely era where getting naked in a hot tub was the thing you did on a Saturday night. There was little risk involved.

Despite the company he kept, there was a sad, empty look in Tom's eyes. I later found out why. Tom was a chemical process engineer who had undersized the pressure relief system on a research reactor. It exploded and killed a lab technician. This had happened ten years before.

I'd like to forget Tom's eyes but I can't. Maybe this is not such a bad thing. It helps remind me how much I don't ever want to look into a mirror and see those eyes staring back.

The failure was not Tom's alone. It was a systematic failure of the design review process. Everybody involved bore some responsibility but Tom had shouldered the guilt – for the rest of his life.

Responsibility: it's a pretty word.

12 On the Helm

I was the first watchman on the helm only once in the voyage. Waking up in the middle of the night was a chore, and finding the courage to take the con[1], the next hurdle. On the question, 'Who's next on the helm?' there was always a pause sometimes extended by a wild sea. Then a volunteer or two, followed by a negotiation on sequence, like a gaggle of boys committed to jumping into the water from a very high cliff.

There was no compulsion to take the con, your reputation would survive a demurral, no harm no foul. But there was also no question that we all would, not driven by some blokey peer pressure but simply the little boy's desire to do dangerous things. A feeling I hadn't had for some time: that inner voice whispering, 'You can't do this,' the tightness in the chest, the foreboding, all in a fog of adrenaline that drew you to it. And for me, a very strong desire not to feel useless.

And when my turn came there was always that shock of transition from presence to responsibility. It would happen in less than thirty seconds. I'd clip my safety harness to the jackstay aft of the helmsman's, grab the stainless steel tubing of the bimini bracket, haul myself out of my seat in the rear cockpit and squeeze past the wheel taking care not to jam it. I'd stand behind the helmsman, gain balance, confirm the heading, then wait for him to reach some steady state, preferably dead on course with the boat well-balanced and the wheel at top dead centre with

---

[1] The position of responsibility and authority for the operation and steering of a sea vehicle.

the rudder running fore and aft of the centre line. For a split second there'd be four hands on the wheel as I took hold and then he'd duck away leaving me in control. No longer just present on deck, but responsible for the boat.

Conditions were different on every watch: the state of the ocean, the wind and the settings of the rig. Not a problem for an experienced helmsman, but for a rank amateur like me in the early days of the voyage, it was a challenge. Especially on ink-black knights when I could not see the bow moving against the horizon and it seemed to Les engineer that the only true thing was a digital read out in the cockpit.

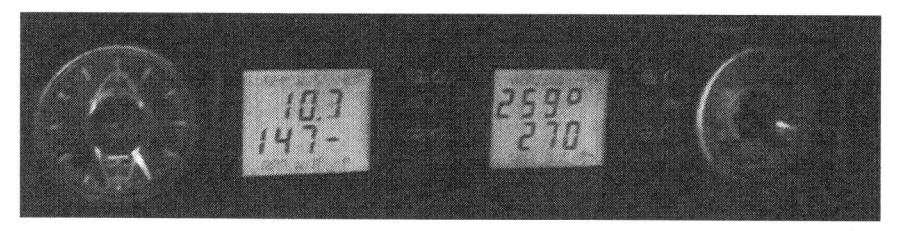

Fig. 5. Cockpit instruments

Unfortunately it lagged the true heading by a fraction of a second, enough to have me off course before I knew it, to over compensate, and then to spend the next fifteen minutes oscillating about the designated course with a snake wake trailing aft.

But finally, after fifteen hours on the helm, I began to feel.

13 Feeling

Night sailing in the mid Atlantic has a rhythm, unique to every session on the helm. Wind gusts peak at the same speed and waves roll in at constant frequency and amplitude. We developed a mentoring discipline. The previous helmsman would stay close and whisper hints in one ear and within five minutes you'd get the beat and fall in sync with nature, a flesh and blood automaton driving west, with the back of your ears sensing the wind, your mind in the stars and your whole visceral self feeling the tempo of ocean's heave.

The Northern Child had a tendency to come up into the wind when hit by a gust.

'Yes wind,' I'd think, 'I can feel you building on the back of my ears. Tonight you'll blow hard. So I'll take her downwind before you arrive and let you blow me back on course.'

And, 'Yes ocean. Tonight you're running a twelve foot swell. So when I feel her transom lift, an eighth turn to port will have us surfing your wave and when we hit the trough, an eighth turn to starboard will put her back on course.'

We humans do love rhythm. It's deep within our nature.

Fig. 6. Uli at the helm in states of no-mind and ultra awareness

14 Transcendence

But best of all were the pre-dawn hours on clear nights when the three stars of Orion's belt pointed due west: a potent metaphor for a boat on course 270. Potent because it's easily understood and translated into action. Some natural wiring in my brain seemed to sense the position of those stars and compute the helming solution, all without conscious thought in a state of no-mind. This in stark contrast to the black nights where the absence of visual cues demanded a mental state of ultra awareness (refer Figure 6). Conversation wasn't possible.

Then on one clear night with a mind free to wander, I thought of Wendy Darling who once asked Peter Pan the way to Neverland, the place of eternal childhood.

'Follow the second star to the right, and straight on till morning,' he said[1].

So there I was, following three stars, and denying Lakoff, for I swear that on that night my mind left the boat and hovered above in the night sky. Disembodied, I looked down on a yacht in full sail, on a moonlit sea. And I saw the planet and the stars and looked back at Earth as a star ship captain will a thousand years from now; his last glimpse of home on a voyage to some outer spiral of the Milky Way. In his reverie he may think of us and the first fifty years of systems engineering and how primitive we were.

---

[1] From the play, Peter Pan or the Boy Who Wouldn't Grow Up (Barrie 1904)

'They were the lost boys (and girls),' he'll think. 'They had no adults to tell them the truth. They had no Wendy Darling to scold them, 'Be nice to your sister, eat your greens and make sure you program run-time stack analysis into your electronic throttle control system, lest you make her sad, stunt your growth, have stack overflow, trigger unintended acceleration and maybe kill someone[1].'

15 What We Know

Systems engineering is a young profession. We should assume that nobody knows anything, at least nothing with the eternal import of the three stars. Maybe Uli was right, we're part of a system for which we don't make the rules. We are yet to discover any grand unified theories (GUTs).

We've found some truth in mathematics but we're weak on the human condition. This ignorance is our existential threat; the thing that could bring down civilisation as we know it. All the more reason for engineers to explore the metaphors that inform man's emotional states, because when sailboat earth hits a squall, humanity will cry for us, not only for our knowledge, but also for our wisdom and we won't find that in a formula. Newton watched an apple fall then derived the inverse square law for gravitational force, not the other way round.

We need to find our metaphors and push them as far as they'll go. For example, I've argued that risk tolerance is a function of fear and desire, but digging deeper we'll find that the state of fear has a short half life (for everyone except the Toms of this world). It's soon forgotten in those transcendent moments of howling down a wave, leaving desire to consume us, unopposed.

So, although risk tolerance may vary between universes, it's a fair assumption that, in any given universe, at any point in time, the levels of risk tolerance are universally too high.

Inappropriate risk tolerance is at the heart of many common mode failures. For example, it's safe to assume that when introducing radically new technology, techno-lust is rampant and you're probably taking risks you are not equipped to manage. This was the truth of the world's greatest oil spill at Deepwater Horizon (Oil Spill Commission 2011). BP had developed deep sea oil drilling technology that did not include a recovery strategy for catastrophic failure of blowout preventers on the seabed. Had a BP hazard analysis looked for accident precursors in human states of consciousness (in this case: all-consuming desire) this disaster could have been avoided.

Where unchecked human desire is involved, probabilistic risk assessment isn't necessary because the probability of an accident approaches 100 percent. History is littered with cases where profit motives have trumped risk assessments with

---

[1] A reference to a ruling against Toyota in a case of unintended acceleration that lead to a death. Central to the trial was dangerous failure of engine control firmware (Dunn 2013).

attendant immoral acts of bastardry. The James Hardie company is a case in point.

> The company knew for decades about the deadly effects of asbestos but they suppressed this information and mounted a campaign of public relations spin while they plotted to escape liability—and would have succeeded but for the union and community campaign for justice for asbestos victims. (Peacock 2009)

16 Safety in Metaphor

So what should we know to save humanity from itself? The Oracle of Rodney Bay was right. People will steal fire from heaven and bring it to earth whether we like it or not.

My solution is to ask a poet. Robert Frost said:

> Unless you are at home in the metaphor you are not safe anywhere. Because you are not at ease with figurative values: you don't know the metaphor in its strength and its weakness. You don't know how far you may expect to ride it and when it may break down with you. You are not safe in science; you are not safe in history. (Frost 1931)

Right now we can cross an ocean in a sailboat but we can't venture further than the moon, because we don't have the metaphors to deal with the mission complexity. It's true, we build massively complex systems, like the global financial markets, but they aren't safe. We don't understand them, they exhibit emergent behaviours that we can't predict, and they fail.

Software engineers have some metaphors: objects, inheritance, proxy, state, thread, queue, tree, layer, relation, semaphore ... but we need higher level concepts, more aligned with the force fields of human experience, more easily imprinted to inform Lakoff's unconscious decision making.

These are not lofty ideas. I've experienced the power of metaphor explaining complex computer control systems to plant operators. They were often poorly educated but smart and dependable and they understood the state engine, quickly and deeply.

David Harel had the same experience. In his paper *Statecharts in the Making: A Personal Account* (Harel 2007) he tells the story of an encounter with an Israeli fighter pilot.

> He was smart and intelligent, but he had never seen a state machine or a state diagram before, not to mention a statechart. He stared for a moment at this picture on the blackboard, with its complicated mess of blobs, blobs inside other blobs, coloured arrows splitting and merging, etc., and asked 'What's that?' One of the members of the team said 'Oh, that's the behaviour of the so-and-so part of the system, and, by the way, these rounded rectangles are states, and the arrows are transitions between states'. And that was all that was said. The pilot stood there studying the blackboard for a minute or two, and then said, 'I think you have a mistake down here, this arrow should go over here and not over there'; and he was right. ... If an outsider could come in, just like that, and be able to

grasp something that was pretty complicated but without being exposed to the technical details of the language or the approach, then maybe we are on the right track.

Thirty years later I'm intrigued that Lakoff agrees. In Philosophy in the Flesh he writes:

> It would be difficult to find concepts more central to philosophy than events, causes, changes, states, actions, and purposes.

This is one example of the nexus of philosophy, psychology and automation resulting in human motor schemas being applied to control systems.

In the most abstract sense, the first ten years of my engineering career were dedicated to preserving the integrity of the state engine model, ensuring it was applied in safety critical process control code - I was Wendy Darling scolding the lost boys. And I was respected, not as the owner of the code, but as the keeper of the metaphor.

Metaphors are a safety issue.

17 Rumours of My Death

We sighted the misty mountains of St Lucia on 12 December. In mobile range I called my three daughters and told them we'd made it without incident (but for a boom that busted on the last jibe before the finish). We crossed the finish line and I thought, 'Well, that's that.' But not so.

The next day Uli, Rolf, Matthias and Johnny insisted on climbing Gros Piton, a 2600 foot peak in the south of the island. I was expecting a stroll on a mountain trail but was handed a challenge, 2000 feet straight up with very large steps cut into the hillside. Lathered in sweat I made it to the top with my fellow watchmen. We rested and listened to the booming reggae music from the Rastermen in the valley. On the return trip we stopped for lunch at a restaurant overlooking a peaceful bay. I ordered a glass of wine.

The food arrived but I'd lost my appetite. I got one sentence away as I slid sideways off my chair,

'Guys I don't feel well ... ,'

I felt life slipping away.

'So this is it,' I thought.

'Sixty-five. That's not so bad. The Bahamas: a good place to die; those warm winds.'

And I could feel loving arms around me as I floated gently to the floor.

'And among friends.'

But then, a heavy sadness.

'I can't say goodbye to the girls,' I thought. 'My darlings. The light of my life.'

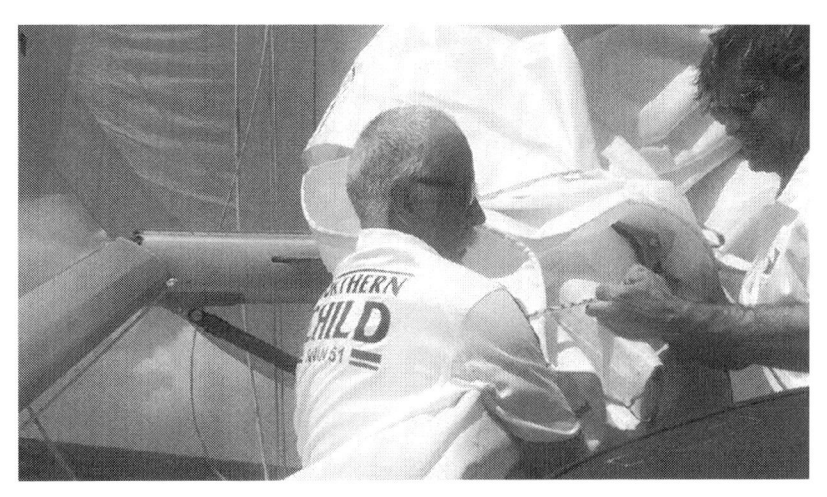

Fig. 7. Johnny and Uli save the main sail on boom failure

If you're going to die, do it at lunch with two anesthetists. Uli and Matthias knew exactly what was happening. Dehydration followed by wine (a diuretic) decreases blood volume and increases its viscosity, thus decreasing the blood pressure in the brain. The subject faints. Uli caught me as I fell. The loving arms were his.

On the way out I bought three necklaces threaded with forest seeds from a Rasterman at the restaurant door. When I told my girls the story, it was the best Christmas present they ever had.

18 Reflections

So in 3000 miles of ocean, forty hours on the helm, one mountain and one death, I learned a lot. Here are some thoughts:

All hazard analyses should commence with a candid review of the balance of emotion in an organisation. Ask questions like, 'Is our unconscious decision making informed by excessive hubris born of desire? Are we fearful enough? Are we too tolerant of risk? Can we learn from other universes?' In particular look for the rush of transcendence, that potent force that draws us into danger. Be aware of triggers such as massive bonuses paid on early delivery and exciting technical breakthroughs. Then take action to counter these urges. And most important of all have someone on the team with bodily experience of catastrophic failure; someone with feeling.

If humanity must break things to advance, systems engineers should be on the edge doing the work. 'Yee har,' is allowed, but:

O humanity
Follow your stars
But safely, safely!

Right now the functional safety specialty advises too much and synthesizes too little. And the boys and girls who perform the hazard analysis and attend the safety design reviews are too often present but not responsible, and there's no harm, no foul, attached to ignoring them. While the boys and girls who ARE responsible don't speak to nature, have never seen a squall and are full of desire. And the education that might give them pause is as cursory as a one hour safety briefing on a level deck at a pontoon in Las Palmas.

So if we're going to educate them it shouldn't be in a classroom. We learn faster when our actions have immediate consequences. We need to find them a helm or a poet to align their thinking with compelling metaphors.

Thinking on design: safe systems are built on good metaphors in sync with the rhythms of human thought. The sign of a good metaphor is a system that's easy to understand and operate in a state of no-mind. There is simplicity and symmetry, with the same metaphor applicable to many application domains. And the metaphor is preserved from requirements to design to code with no cognitive discontinuity, as is the case with the finite state model.

The sign of a bad metaphor is a system that requires the massive concentration of a helmsman on a black night in a stormy sea with the experiential wisdom of eight Atlantic crossings (refer Figure 6 - right-hand image). Bad metaphors are our primary existential threat. It's possible to crash jibe a civilisation with lack of understanding.

And lastly, dying is a profoundly disappointing experience, worse if you visit it on someone else. In contrast loving is a wonderful thing. Spend time, engineers. Put your arms around humanity and know how comforting that is. And part of loving is understanding. Reading the thought patterns that drive our actions and designing better metaphors to serve and protect the lives of those who depend on us.

Have I spoken the truth? I don't know. But just the act of reflecting on these things is to know your parents, so no one will ever call you a bastard, and a millennia from now, when we are judged, they may even say,

'They were professionals.'

Acknowledgements The thoughts expressed in this paper would not have existed without my fish-out-of-water exposure to the men and women of the bluewater sailing community. I am particularly grateful to Mark Burton, the skipper of the Northern Child, for the experience of a lifetime. Mark, your seamanship and problem solving abilities were a pleasure to behold. I have never felt more safe in anyone's hands. Thanks also to the men of the German watch for your fellowship, conversation and insights on those long nights in the rear cockpit, especially to Uli who caught me when I fell. Fair winds and following seas to you all.

References

Barrie JM (1904) Peter Pan or The Boy Who Would Not Grow Up. Project Gutenberg of Australia. http://gutenberg.net.au/ebooks03/0300081h.html. Accessed 10 November 2014

Dunn M (2013) Toyota's killer firmware: *Bad design and its consequences*. EDN Network. http://www.edn.com/design/automotive/4423428/1/Toyota-s-killer-firmware--Bad-design-and-its-consequences. Accessed 30 October 2014

Frost R (1931) Education by Poetry. A talk Frost delivered at Amherst College and subsequently revised for publication in the Amherst Graduates' Quarterly of February 1931. http://www.en.utexas.edu/amlit/amlitprivate/scans/edbypo.html. Accessed 30 October 2014

Harel D (2007) Statecharts in the Making: A Personal Account. The Weizmann Institute of Science http://www.wisdom.weizmann.ac.il/~harel/papers/Statecharts.History.pdf. Accessed 30 October 2014

Lakoff G, Johnson M (1999) Philosophy In The Flesh: *The embodied mind and its challenge to Western thought*, Basic Books

McKee R (1999) Story: *Substance, structure, style, and the principles of screenwriting*, Methuen, Publishing

Oil Spill Commission (2011) Deep Water: Gulf Oil Disaster and the Future of Offshore Drilling. http://www.gpo.gov/fdsys/pkg/GPO-OILCOMMISSION/pdf/GPO-OILCOMMISSION.pdf. Accessed 30 October 2014

Peacock M (2009) Killer Company: *James Hardie Exposed*, ABC Books

Plato (360 BCE) The Republic. MIT. http://classics.mit.edu/Plato/republic.html. Accessed 10 November 2014

Assessing the Safety Risk of Collaborative Automation within the UK Aerospace Manufacturing Industry

Amira Hamilton, Phil Webb

Cranfield University

Cranfield, UK

Abstract *The ever increasing demand to improve efficiency, raise production rates and reduce costs, whilst maintaining or improving quality, is driving the need for the automation of many manufacturing processes. Traditionally speaking, the default control measure for the risk management of automation in manufacturing has been the use of physical guards. Robot cell installations with physical guards can be difficult to deploy on some manufacturing assembly line. Given that these physical safeguarding measures separate the human operator from the moving robot, the work performed by robots is effectively detached from that performed by human operators. This can be a limitation on some flow lines where the human tasks cannot be completely separated from those of the robot. From a safety perspective, physical guards do not completely mitigate against the risk of injury due to a human trespassing the guarded area with the locks and interlocks engaged, or deliberately limiting their effectiveness through local process modifications.*

Aerospace manufacturing processes typically involve a complex mixture of high and low skilled tasks, and therefore are a prime example of a system that cannot be completely automated. Hence, it is proposed that a collaborative human/robot work cell would allow the lesser skilled manufacturing tasks to be allocated to a robot while the human operator carries out the more skilled tasks. This would in effect result in the design of a semi-automated flexible assembly line where the human and the robot work alongside each other and collaborate with each other within a predefined, shared and safeguarded workspace. Work has been undertaken at Cranfield University to investigate the potential for using a collaborative robotic system in aerospace manufacturing. The aim of this research is to offer an example of a truly collaborative, inherently safe, robot work cell that integrates readily available robotic systems with affordable, off-the-shelf, safety-rated electro-sensitive protective equipment.

© Amira Hamilton and Phil Webb 2015.
Published by the Safety-Critical Systems Club. All Rights Reserved

1 Introduction

The production of modern commercial aircraft involves many processes that could potentially be automated in order to improve efficiency and quality. The use of readily available robotic systems could provide a means of achieving such automation in a cost effective manner. However, there remain many processes that will not be possible or desirable to automate, leading to the requirement for both automated and manual processes on the same production line. Current safety regulations require the separation of human operators from automated equipment, typically with fixed safety fencing. Such installations are difficult to arrange on a flexible assembly line, as is typically used for aircraft wing equipping, restricting the efficiency of the overall assembly process.

A combination of improving technology and changing regulations has the potential to allow for closer interaction between human operators and robotic equipment. Improving sensor technologies, combined with high speed computer processing, could allow the real-time monitoring of the environment around automated equipment and thus remove the need for fixed guarding.

Work has recently been undertaken at Cranfield University (Walton et al. 2011) to investigate the potential for using a robotic system to assist in aerospace manufacturing processes. The work proposed in this project refers to the fitting of an assembly item to a counterpart as a process demonstrator. The work has been conducted with the view of developing a generic safety solution for use with any automated equipment on an assembly line.

This project does not only consider the robotic technical features that could be introduced to increase the efficacy and flexibility of aerospace manufacturing processes, but also considers the safety aspects of close proximity collaborative automation in depth. State of the art off-the-shelf safety systems have been integrated into the test facility with the view of demonstrating their use in a flexible and cost effective manner that does not compromise safety.

1.1.   Literature Review

Some research has been carried out in the general field of human-robot interaction. In fact, most authors recognise that a flexible assembly design where robots and humans can cooperate together directly can improve efficiency. For example, a comprehensive survey was written about the different forms of human-robot interaction in assembly lines and the available technology to allow for this to happen (Kruger et al. 2009). It was stated that one way of making the most of both robot and human capabilities is to design the assembly line in such a way so as to allow for sequential division of tasks - known as a hybrid assembly line. The idea being that the simpler parts of the assembly could involve the use of a robotic device upstream of the line, whilst the more complex part that requires human skill could be set downstream of the line. The main drawback of this type of assembly design is that it assumes that one can clearly separate the two processes. There are some aerospace manufacturing processes that do not have a clear boundary where the robot's work ends and the human's work starts, therefore research that considers hybrid assembly lines of this nature would be of limited value in certain areas of aerospace manufacturing.

An alternative assembly design is referred to as 'workplace sharing systems' in the survey (Kruger et al. 2009). In this assembly design, the robot and human are both working in the same workplace, whereby one of them is performing a handling task (i.e. lifting and manoeuvring items), whilst the other performs an assembly task (e.g. fitting items together). In this type of assembly design, the interaction of the robot and the human is limited to the avoidance of collision. Again this assumes that one can distribute the task clearly between the robot and the human operator and some aerospace manufacturing processes do not fall within this category.

Due to the need to reduce the risk of occupational related injuries and to meet ergonomic targets in certain assembly lines, intelligent assist systems (IAS) or intelligent automation devices (IAD) have been designed. Therefore, a third category of assembly line design was invented, one known as the 'workplace and time sharing system' (Kruger et al. 2009). In this category, the human operator and the robot are able to jointly perform a handling or assembly task at the same time which is more within the context of the type of manufacturing processes that this project is considering.

An example of a 'workplace and time sharing system' is the PowerMate robot (Scharft et al. 2005). The PowerMate robot would handle an item for assembly, and would move at maximum speed for parts of the task that took place in an area where the human could not access. Once in the cooperation area, the robot would come to a stop and change to cooperation mode. The human operator would then be able to move the robot by pushing, pulling or turning a handle which was installed on the robot gripper. This functionality was made possible through a force-torque-sensor on the robot gripper that allowed the robot to be moved by the human operator. This solution, although effective for certain applications, could not be practically installed on certain rigging processes. For example, if the assembly item is particularly large, the operator may not be able to reach over to the gripper to push and pull the robot.

The literature review on human-robot interaction further unearthed some more complicated solutions. For example, one piece of research involved using a robot without range sensors for the detection of the position of the assembly item (Wojtara et al. 2009). Hence, the robot had to rely on the force and displacement information gained from the direct interaction with the human. To allow for this, the system was developed using some complex algorithms to provide the robot with the functionality to distinguish when the human wanted to rotate the assembly item or to translate it laterally. In order to assist the human with the high mental workload required in the scenario, given that the robot did not make any decisions, the researchers proposed a concentration aiding scheme. The author claims that the research was successfully tested at one of Toyota Motor's plants, where the robot attached a windscreen to the front and rear of the car in cooperation with a human. The robot in this solution can be seen as a passive machine, in that it responds to the input provided directly by the operator. From a safety perspective, it could be argued that the hazards relating to such a robot colliding and causing harm to a human are negligible. However, the high mental workload required of the operator may in itself be a source of error and may not be a practical solution that can be used within certain manufacturing processes.

A number of studies in this area have also been dedicated to providing bespoke solutions for collision detection and collision avoidance. For example, an approach was considered for collision avoidance based on connecting virtual 3D models to a set of vision and motion sensors in order to monitor and detect collision in real-time in an augmented reality (Schmidt et al. 2013). It is accepted that collision detection and avoidance could contribute to a safer work environment when humans and robots collaborate together, however, this bespoke technology is relatively new and under development. It could be said that this innovation introduces a level of complexity to a safety-critical system and therefore could undermine its ability to protect operators sufficiently. Furthermore, an assessment was carried out of a collision avoidance strategy that involved the robot modifying its path once potential collision was detected (Schmidt et al. 2013). This strategy does not take into account the existence of other operators that could be in the modified pathway and at risk of injury due to collision. It also does not take into account the existence of high value assembly items that could be damaged in the process.

Collision avoidance as the means of controlling the risk of injury with a moving machine is not a relatively recent interest in research. A paper was written relating to the fuzzy logic control approach used to assist in collision avoidance for an automated guided vehicle (AGV) (Lin et al. 1997). This research proposed a sensor modelling method which allowed for the use of an optimal number of sensors and arrangement on an AGV to provide views at all angles. It was concluded that the fuzzy logic control was successful in navigating the AGV to get out of traps as well as to avoid collision with moving objects. This type of research shows that the theory of applying fuzzy logic control to an industrial robot to avoid collision is viable, however, it can be said that it would be more difficult to construct a safety argument with a novel fuzzy logic control method. It is more feasible to integrate an industrial robot with well tested and tried safety monitoring and control systems to provide the required evidence for deployment.

Other efforts in this area relate to increasing human's acceptability of robots by having on-board adaptive learning mechanisms that can intelligently update a human model for effective human-robot interaction (Sekman et al. 2013). The robot used in this research was a small mobile service type. Although this research is effective for future service robots, its readability across to industrial robots is limited. The risk profiles involved differ greatly between the two, and the errors that can be introduced during the teaching phase of the robot could be numerous and largely unidentifiable before operational use.

1.2. ISO 10218

In order to demonstrate the viability of the proposed safety monitoring system for integration in a collaborative automation cell, a review of the applicable standards was carried out. Specifically, those sections that relate to collaborative automation were considered. In order to establish a clear understanding of the definition of collaborative automation, a description in the ISO 10218-1:2011 standard "Robot and robotic devices – Safety requirements for industrial robots" was used for reference. In this standard, collaborative operation is the 'state in which purposely designed robots work in direct cooperation with a human within a defined workspace'. This definition is in line with the basis of this project, in that it considers the safety aspects of an industrial robot collaborating directly with an operator within a safeguarded working envelope. Furthermore, ISO 10218-1:2011 standard provides a clear set of safety requirements for collaborative automation as follows:

- visual indication shall be used;

- robot shall stop when human is in the collaborative workspace;

- robot shall maintain a determined speed and separation distance from the operator;

- Standstill Monitoring shall be used;

- a protective stop shall be issued if robot exceeds any parameter limit;

- and, collaborative operation shall be determined by a risk assessment."

With regards to the use of sensitive protective equipment instead of fixed guarding, requirement 5.10.5.1 of ISO 10218-2:2011 states:

Sensitive protective equipment is typically selected when an application requires frequent access, personnel interaction with a machine, good visibility of the machine or process, or when it is not ergonomic to provide fixed guarding.

Based on the above, it is clear that the applicable standard does allow for the use of collaborative automation robot cells without fixed guarding. This is on the provision that the risk assessment underpinning the design of the robot cell does not highlight process related hazards that require fixed guarding as a risk control measure. Requirement 5.10.5.1 of ISO 10218-2:2011 provides the following examples of application hazards that would preclude the sole use of sensitive protective equipment:

- Ejection of materials.

- Thermal or radiation hazards.

- Noise hazards.

- Environment (workspace or process) induced hazards whereby functionality of the sensitive protective equipment would be degraded.

-

2 System Specification

This section describes the top level view of the proposed safety system that complements the collaborative automation demonstrator at the Cranfield University test facility. The integrated system consists of a high payload industrial robot, various off-the-shelf safety monitoring systems, a bespoke end-effector and representative assembly installation parts.

2.1 Robot

The system includes a foundry variant of the KUKA KR240-2 industrial articulated robot with 6 degrees of freedom and a 240Kg payload capability. It can achieve repeatability of ±0.06mm and has a reach of 2700mm. The KR240-2 is equipped with a bespoke end-effector which is used to perform the scanning and manipulation of the assembly component.

2.2 Safety Monitoring System

The demonstrator integrated safety monitoring system consists of two state of the art sensitive protective and monitoring devices, consisting of a Pilz SafetyEYE and a SICK Safety Laser Scanner integrated via a Pilz Safety Relay.

2.2.1 Pilz SafetyEYE

The Pilz SafetyEYE is a 3D stereo vision system that uses a trinocular sensor suspended over the collaborative workspace. It is used to monitor the safeguarded space, the dimensions of which are based on a calculated separation distance, as stipulated by EN ISO 13855:2010, between the operator and the anticipated maximum reach of the moving robot and the edge of the assembly item. The Pilz SafetyEYE is safety rated at a maximum monitoring area of $72m^2$ (as shown in figure 1) with a resolution of 210mm for body movement detection (PILZ SafetyEYE, Safe Camera Systems Manual).

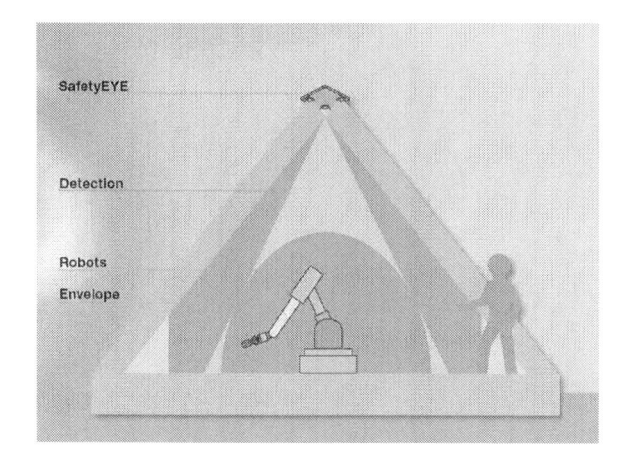

Fig. 1. The Pilz SafetyEye

The monitored safeguarded space is pre-programmed in the Pilz SafetyEYE using a number of drawn zone arrangements. These zone arrangements are designed based on the knowledge of the programmer of the expected automated movements of the robot and the calculated separation distance that must be maintained. The programmer can draw warning zones which cause the robot movement to slow down if they are violated, either by an operator or an object entering their field. The programmer ought to also draw detection zones that initiate a protective stop of the robot if they are violated. The area would then have to be cleared before the system can be re-started after re-setting the safeguards. The integrity of the overall programme is verified (preferably by an independent, trained programmer) once configuration is complete.

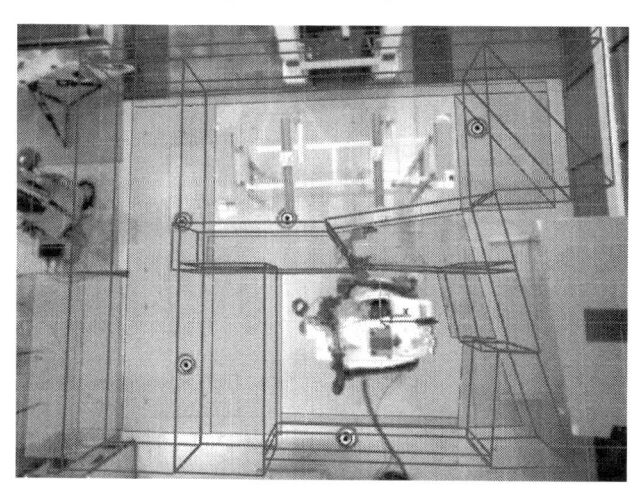

Fig. 2. Pilz SafetyEYE zone arrangements.

2.2.2 SICK Safety Laser Scanner

A SICK Safety Laser Scanner is used to cover the safety monitoring of areas where the Pilz SafetyEYE cannot detect object(s). The Pilz SafetyEYE operates from an overhead view and is therefore unable to monitor movements under areas such as fixtures or rigs. The SICK Safety Laser Scanner has a 190° field of view and 7m range, which is sufficient to cover most of the demonstrator cell (S3000 Safety Laser Scanner, Operating Instructions). It also has a resolution of 30mm.

2.2.3 End-effector

The end-effector is a tool attached to the end plate of a robot arm. In this case, the end-effector consists of a bespoke suction device for holding the assembly item – see figure 3.

Fig. 3. Bespoke end-effector for picking up assembly item

2.2.4 Push Buttons

A set of push buttons have been installed into the integrated system, see figure 4. These buttons include a green button for the start/re-start of the robot motion, a blue button for the re-setting of the safeguards as well as a red emergency button. The selection of these push buttons is in accordance with the guidance in BS EN 60204-1:1993. Furthermore, two emergency stop buttons have been installed on the assembly rig itself.

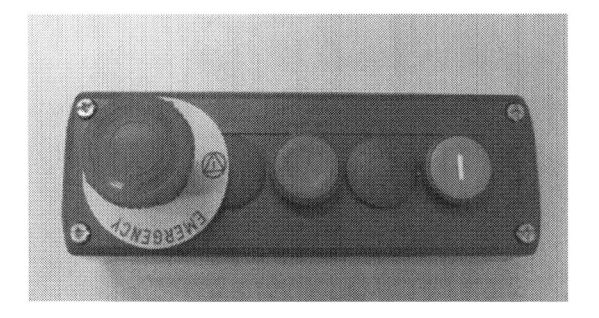

Fig. 4. Push buttons for robot control

2.2.5 Facial Recognition System

Cranfield University is prototyping a Facial Recognition System that has been integrated into the test facility. The aim of this is to provide an extra layer of security and safety during start up and re-start of the robot arm motion. The Facial Recognition System ensures that only authorised and suitably trained personnel provide clearance for the robot system to initialise its hazardous movement.

2.2.6 Projection of a Warning Sign

During the movement of the robot arm, the operator is expected to remain outside of the safeguarded workspace. In order to facilitate for an efficient use of the system, it is important to clearly mark out the safeguarded workspace and communicate to the operator when to keep out and when it is appropriate to approach the robot arm. Two standard projectors have been installed either side of the SafetyEYE to allow for projection across the robot cell if required.

2.2.7 Audio and Visual Warning System

An audio message is emitted to caution the operator when the warning zone has been violated. It also emits a message to indicate to the operator that the robot is in collaborative mode during standstill. In addition, a traffic light system is used to indicate when the robot is in motion or in standstill. The traffic light system also allows for a visual indication of the warning zone having been violated or a protective stop having been initiated.

2.2.8 Test components

As mentioned previously, the aerospace manufacturing application being used for the demonstration is based on the installation of a moveable assembly item. Test components include a structure which holds the moveable assembly item in the initial position as well as a jig with 4 brackets which represent the counter part with its relevant mountings (approximately 1.6m high), as well as an assembly item of a length of 2.3m.

3 Robot Cell Design

3.1 Description of Simulated Manufacturing Process

As already mentioned, one of the aims of the collaborative automation robot cell is to demonstrate the use of an integrated safety monitoring and control system in the context of a generic and typical aerospace assembly process. In order to achieve this, a handling task was used as a basis for the cell design. It was decided that a medium payload KUKA robot would be programmed to pick up an example moveable assembly item from one point to another, with a human approaching the robot at appropriate points to carry out the assembly task.

Figure 5 provides an aerial view of the robot cell. The robot arm is programmed to pick up the item from point A and place it at point B. The operator would be expected to stand outside of the robot's working envelope, as bounded by the detection and warning zones configured in the SafetyEYE, during the motion of the robot arm. The operator would have easy access to a control station which includes the push buttons and the Facial Recognition System.

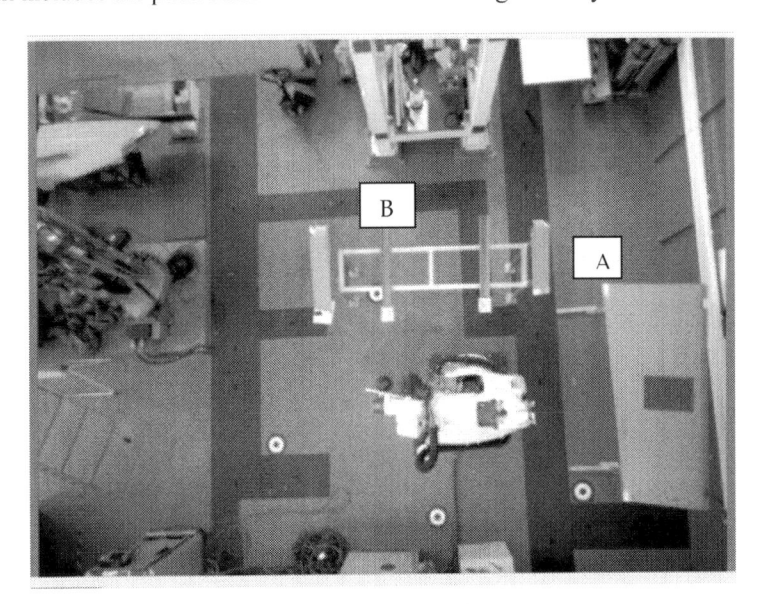

Fig. 5. Aerial view of the collaborative automation robot cell

Once the robot has manoeuvred the item to point B, it comes to a standstill. During this movement, a projected warning sign is shown on the ground – see figure 6. Once in standstill, the projection of the safeguarded zones switches to a 'go' sign, and an audio message is emitted to prompt the operator to approach the robot. The operator is then expected to insert a set of pins through the bracket, thereby joining the moveable assembly item to the brackets attached to the track.

Fig. 6. Projection of a 'Keep Out' warning sign during the movement of the robot arm

The operator then returns to a location outside of the projected area and initiates a re-start of the robot motion via the Facial Recognition System at the control station. The robot then moves the moveable assembly item up the track and comes to another standstill at the second pinning position. During this movement, a projected warning sign is shown on the ground. Once in standstill, the projection of the safeguarded zones switches to a 'go' sign, and an audio message is emitted to prompt the operator to approach the robot for the second pinning. Once this is done, the demonstration is complete.

4 Safety Analysis

In accordance with the principles outlined in BS EN ISO 12100:2010, a risk assessment was carried out on the proposed collaborative automation robot cell. This standard provides the principles of risk assessment and risk reduction in the design of machinery. To assist in the identification of the hazards, a Functional Failure Mode and Effects Analysis was carried out. A Risk Priority Number (RPN) was generated for each failure mode. This analysis showed that the failure modes relating to the bespoke end-effector system had the highest RPN. Whilst the corresponding RPNs for failure modes relating to the Pilz SafetyEYE and SICK Safety Laser Scanner were very low. Fault Tree and Event Tree Analyses was also carried out and these demonstrated that the probability of death or harm as a result of a system failure is negligible and therefore the residual risk is very low.

Future work in this subject will be focused on two main areas; carrying out various qualitative and quantitative risk assessment methodologies using the same case study in order to compare and contrast them, as well as identifying the tolerability and acceptability of risk within the UK aerospace manufacturing industry.

5 System Testing

The capability and flexibility of the integrated safety monitoring system has been demonstrated at the Cranfield University test facility. In order to validate the system, a number of test scenarios were designed and the behaviour of the system was logged. The test scenarios were chosen to simulate the most probable and foreseeable hazardous situations that could occur within the context of the test facility. The test scenarios and their actual outputs were as follows:

Scenario 1: The operator stands next to the robot during initiation. The SICK Safety Laser Scanner and Pilz SafetyEYE are active and the robot programme is initialised.
Desired output: Safety relay stops the robot from moving.
Actual output: Safety relay stops the robot from moving.

Scenario 2: SICK Safety Laser Scanner and Pilz SafetyEYE are active. The operator moves towards the robot which is moving the assembly item towards the track.
Desired output: Protective stop triggered.
Actual output: Protective stop triggered.

Scenario 3: SICK Safety Laser Scanner and Pilz SafetyEYE are active. A protective stop had been initiated. The safeguards are reset and the start button is pressed in order to re-start the motion of the robot. An operator is located near the robot arm on re-start.
Desired output: Robot does not re-start.
Actual output: Robot does not re-start.

Scenario 4: SICK Safety Laser Scanner is deactivated (simulating failure) but Pilz SafetyEYE is active. Operator presses the button to start the motion of the robot arm.
Desired output: Robot does not start.
Actual output: Robot does not start.

Scenario 5: SICK Safety Laser Scanner is active but Pilz SafetyEYE is deactivated (simulating failure). Operator presses the button to start the motion of the robot arm.
Desired output: Robot does not start.
Actual output: Robot does not start.

Scenario 6: SICK Safety Laser Scanner and Pilz SafetyEYE are active. The operator attempts to re-start the robot motion without re-setting the safeguards after they have been triggered.
Desired output: Robot does not start.
Actual output: Robot does not start.

Scenario 7: The operator presses the emergency stop button near the robot arm.
Desired output: Emergency stop triggered.
Actual output: Emergency stop triggered.

Scenario 8: The operator attempts to start the robot arm motion without being authorised by the Facial Recognition System.
Desired output: Robot does not start.
Actual output: Robot does not start.

6 Discussion

It is clear from the literature review that within the subject area of human-robot interaction, most research has been carried out in developing bespoke robotic systems, some of which include complex algorithms. The review did not find examples of research focussed on a human interacting with an unmodified, off-the-shelf, industrial robotic system. From a safety perspective, it can be argued that the deployment of bespoke robotic arms with complex algorithms could introduce some unacceptable risk. There is a more sound business case to be made, along with a more robust safety argument, if standard and safety-rated robotic devices and control measures are used for close human-robot interaction.

In addition, innovative collision avoidance systems would require some effort to be validated and verified, as well as sufficiently demonstrated in a factory environment, before they can be deployed. Some significant research seems to have been dedicated in the area of collision avoidance systems, and it is expected that their future deployment would be inevitable. However, the literature review has not shown evidence of effort in the assessment of bespoke integration of standard, off-the-shelf, safety-rated monitoring systems with standard industrial robotic devices without physical guards. Such a solution could be seen as a more viable solution to industry which could be deployed in the nearer term.

The aim of this project is to investigate the viability of truly collaborative automation robot cell from a safety perspective. It is evident from a review of the applicable safety standard that sole use of sensitive protective equipment is permitted in cases where frequent access is required or where a fixed guard would not be practical. Many aerospace manufacturing processes would fall under this category given the large size of assembly items.

The risk assessment that was carried out did not identify hazards that require fixed guarding as a risk control measure. The process of using Functional Failure Mode Effects and Analysis with Priority Risk Numbers did identify that the highest risk within the current proposed system lies with the design of the end-effector. In theory, the suction system used to lift the moveable assembly item from above could lose pressure or eject the item due to operator error or an internal failure. This indicates that future work could be to develop an enhanced version of the current system, or design a new end-effector. During the validation process of the collaborative automation robot cell, there has been no incidence of failure of the suction system. However, it is worth noting that the end-effector would undergo different design modifications depending on the application of the robot cell. For example, a cradle design could be used as alternative. Alternatively, extra suction pads could be added to cover a wider area of the assembly item. This risk assessment activity ultimately demonstrates that the proposed collaborative automation robot cell is indeed viable from a standard compliance point of view. Note that the risk assessment was carried out only for the 'operation' phase of the robot cell lifecycle.

The testing of the integrated system further demonstrates that the system provides a practical and an inherently safe solution to the problem of automating aerospace manufacturing processes. Considering possible future iterations of the robot cell design, it is recommended that a second SICK Safety Laser Scanner is located on the opposite side of the robot base to allow for a full 360 degree coverage.

Once different risk assessment methodologies have been researched and applied to the current case study, with the possibility of the development of a new risk assessment framework, different applications will be considered with proposed robot cell designs. Future considerations to develop this research further will be to propose a safety system design for a robot arm that manoeuvres along a track, and is therefore not in a fixed position. In addition, the research into the tolerability and acceptability of risk across the UK aerospace manufacturing industry will be taken into account. This will assist in the formation of a view regarding the viability of deployment of a generic proposed collaborative automation safety system design, with an accompanying proposed risk assessment framework, into different assembly lines and process applications.

7 Conclusions

A collaborative automation robot cell has been designed, installed and tested at Cranfield University. This cell is designed to simulate a typical aerospace manufacturing assembly process, with an industrial robot collaborating directly with an operator/participant without the use of fixed guarding as a risk control measure. Safety-rated sensitive protective equipment was used to provide the required non-physical safeguarding.

Safety analysis was carried out on the cell design using a traditional probabilistic risk assessment methodology. The residual risk was found to be very low. Furthermore, the integrated safety system within the large scale human/robot collaboration cell has been exposed to a number of hazardous situations and its behaviour has been recorded and reported within this document. The most likely scenarios were used in this validation work and the system was shown to repeatedly behave in a predictable and in an inherently safe manner. The robot arm has been shown to not move if an object is located within the safeguarded workspace on initiation. It will also come to a stop if an object approaches it during movement. In addition, the robot arm will not re-start if the safeguards have not been re-set.

Acknowledgments The project is supported by the Engineering and Physical Sciences Research Council Centre for Intelligent Automation, as well as Airbus UK. Particular thanks goes to John Thrower and Seemal Asif, whose collaboration and input into this work has been crucial to its success. Thanks also to Gilbert Tang and the Human Factors team at Cranfield University, led by Dr Sarah Fletcher.

References

BSI Standards Publication (2011), *Robots and robotic devices – safety requirements for industrial robots.* Part 1: Robots (BS EN ISO 10218-1:2011). Switzerland: BSI Standards Publication

BSI Standards Publication (1993), *Safety of machinery. Electrical Equipment of Machines – Part 1 Specifications for general requirements.* (BS EN 60204-1:1993). Switzerland: BSI Standards Publication

BSI Standards Publication (2010), *Safety of machinery. General principles for design. Risk assessment and risk reduction.* (BS EN ISO 12100:2010). Switzerland: BSI Standards Publication

BSI Standards Publication (2010), *Safety of machinery. Positioning of safeguards with respect to the approach speeds of parts of the human body.* (BS EN ISO 13855:2010). Switzerland: BSI Standards Publication

Kruger, J., Lien, T.K., Verl, A. (2009) 'Cooperation of human and machines in assembly lines'. *CIRP Annals – Manufacturing Technology* 58, pp. 628 – 646.

Lin, C., Wang, L. (1995) 'Intelligent collision avoidance by fuzzy logic control', Institute of Computer Science, National Tsign Hua University, Hsinhu, 30043 Taiwan, Republic of China, accepted 15th April 1996, *Robotics and Autonomous Systems* 20 (1997) pp. 61-83.

PILZ SafetyEYE, Safe Camera Systems Manual, Nr. 21 743-EN-11

S3000 Safety Laser Scanner, Operating Instructions, 8009942/V430/2011-06-03

Scharft, R.D., Meyer, C., Parlitz, C., Helms E., (2005) 'PowerMate – A safe and Intuitive Robot Assistant for Handling and Assembly Tasks', *Proceedings of the 2005 IEEE, International Conference on Robotics and Automation,* April 2005, pp. 4074 – 4079.

Schmidt, B., Wang, L. (2013) 'Contact-less and Programming-less Human-Robot Collaboration', *Procedia CIRP 7, Forty Sixth CIRP Conference on Manufacturing Systems* 2013, pp. 545 – 550.

Sekman, A., Chall, P. (2013) 'Assessment of adaptive human-robot interactions', *Knowledge-Based Systems* 42 (2013), pp. 49 – 59.

M. Walton, P. Webb and M. Poad, "Applying a concept for robot-human cooperation to aerospace equipping processes", *Proc SAE AeroTech Conference*, Toulouse, 2011.

Wojtara, T., Uchihara, M., Murayama, H., Shimoda, S., Sakai, S., Fujimoto, H., Kimura, H., (2009) 'Human-robot collaboration in precise positioning of a three-dimensional object'. *Automatica* 45 (2009) pp. 333 – 342.

Uncertainty in Demonstrating Requirements

Clive Lee

ERA Technology Ltd
Leatherhead, UK

Abstract *Deterministic requirements are often used to express criteria for system properties such as system safety. However, the evidence presented to meet these requirements or targets is often uncertain, embodying both a lack of detail and inherent variability. Consequently it is difficult to establish definitively that these requirements have been satisfied. This paper presents a Bayesian approach that provides a framework for consideration of the 'epistemic' uncertainty (lack of knowledge) together with the 'aleatory' uncertainty (random variability) in system and component estimates. The 'classical' statistical approach is presented for some common scenarios each followed by the tool-assisted Bayesian inference approach. Finally, an example is presented of the uncertainties in the individual steps of an accident sequence that accumulate to the uncertainty in the likelihood of the accident itself. The main finding is that the use of a Bayesian inference tool such as OpenBUGS (Lunn et al. 2009) based on Markov Chain Monte Carlo (MCMC) techniques (Gelfand and Smith 1990) enables the burden of the computation to be lifted so that the engineer can then concentrate on the interpretation of the uncertainty in relation to the requirement.*

1 Bayesian Statistics

In this paper we report the use of a tool for the modelling of uncertainty in the estimation of the continuous variables used to express required reliability and safety properties. We have found that the use of a Bayesian tool that uses algorithms based on MCMC (Markov Chain Monte Carlo) techniques to update the uncertainty distributions of these variables has been effective and its application fairly straightforward. The history of development of Bayesian techniques together with the MCMC algorithms used by the tool has been separated into Annex A.

We compare the classical and Bayesian approach to simple scenarios that we regularly meet in our safety work in defence and railways. The underlying statistics has been separated into Annex B as far as possible.

Section 2 presents the Bayesian approach - with Bayes' theorem in its most general form in equation (2).

© ERA Technology 2015. Published by the Safety-Critical Systems Club. All Rights Reserved

Section 3 introduces a very common statistical problem based on the Binomial distribution that is familiar to most engineers. Its application occurs regularly in our safety work in railways and defence projects to present safety evidence from trials in Safety Cases for the introduction of new equipment.

The classical approach and the Bayesian approach are presented in detail. The problem is then modelled with a Bayesian tool and results are generated to demonstrate both the accuracy of its calculations and its ease of use.

Sections 4 and 5 provide further examples of statistical models that we use often in our work in safety engineering concerning failure rates of components and their cumulative effect in accident sequences. The example in Section 5 has no classical statistical approach as the uncertainty inherent in engineering judgement is modelled and combined in an accident sequence together with random failures.

The conclusions in Section 6 outline a follow-on paper that will present work we have done on the application of the framework to estimate risk when the 'safe life'[1] exposure of a safety-related mechanical structure is exceeded. This project included the modelling of uncertainty in the exposure or "consumed life" of the structure. It was felt that this initial paper is necessary to present the power of Bayesian techniques and tools as they are not widely appreciated.

1.1 Bayesian Methods - Background

Bayes' Theorem has been known, accepted and used in statistical work for centuries (McGrayne 2011). The Bayesian interpretation of probability has similarly been known and used but the validity and appropriateness of its application is still not universally accepted.

The main controversy concerns the validity of the modelling of epistemic uncertainty and its combination with more objective evidence from experimentation and analysis.[2] The Bayesian statistician argues that the Bayesian interpretation of probability extends the classical approach and enables combination of random and epistemic uncertainties.

The development of Bayesian Methods and their relationship to classical statistics is summarised in Annex A.1

---

[1] Safe-life in structural integrity is defined as the life (in appropriate measure of exposure such as hours of operation or landings etc.) such that the largest flaw that could remain undetected after non-destructive examination would not grow to failure during this exposure.

[2] There is extensive literature on the applicability of Bayesian statistical methods summarised by the 'Objections to Bayesian statistics' (Gelman 2008). He argues that controversy has moved from theory (settled) to practice - the appropriateness of Bayesian methods.

1.2 Algorithms for Bayesian tools

The incorporation of evidence to update the Bayesian probability distribution of a statistical variable requires the numerical calculation of an integral[1] over its probability distribution as shown in equation (2) below. In the late 1980s Bayesian statisticians began to exploit the MCMC algorithms developed in the physical sciences to enable the numerical integration to be performed by computer tools running on UNIX / PC platforms.

The development of the algorithms which are used by the OpenBUGS tool is described in Annex A.2.

2 Bayesian Statistics

2.1 Informal Introduction

'Bayesian probability' is an interpretation of probability as a quantity that is assigned to represent our 'degree of belief' or 'state of knowledge' in a hypothesis or proposition such as the value or range of possible values of a variable.

Bayes' theorem[2] is used to describe how 'uncertainty' or more accurately 'degree of belief' is modified in the light of evidence such as observations and analyses. Finally, 'Bayesian statistics' refers to the statistical methods that derive from the Bayesian interpretation of probability together with use of Bayes' theorem.

An important application is Bayesian inference which is used to estimate the value of a parameter of a population from a sample. The approach starts by modelling the initial or prior knowledge (i.e. before the sample is taken or values observed) by a probability distribution. This prior distribution may represent previous samples or experiments embodying random uncertainty or epistemic uncertainty based on engineering judgement or indeed a combination of both from a previous Bayesian analysis.

Consider the tossing of a possibly biased or 'unfair' coin several times to estimate the 'probability of heads', θ assumed to be a constant for each throw.

In Bayesian statistics the initial knowledge of the 'probability of heads' θ is modelled as a continuous variable with a probability distribution over the range $0 \leq \theta \leq 1$. Two important points to note are that:

1. In classical statistics, a parameter such as θ is always regarded as a constant (albeit unknown).

---

[1] For a discrete variable, the integration is replaced by summation over the distribution. The integration is often over several variables i.e. a multidimensional sample space.

[2] Bayes' theorem applies to both epistemic and random probabilities.

2. Continuous variables (variables over a continuous range) occur frequently in practice; a tool should support the use of continuous (as well as discrete) variables.

If nothing is known about the extent to which the coin is biased, the initial knowledge of the probability of heads may be modelled as a continuous probability density function, $\pi(\theta) = c$ (i.e. a constant density) over the range $0 \leq \theta \leq 1$. It is customary to use Greek letters for the distribution $\pi(\theta)$ to distinguish from a random variable and its probability distribution such as p(x) [or f(x) if a probability density i.e. x is continuous]. As the total probability $\int_0^1 \pi(\theta) \, d\theta$ must be 1 then then $\pi(\theta) = 1$ over range.

The prior knowledge of the parameter θ can be modelled by other distributions. For example, it might be modelled as a normal distribution centred on 0.5 and given a standard deviation to reflect the degree to which we believe the coin may be biased. The distribution must be truncated at 0 and 1 and normalised.

This simple example demonstrates one of the seemingly most complex steps of the Bayesian method, the modelling of initial uncertainty by the selection of the prior distribution. Nevertheless, in many cases, the evidence "speaks for itself"; that is the prior distribution becomes less significant in its influence on the posterior distribution as more and more observations are taken into account.

2.2 Formal Presentation

We start from a general statistical model or distribution:

$$f(x|\theta) =' \text{probability distribution}' \quad e.g. \quad f(x|\theta) = \theta \exp(-\theta x) \qquad (1)$$

- f in (1) is the probability density or probability p (depending on whether the model is continuous or discrete) of observing x given parameter θ.
- x and θ can be vectors to represent a series of observations or several parameters - sometimes **bold** to emphasise vector e.g. $x = (x_1, x_2...x_n)$
- '|' in $f(x|\theta)$ can be read as the probability [density] of x 'given' θ.

In classical statistics, θ is an unknown constant and the observations $x = (x_1, x_2...x_n)$ can be used to estimate the parameter. For example, the mean μ is a parameter of many models which can be estimated from the sample mean written $\bar{x}$ where $\bar{x} = \sum_{j=1}^{j=n} x_j / n$.

Bayesian probability is introduced by modelling the epistemic uncertainty in the parameter θ by a 'prior' probability distribution $\pi(\theta)$. The observed data x is then used to update the prior distribution $\pi(\theta)$ to become the 'posterior' distribution $\pi'(\theta|x)$ of the parameter θ, using the general form of Bayes' theorem to obtain Bayes' formula (2):

$$\pi'(\theta|x) = \frac{f(x|\theta)\pi(\theta)}{\int f(x|\theta)\pi(\theta)d\theta} \quad (2)$$

The distribution, $\pi(\theta)$, encapsulating prior knowledge, is transformed into the posterior distribution written $\pi'(\theta)$ or $\pi'(\theta|x)$ to emphasise that observation(s) x have been taken into account. A primed character such as π' represents the updated distribution of π.

The numerator in (2) gives the relative contribution of the observations to the posterior probability density at θ by the prior probability $\pi(\theta)$ weighted by the probability density, $f(x|\theta)$, of the observation(s) x at this θ. The integral in the denominator is required to normalise these contributions i.e. the probability of the posterior distribution must integrate to 1.

The Bayes' formula above is the most general form (see Annex B.1 for other formulations).

3 Simple Application: Inference from Binomial distribution

A common problem in safety and system engineering concerns the inference from a series of independent trials such as tests and field observations, each assumed to have the same constant (but unknown) probability p of success

If r successes are observed in the n trials, what can be inferred about p and how confident are we in the result? We regard n and p as constants and r as the observed variable.

The probability distribution of the number of successes is given by the well-known Binomial distribution where $_nC_r$ is the number of combinations or ways the r successes can occur in the r trials:

$$f(r|p,n) = \,_nC_r p^r q^{n-r} \text{ where } q = 1-p \quad (3)$$

3.1 Classical approach

The unknown p is selected to fit the observed variable r. For example, under the assumption that we have a typical set of trials, then a reasonable estimator of p is the mean number of successes i.e. $p = r/n$. For 2 successes in 8 trials, we obtain $p = r/n = 2/8 = 0.25$.

For the estimation of the upper and then lower limit of p, we assume we have an unusually low and then high number of successes respectively. The details of the significance level (5%) and the associated confidence region (95%) in the interpretation of "unusual/significant" are provided in Annex B.1.

The example of 2 successes in 8 trials is given in Annex.B1 where the 95% confidence region is determined to be the range: " $0.03 \leq p \leq 0.65$ ".

Zero successes

If we observe zero successes (all failures) in the series of trials, the one-sided 95% confidence region is used as described in Annex B.1. For 8 trials, the 95% confidence region is determined to be the range : " $0 \leq p \leq 0.31$".

3.2 Bayesian Approach

The Bayesian approach is based on the same Binomial distribution (3) but the parameter θ (i.e. p) is not a constant but has a prior uncertainty distribution representing our incomplete knowledge. If nothing is known about this distribution (starting from "a blank sheet"), the initial uncertainty can be modelled as an "uninformative" Bayes probability distribution $\pi(\theta) = 1$ for range of θ in $0 \leq \theta \leq 1$ i.e. equal uncertainty density at any value between 0 and 1 as before.

Note that we have a Bayesian probability density function $\pi(\theta)$ (measuring uncertainty) of a parameter which happens to be a probability θ (but unknown) in the model. Substituting (3) in (2) gives the following updated distribution (4):

$$\pi'(p|r,n) = {}_nC_r(n+1)p^r q^{n-r} \qquad (4)$$

As an example, r=2 successes in n=8 trials gives the following posterior distribution obtained from a spreadsheet table.

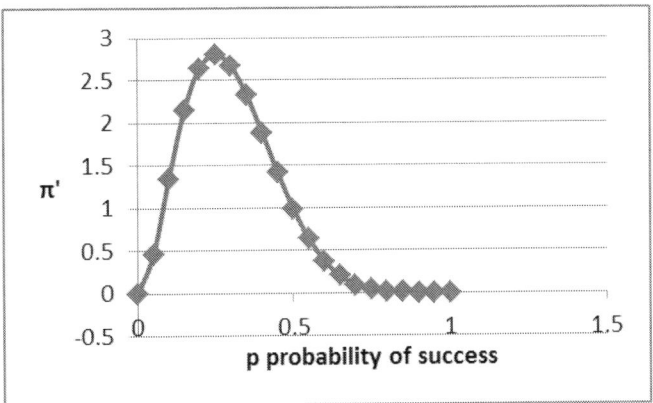

Fig. 1: Posterior Probability $\pi'(p)$

Knowledge of the form of the posterior distribution $\pi'(p)$ in (4) enables the forms of statistics to be derived such as (5) for the mean of the distribution $\pi'(p)$ of p as derived in Annex B.1:

$$mean\ of\ posterior\ distribution = \frac{(r+1)}{(n+2)} \qquad (5)$$

However, even in this simple example, the integration required to obtain formulae is starting to become difficult. For more complex examples, there may be no theoretical analytical formula (i.e. a function finitely expressible in terms of known functions) for the required integral.

3.3 Tool assistance

One of the open source tools started and developed from the 1980s is OpenBUGS (Lunn et al. 2009) which uses the Gibbs' sampling algorithm (Gibbs 1902 [1960]) to perform the calculations almost entirely transparently[1].

The OpenBUGS model for Binomial distribution example is shown below.

```
model {
    r ~ dbin(p, n)              1.
    p ~ dunif(0, 1)             2.
    n <- 8                      3.
}
list( r=2)                      4.
```

The model is essentially the 4 lines labelled for reference – **model, list** are keywords. Each line represents the following information:

1. Variable r is distributed binomially '~dbin' with parameter p for n trials.
2. Variable p is distributed uniformly '~dunif' from 0 to 1 - 'an uninformative prior'.
3. The variable n is assigned the value 8 (trials).
4. The variable r is observed as 2 (successes).

The distribution of r is not fully defined as its parameter p is not a constant but a variable with a distribution. In this case, the observed value of r is used to infer the posterior distribution of p.

The following posterior distribution $\pi'(p|r)$, written P(p) in Figure 2 was produced by the tool in 6 seconds for a sample of 10,000 points.

---

[1] There are options such as the selection the initial point of the random walk (Markov chain) through the variables of the distribution although this can be generated automatically.

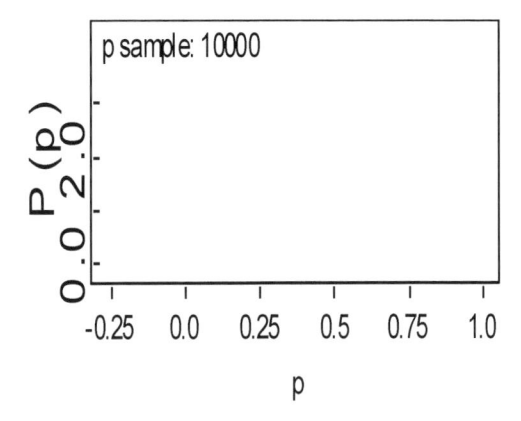

[Figure: plot area with axes labeled P(p) on y-axis (0.0 to 2.0) and p on x-axis (-0.25, 0.0, 0.25, 0.5, 0.75, 1.0), titled "p sample: 10000"]

Fig. 2: Posterior P(p) i.e. $\pi'(p|r)$

The tool generates the following basic statistics (mean and percentiles) for the posterior distribution P(p) (other statistics can be generated):

mean	val2.5pc	median	val97.5pc	sample
0.2996	0.07754	0.2851	0.6047	10,000

Accuracy of tool calculation

The output of the OpenBUGS tool has been verified on many examples at NASA (Dezfuli et al. 2009) and other sites varying from simple very complex applications but its use still requires care (see OpenBUGS manual).

In this example, the mean 0.2996 compares well with value 0.3 calculated from (5) above. As a further test, the % point at 0.09810 was calculated from (4) using direct integration and found to be 5% (the value generated by the tool) as shown in Annex B.1. Furthermore, it can be seen by comparing Figures 1 and 2 that the same posterior distribution shape is generated.

Comparison with classical result:

The Bayesian range of '$0.08 \leq p \leq 0.60$' is contained entirely within the classical range of '$0.03 \leq p \leq 0.65$' derived in Annex B.1 under certain conservative assumptions. We contend the Bayesian result is easier to understand and generate.

4 Poisson Probability Example

The Poisson process has been applied to the estimation of the reliability of electronic components and has also been found useful in modelling the loss rate of many systems in the random failures part of the "bathtub" curve after early failures but before wear-out failures. It is characterised as a constant failure rate (i.e. "no memory") and therefore the times of different components or systems operated in parallel can be combined as if one component or system were operated for the cumulative time with immediate replacement after each failure.

The failure rate λ of the component or system can be estimated by observing the number of failures n in a cumulative time T where n is regarded as variable and T fixed. Alternatively, the components/systems can be operated until the nth failure when the cumulative T is regarded as variable and n fixed.

4.1 Classical approach

Consider operating components/systems for fixed time T. The number of failures is the well-known Poisson distribution and the probability of 'n failures or less' is given by the following formula:

$$Prob\ (n\ failures\ or\ less) = \sum_{r=0}^{r=n} \frac{\exp(-\lambda T)(\lambda T)^r}{r!}$$

T is fixed and known, λ is an unknown constant, n is observed and r is used to express the summation. In summary, the two-sided 95% confidence limits are given by formula (6) as described in Annex B.2.

$$\frac{\chi^2_{2n}(2.5\%)}{2T} \leq \lambda \leq \frac{\chi^2_{2n+2}(97.5\%)}{2T} \quad (6)$$

As a concrete example, 8 failures in 10 million hours gives the two sided 95% confidence range for λ as " $3.45 * 10^{-7} \leq \lambda \leq 1.58 * 10^{-6}$ " failures per hour from χ^2 tables.

4.2 Bayesian approach

We present the following OpenBUGS model:

```
model{
    T <-10000000
    parameter <- lambda * T
    n ~ dpois(parameter)              1.
    lambda ~ dunif (0,1)              2.
}
list
    ( n=8)
```

Line 1 is the Poisson distribution for n and Line 2 is a fairly uninformative prior distribution for lambda; assumption that the failure rate is not be greater than 1 per hour. The following distribution (Figure 3) was obtained for the posterior distribution for lambda (λ).

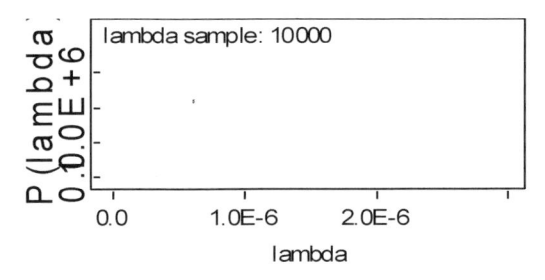

[Graph: lambda sample: 10000, x-axis lambda from 0.0 to 2.0E-6, y-axis P(lambda) 0.0E+0 to 0.0E+6]

Fig. 3: Posterior P(lambda) i.e. $\pi'(\lambda)$

A sample of the statistics generated is shown below.

mean	val2.5%	median	val97.5%	sample
8.98E-7	4.131E-7	8.661E-7	1.577E-6	10000

The 2.5% cumulative distribution of Bayesian variable lambda at 4.131E-07 ($4.13 * 10^{-7}$) is slightly different from the classical result $3.45 * 10^{-7}$ but the upper limits align exactly 1.577E-06 ($1.58 * 10^{-6}$) compared with $1.58 * 10^{-6}$.

The Bayesian model can be set up and run in minutes and the engineer can start to consider the uncertainty in the evidence in Figure 3 against the requirement. . In this example, the requirement may be that the failure rate is not greater than 1E-06 per hour (1 in a million hours). It should be noted there is no corresponding diagram in classical statistics to Figure 3.

5 Accident Sequence example

The uncertainties in the estimates of probabilities in accident sequences propagate and accumulate in the uncertainty of the resultant accident. Where two steps are independent, their combined probability p is the product $p_1 * p_2$ of the probabilities of individual steps p_1, p_2. Assuming their uncertainty distributions $f_1(p_1)$, $f_2(p_2)$ are independent, the uncertainty distribution $f(p)$ of the $p = p_1 * p_2$ is given by the following formula:

$$f(p) = \int_p^1 \frac{f_1(p_1) f_2\left(\frac{p}{p_1}\right)}{p_1} dp_1 \tag{7}$$

The formula is applicable to the combination of epistemic probabilities $\pi(\theta)$ as well as $f(p)$ (random probabilities). Indeed the two types can be mixed

Derivation of uncertainty of product of independent probabilities

The result is a special case of the well-known formula for the distribution of the product of independent random variables. One mathematical derivation is supplied in Annex B.3 but there are many ways to show this result.

The mathematical techniques we have tried to use this formula directly are presented in Annex B.4. These techniques are difficult to master and are tractable only in fairly simple cases. However, no knowledge of formula (7) is required to use OpenBUGS ; the formula will be automatically applied by the tool.

Example of uncertainty of accident sequence

The following model represents an accident sequence of 3 independent events with probabilities pe1, pe2, and pe3 to give the probability of accident pa. The epistemic uncertainties in pe1 and pe3 are modelled with the uncertainty uniformly distributed in ranges derived from the judgement of experts. The random variability of pe2 is derived from physical theory and its normal distribution validated by experiment. What is the uncertainty in pa?

Our Bayesian model is shown below

model{
    pa <- pe1 * pe2 * pe3

    pe1 ~ dunif(1.0E-04, 1.0E-03)   judged *between 0.0001 and 0.001*

        sd <-0.01   *standard deviation is 0.01*
        tau <- 1/pow(sd, 2)   *tau=1/variance*

        pe2 ~ dnorm (1.0E-01, tau)   *tau required for Normal by tool*

        pe3 ~ dunif (1.0E-03, 1.0E-02)   judged *between 0.001 and 0.01*
}

The uncertainty in the probability of the accident pa is shown in the graph (Figure 4 below) with the graphs of the uncertainties of probabilities of events pe1, pe2, and pe3 shown in the following figures.

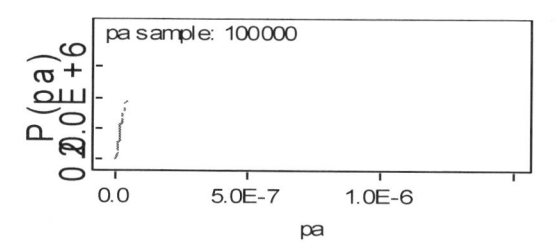

Fig. 4: Uncertainty in probability of accident

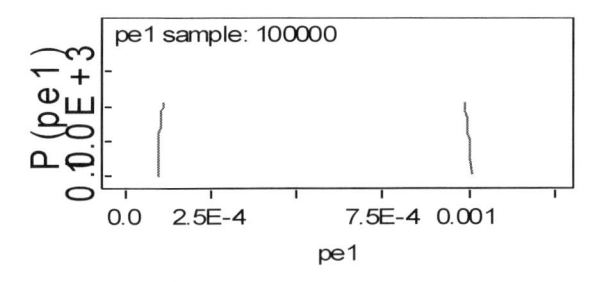

Fig. 5: Uncertainty in probability of event e1

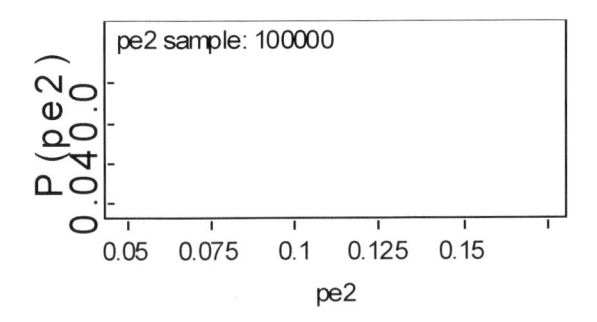

Fig. 6: Uncertainty in probability of event e2

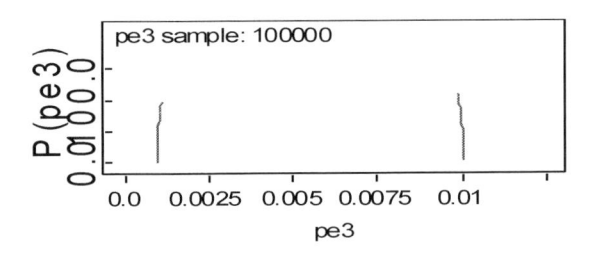

Fig. 7: Uncertainty in probability of event e3

From the graph generated by the tool in Figure 4 and associated statistics below, we can consider whether we have met a requirement such as the probability of the accident is not greater than 10^{-6} per hour.

mean	sd	val2.5pc	median	val97.5p
3.004E-7	2.162E-7	3.411E-8	2.451E-7	8.211E-7

For example, the 95% confidence region for the probability of the accident (pa) is $2.2 * 10^{-7} \leq pa \leq 8.2 * 10^{-7}$ which is less than 10^{-6} per hour.

The worst case calculations taking the top of the range for pe1 and pe3, and for pe2 taking standard deviations above the mean would give the following inconclusive result.

$pa = pe1*pe2*pe3 = 10^{-3} \times 1.02 \times 10^{-1} \times 10^{-2} = 1.02 \times 10^{-6}$ which is greater than 10^{-6}.

6 Conclusions

This paper has presented the use of continuous distributions to model uncertainty (both epistemic and aleatory) in basic evidence and its propagation and accumulation in the uncertainty of system properties. We have found that a tool assisted Bayesian approach can help engineers to avoid the details of the mathematics and the computations required so they can concentrate on interpretation of the results.

The Bayesian modelling does require familiarity with basic probability distributions of continuous and discrete variables such as the uniform, normal, Poisson and Binomial distributions for the prior distribution but the accuracy of this modelling becomes less important as the data is taken into account. We have found that significant results can be achieved with the least informative of prior distributions such as the uniform distribution.

Our intention is to produce a following paper on work we have undertaken on the modelling of the uncertainty in the exposure of a structure to stresses (due to recording practices). This work introduces the further complication of uncertainty in the underlying statistical model. For example, the failure rates for structures start to increase dramatically with exposure around an unknown point, sometimes referred to as the "knee joint" from its form on the graph of failure rate against exposure.

A structure is subject to exposure until failure occurs and the "safe-life" exposure is set at a fraction of this value to ensure the "knee joint" is not encompassed within the "safe-life" period. The "safe-life" is intended never to be exceeded but can the increased risk be judged if there is a possibility that the safe-life will be exceeded?

314   Clive Lee

Annex A - Bayesian statistics

A.1 Bayesian Methods - Background

Bayes' Theorem has been known, accepted and used in statistical work for centuries (McGrayne 2011). The Bayesian interpretation of probability has similarly been known and used but the validity and appropriateness of its application remains controversial mainly because of the modelling of subjective judgement of uncertainty and its combination with objective probabilities from experimentation and analysis.[1]

The rigorous basis of classical statistics was developed initially for the 'frequentist' interpretation of probability in the first half of the 20th century.

With these classical foundations established, the development and application of Bayesian inference was then revived in the 1950's and 1960's and reported in classic works such as "Bayesian Inference in Statistical Analysis" (Box and Tiao 1963). The work was undertaken in Universities and industrial research centres particularly in statistical decision analysis to aid management in the selection of optimal decisions under uncertainty. By the 1970's, the theory and practice of Bayesian decision analysis had been established in many fields such as management sciences (e.g. Operational Research), experimental design and economic modelling but only peripherally in hard science and engineering disciplines such as system and safety engineering.

During the 1980's research in Bayesian methods continued in management sciences, artificial intelligence, IT applications, search techniques, data visualisation and experimental design for medical, psychological and sociological sciences. The properties of Bayesian (belief) nets were described by Judea Peal (Pearl 1988) and Richard Neopolitan (Neopolitan 1989). At this time, the MCMC algorithms became known to statisticians and useful Bayesian inference tools were developed.

Despite these advances, the use of Bayesian techniques and tools is still considered to be a research area as they are still not widely known or applied outside the statistical community.

A.2 Algorithms for Bayesian tools

---

[1] There is extensive literature on the applicability of Bayesian statistical methods summarised by the "Objections to Bayesian statistics" (Gelman 2008). His conclusion is that controversy has moved from theory to practice , the appropriateness of Bayesian methods compared with alternative methods.

The incorporation of evidence to update the Bayesian probability distribution of a statistical variable requires the numerical calculation of an integral[1] over its probability distribution. [see equation (2)].

The MCMC algorithms are based on the generation of a 'random walk' through the possible values of the variable to represent a sample from the distribution. From a particular value in the sample, a random number is generated (hence Monte Carlo after the casino) to determine a candidate for the next value which is selected or rejected according to defined criteria. The next value depends only on the current value (and no other values in the sample) and hence the values constitute a Markov chain (Markov 1906). The Markov chain generates a representative sample from the distribution to enable integrals to be approximated.

The use of MCMC (Monte Carlo Markov Chain) algorithms was first reported for applications in statistical physical sciences (Metropolis et al. 1953) and simplified by Hastings (Hastings 1970) to become known as Metropolis-Hastings algorithms. Specific Metropolis-Hastings algorithms were described by work (Geman and Geman 1984) based on ideas from statistical physics by Gibbs (Gibbs 1906) and further extended at the University of Nottingham (Gelfand and Smith 1990) to become known as known as Gibbs sampling.

The Bayesian tool used at ERA is OpenBUGS (Lunn et al. 2009) where BUGS stands for "Bayesian inference using Gibbs sampling". The BUGS project was developed from artificial intelligence work in the 1980s on the propagation of uncertainty in graphical structures using simulation methods and object-oriented programming techniques. It was started at the MRC Biostatistics Unit at Cambridge with the chief programmer Andrew Thomas working under David Spiegelhalter.

Annex B - Mathematics

This section separates mathematical details to avoid distraction from the main presentation.

B.1 Forms of Bayes' Theorem

The continuous form of Bayes' Theorem is given by equation (2) above. In its discrete form, Bayes' theorem is usually presented where there are n mutually exclusive and comprehensive possibilities $A_1...A_n$ with probabilities $p_1...p_n$. This context gives perhaps the more familiar form of Bayes' theorem to provide the updated probabilities p'_i. The '|' notation is used for conditional events and their

---

[1] For a discrete variable, the integration is replaced by summation over the distribution. The integration is often over several variables i.e. a multidimensional sample space.

316 Clive Lee

probabilities and '$A_i |x$' can be read as 'the event A_i given (or conditional on) the event of obtaining observations x.

$$p'_i = prob(A_i|x) = \frac{prob(x|A_i)p_i}{\sum_{j=1}^{j=n} prob(x|A_j)p_j} \tag{8}$$

Probability is shortened to *prob*. The equation (2) is derived from the elementary identities for conditional probability:

$$prob(A|B) = \frac{prob(B|A)p(A)}{p(B)}$$

$$where\ p(B) = \sum_{j=1}^{j=n} prob(B|A_j)p(A_j)$$

Equation (2) is a natural generalisation of (8) where the mutually exclusive and comprehensive events A_j are interpreted as all of the different possible values of the parameter θ and the summation over j becomes integration over θ.

B.1 Classical Approach- Significant Levels/Confidence Regions

A level of probability is selected (commonly 5%) as the criterion for an unusual or 'significant' event that would not normally be expected under the assumptions of the model. An observation is therefore significant at the 5% level if it lies outside the 100% - 5% = 95% confidence region in which we expect 95% of its values.

For a simple variable, the complementary 5% significance region is split into two 2.5% regions considered to be 'too high' and 'too low'. The limits of the 95% confidence region are then determined so that the probability of obtaining a value in this upper or lower interval respectively is 2.5%.

Example of 2 successes in 8 trials for Binomial distribution

The Upper Limit of p is chosen so that probability of '2 or fewer' is only 2.5%

$$prob(0) + prob(1) + prob(2) = 0.025\ (i.e.\ 2.5\%) \tag{9}$$

$$q^8 + 8q^7p + 28q^6p^2 = 0.025\ \text{Substituting (3) in (9)}$$

Solving this numerically we obtain $p \approx 0.65$.

Lower Limit

In the same way, the Lower Limit is chosen so that probability of the upper tail '2 or greater number of successes' is 2.5% as follows:

$$prob(2) + prob(3) + \cdots = 0.025\ (i.e.\ 2.5\%)$$
$$prob(0) + prob(1) = 0.975\ (i.e.\ 97.5\%)$$

$$q^8 + 8q^7 p = 0.975$$

and solving numerically p ≈0.03

Two sided confidence limits

From above, the two sided 95% confidence limits are approximately 0.03 and 0.65 i.e. the probability that p is in range: $0.03 \leq p \leq 0.65 = 95\%$.

Estimator of p.

The mean number of successful trials 'r/n' is often used as an estimator of p. This gives 2/8 = 0.25 for 2 successes in 8 trials. The median is also used as an estimator. In this case, the calculation is similar to the equations above set to 0.5 (50%) which is solved to obtain p ≈ 0.32

Zero successes

In this case, we work with the one sided 95% confidence region based on the upper limit. The lower limit of p is 0 and all possible significant observations are in the upper tail of the distribution.

As a concrete example, if we observe 0 successes in 8 trials then a simple calculation ($q^8 = 0.05, q \approx 0.69$) gives the upper limit $p \approx 0.31$.

A reasonable estimator of p is the value for which the median of the distribution of the number of successes is the observed value. In 8 trials, the probability of observing 0 successes i.e. $q^8 = 0.5$ (50%) when $q \approx 0.92$ i.e. $p \approx 0.08$.

B.2 Binomial Distribution- Bayesian Approach

If we substitute (3) in (2) we obtain the updated probability distribution

$$\pi'(p|r) = \frac{{}_nC_r p^r q^{n-r}}{\int_0^1 {}_nC_r p^r q^{n-r} dp} = \frac{p^r q^{n-r}}{\int_0^1 p^r q^{n-r} dp} = {}_nC_r(n+1)p^r q^{n-r} \quad (10)$$

The denominator has been simplified by repeated integration parts to give the factorials shown in (11) below or equivalently using the fact that the integral is the definition of the Beta function (**B** below) which is known to reduce to these factorials.

$$\int_0^1 p^r q^{n-r} dp = B(r+1, N-r+1) = \frac{r!(n-r)!}{(n+1)!} = \frac{1}{{}_nC_r(n+1)} \quad (11)$$

Formula Mean

Using (10) to calculate the general formula for the mean and (11) with n replaced by n+1 to take account of the extra p in the formula for the mean:

$$mean = \int_0^1 \pi'(p|r) p \, dp = \frac{(r+1)}{(n+2)}$$

Verification of calculation of 5% point by tool

The 5% point is selected at random to demonstrate correct calculation of integral. The tool gives 0.09810 as the 5% point. This is verified by integrating :

$$\int_0^{0.09810} \pi'(p|r) \, dp = \int_0^{0.09810} {}_nC_r (n+1) p^r (1-p)^{n-r} \, dp$$

$$= \int_0^{0.09810} 252 p^2 (1-p)^6 \, dp$$

≈ 5.0%

B.2 Poisson Distribution

The probability of 'n or fewer failures' on the LHS is given the integral below (established by repeated integration by parts of the RHS) which can be used to estimate the upper and lower limits:

$$\sum_{r=0}^{r=n} \frac{\exp(-\lambda T)(\lambda T)^r}{r!} = \int_{\lambda T}^{\infty} \frac{e^{-x} x^n}{n!} dx \qquad (12)$$

The integrand is a Gamma distribution with parameter (n+1) so the inverse of its cumulative distribution can be used to find λT using Excel function GAMMA.INV to give. the 95% confidence range for λ by the following formula

$$\frac{GAMMA.INV(0.025, n, 1)}{T} \leq \lambda \leq \frac{GAMMA.INV(0.975, n+1, 1)}{T}$$

Most texts note that integrand is related to the χ_k^2 distribution i.e. with density $(e^{-x/2} x^{k/2-1})/\Gamma(k/2)$ for which statistical tables are widely available. After some manipulation, the two-sided confidence limit of 95% results in following formulae (see Table 8.13 of MIL-HDBK-338B [DoD 1998]) where

$$\frac{\chi_{2n}^2(2.5\%)}{2T} \leq \lambda \leq \frac{\chi_{2n+2}^2(97.5\%)}{2T}$$

B.3 Product of probabilities with independent uncertainties

There are several different methods of proof of this result.

Using F for the cumulative probability function of probability density f:

$F(p)$ = Probability of $\{(p_1, p_2) \bullet p_1 * p_2 \leq p \}$
       = Probability of $\{(p_1, p_2) \bullet p_2 <= p/p_1 \}$
       = Sum over p_1 of [Probability $\{p_1\}$ * Probability $\{ p_2 \bullet p_2 <= p/p_1 \}$]
       = Sum over p_1 of $[f_1(p_1) * F_2(p/p_1)]$
       = $\int_0^1 f_1(p_1) * F_2(p/p_1) \, dp_1$ *generalising to integral*

Differentiating $F(p)$ wrt p to obtain density function f(p)

$$f(p) = \int_p^1 \frac{f_1(p_1) f_2(p/p_1)}{p_1} dp_1 \qquad (13)$$

N.B. Probability $p/p_1 \leq 1$ so $p \leq p_1 \leq 1$ for limits of integral

B.4 Uncertainty in product of probabilities

We have spent a lot of effort in trying to perform the integration in (7) for the propagation of the uncertainty in the product of uncertain probabilities. Direct integration to produce an analytical formula was only successful in simple cases due to its complexity.

    We did notice that the formula is a multiplicative convolution of functions and techniques from signal processing for additive convolution were applicable. Indeed, it has long been established that the Mellin transform (Epstein 1948) replaces the role of the Fourier transform so that the transform of the multiplicative convolution is the product of the Mellin transforms of the functions themselves. The problem we faced is that the Mellin transform and its inverse are integrals themselves and libraries of Mellin transforms are not large.

    A particular advantage of the transform approach is that we can derive the distribution of the product of many probabilities by simply multiplying their transforms before inversion. We were partially successful in deriving formulae for the distribution of the uncertainty in the product of many probabilities with independent distributions of similar form (e.g. uniformly distributed uncertainties over different ranges).

    Finally, we also used the Fourier transform itself by working with logarithms of the probabilities when the multiplicative convolution reduces to additive convolution. The extra steps in moving to the distribution of the logarithms at the start and reversing at the end of the procedure are fairly straightforward and worthwhile in that the full theory and library of Fourier transforms are available.

Acknowledgments We would like to thank both NASA for making their work on Bayesian Inference publicly available (Defulzi et al. 2009) and similarly the OpenBUGS project (Lunn et al. 2009) for the publicly available tool.

References

Box,G E P , Tiao G C (1963). Bayesian Inference in Statistical Analysis. NY: Wiley Classics.

Dezfuli H, Kelly D, Smith C, Vedros K, Galyean W (2009) Bayesian Inference for NASA Probabilistic Risk and Reliability Analysis, NASA/SP-2009-569

DoD (1998) Military Handbook, Electronic Reliability Design, MIL-HDBK-338B, DoD

Epstein (1948) Some applications of the Mellin transform in statistics, Coal Research Laboratory, Carnegie Institute of Technology

Gelfand A E, Smith A F M (1990) Sampling-Based Approaches to Calculating Marginal Densities, Journal of the American Statistical Association, Vol. 85, No. 410. , pp. 398-409.

Gelman A (2008) Objections to Bayesian statistics. International Society for Bayesian Analysis, Number 3, pp. 445–450

Geman S, Geman D (1984) Stochastic relaxation, Gibbs distributions, and the Bayesian restoration of images. IEEE-PAMI, 6, 721-741.

Gibbs J W , Elementary Principles in Statistical Mechanics, developed with especial reference to the rational foundation of thermodynamics, (New York: Dover Publications, 1960 [1902])

Hastings W K(1970) Monte Carlo sampling methods using Markov Chains and their applications, Biometrika 57, 97-109

Lunn, D., Spiegelhalter, D., Thomas, A. and Best, N. (2009) The BUGS project: Evolution, critique and future directions (with discussion), *Statistics in Medicine* **28**: 3049--3082. "OpenBUGS": (http://www.openbugs.net) Accessed December 2014

McGrayne S B (2011), The Theory That Would Not Die : How Bayes' Rule Cracked the Enigma Code Hunted Down Russian Submarines, & Emerged Triumphant from Two Centuries of Controversy. New Haven: Yale University Press. 13-ISBN 9780300169690/10-ISBN 0300169698; OCLC 670481486

Markov A A (1906) "Extension of the limit theorems of probability theory to a sum of variables connected in a chain". reprinted in Appendix B of: R. Howard. Dynamic Probabilistic Systems, volume 1: Markov Chains. John Wiley and Sons, 1971.

Metropolis N, Rosenbluth A W, Rosenbluth M N, Teller A H and Teller E (1953) Equations of State Calculations by Fast Computing Machines, Journal of Chemical Physics 21 (6): 1087-1092

Neapolitan R E (1989). Probabilistic reasoning in expert systems: theory and algorithms. Wiley. ISBN 978-0-471-61840-9.

Copernic Safety

José Miguel Faria

>Educed Lda
>
>Coimbra, Portugal

Abstract *The difficulty in explaining observations of Saturn's motion in the context of the geocentric model was a crucial stepping-stone that eventually led Copernicus to propose the heliocentric model. This paper analyses approaches for building safer control systems related to this analogy: detecting mismatches between a controller's view of the environment and its actual state can greatly prevent accidents. We survey existing work on the topic and look for ways of systematically comparing, assessing, and devising effective monitoring solutions, from the detection of a simple component failure to the circumvention of flaws in the system requirements specification.*

1 Science Goes by Data – So Should Safety

A scientific theory provides an explanation of some aspect of the natural world. For a theory to be validated and accepted, it must comply with the observations and experimentation. If the collected data does not match the expected results, there must be some defect in the theory. Taking an example from the history of humanity: in the ancient times, people believed that Earth was the centre of the Universe, and the Moon, the Sun, Mercury, Venus, Mars, etc., all circled around the Earth. But astronomers found a difficulty: Saturn seemed to move back and forth, which could not be possible if it were simply revolving around the Earth. This mismatch between the (geocentric) theory and the data collected (observation of Saturn) meant that something had to be wrong in the theory. Long after Ptolemy's intermediate tentative explanation with Epicycle movements (abandoned for similar caveats), Copernicus[1] presented the heliocentric model, which finally complied with all the data.

The analogy to safety comes obviously: the same way that observations of Saturn showed the geocentric model had to be flawed, eventually leading Copernicus

---

[1] Historically, the notion that the Sun was the centre of the (then known) Universe and the Earth revolved around it goes back to Aristarchus of Samos, in the 3rd century BC, but the idea was either neglected or rejected until Copernicus (16th century).

© José Miguel Faria 2015. Published by the Safety-Critical Systems Club. All Rights Reserved

to devise Heliocentrism; a system's automated controller can recognize potential problems by detecting inconsistencies among data, and consequently take appropriate safety measures. While not a novel idea – it has been explored in multiple ways – the analogy placed emphasizes a powerful basic principle for developing and verifying safety-critical systems.

Arguably, the importance of identifying inconsistencies between an actual process state and the controllers' internal view of the process has best been exhibited by N. Leveson's STAMP accident model (Leveson 2012). STAMP (Systems-Theoretic Accident Model and Processes) accident model of causation proposes the evolution from traditional chain-of-events causality models to *systemic thinking*, where safety is an emergent property that may or may not arise from the interaction of the system components within a larger environment. In this paradigm, N. Leveson proposes the analysis of the different *control loops* identifiable at every level, including management and organizational. Every controller, either automated or human, keeps an internal model of the current state of the system it is controlling, in order to decide its control actions (Fig. 1). Very often, it is the mismatch between the process model used by the controller and the actual state of the process that results in unsafe or missing control action that eventually lead to an accident.

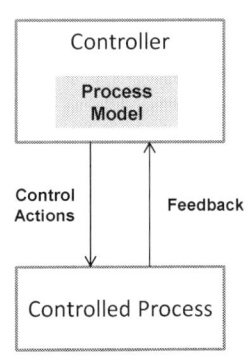

Fig. 1. Control flow example (adapted from Leveson 2012)

This paper surveys and argues in support of solutions where the data collected by a controller is essential to detect inconsistencies and potential problems in a system, be it the runtime detection of simple component failures or the prevention of flaws in the requirements and design specification stage. It cannot be too often emphasized that safety and reliability are two very different concepts. In that respect, despite the paper title, the solutions herein presented do not strictly concern safety issues, but dependability in a more general sense.

2 Execution Monitoring

In any field, the first step towards the correction of an error is its proper identification and diagnosis. This sections starts with an overview of traditional redundancy architectures where the presence of multiple similar components plays a fundamental role to circumvent faults at execution time. We progressively move into asymmetric architectures and to the core topic of the paper, efficient monitoring solutions.

2.1 Traditional Redundancy Architectural Patterns

Mechanical and simple EEE (Electrical, Electronic and Electromechanical) components typically fail due to vibration, physical damage, humidity, and thermal stress after a number of hours in operation. One of the most effective solutions to tolerate such degradation faults in a system is to rely upon redundancy, be it *static* or *dynamic*. Static redundancy denotes solutions where all devices operate all the time; whereas in dynamic redundancy, redundant devices are only activated when needed, the main rationale being that keeping the redundant devices powered down lowers the thermal stress they are subject to, potentially extending their life. Yet, this is not a guarantee for longevity; failure rates at restoration of power are relatively high. Nor is it a guarantee of higher safety or reliability: knowledge of the current process status and restoration of service may be quite complex.

Two traditional static redundancy architectural patterns are *dual redundancy* and *N-modular redundancy*. In a dual-redundant architecture, two identical components operate in parallel, with synchronized clocks, and their results are passed to a comparator, which checks their equality. If they do not match, then some error must have occurred, either in one of the two identical components or in the comparator (or some other complementary element such as the input distribution mechanism). In the N-modular redundant architecture, a set of N identical components operates in parallel, with synchronized clocks, and their results are passed to a voter. To validate the result, the voter must receive a majority M (lower than or equal to N) of identical values. This solution tolerates the failure of N-M components.

The two solutions are quite effective preventing random hardware faults. Yet, some intricate scenarios can still go undetected: for example, the failure of some unit(s) combined with the silent failure of the comparator or voter that bypasses a wrong result; or the simultaneous failure of the units coincidentally producing the same (erroneous) result. Sound engineering analysis should guarantee that the likelihood of occurrence of such scenarios is sufficiently low for the given application.[1]

Dual redundancy and N-modular redundancy constitute an initial approximation to the analogy explored in this paper: it is the observation of inconsistent data that allows the detection of faults.

In general, the effectiveness of hardware redundancy relies on the principle that random/degradation faults of hardware components tend to be statistically independent. Unfortunately, this is not true for design faults. In addition, a similar approach is not effective for detecting software faults. These topics are explored further in the next sections.

2.2 From Replication to Asymmetric Diversity

In the hardware redundancy patterns described above, identical components are used as the main computation units. And, as said, the patterns' efficiency is largely based on the fact that random hardware faults show to be statistically independent. In software, or in logically complex hardware like application-specific integrated circuits, however, independence of failure is much harder to attain: If the software program is the same and runs deterministically, it should always return the same results, regardless of these being the expected ones or not. To overcome this issue, *N-version programming* (Chen and Avizienis 1978) was introduced in the 1970's.

In N-version programming, two or more functionally equivalent programs are independently developed from an initial common specification. During execution, the results of each version are compared, in an analogous fashion to hardware redundancy. The underlying assumption in the method proposal was that the independence of programming efforts should greatly reduce the likelihood of identical software faults in different version of the program.

A number of years after the method proposal, a thorough experiment to examine the core hypothesis of statistical independence of failure in N-version programming was conducted (Knight and Leveson 1986). The results revealed that the observed number of coincident failures was 'substantially more than would be expected if they were statistically independent'. Analysing the causes, most notably, it was found that programmers ended-up making equivalent logical errors in some more complex parts. In other cases, for some particular inputs, apparently different logical errors resulted in faults that caused statistically correlated failures.

The debate on the benefits of N-version programming has been long (Knight and Leveson 1990). Above all, it is very difficult to evaluate whether or not the

---

[1] Many more considerations can be made about these two solutions. Nevertheless, an exhaustive overview of redundant architectural patterns is beyond the scope of this paper – for that we refer to reader to, e.g., (Armoush 2010, Knight 2012).

gains compensate for the extra effort in development. On the positive side, is should be noted that the fact that independence of failure cannot be simply assumed does *not* imply that there cannot be significant benefits. Also, if more than error detection is wanted, N-version programming can provide unique benefits in software fault masking for configurations of N equal or greater than three, with majority voting.

This last point leads to an important consideration: very frequently, N-version systems are built into a logical configuration where a mismatch in the results of the different versions – of similar complexity and implementation effort – is used "solely" for error detection, leading the systems to be placed into some form of fail-safe state (Hampton 2012). In many cases, error detection can be achieved with significantly simpler versions. In our central analogy: all that Copernicus knew to start with was that the geocentric model was flawed; not what was the correct answer. Such is the underlying principle of redundancy solutions as simple as *memory parity* – where a single extra (parity) bit added to a set of data bits is sufficient to detect an odd number of bit inversions – or *cyclic redundancy checks* (CRCs), used mainly in data transmission.

While memory parity or CRCs are mainly useful to detect hardware related faults, like bit flips, bit losses, or noise in transmission channels, the reasoning of using a redundant much simplified element can equally adopted in software. Put into other words: if the invalidation of independence of failure in N-version programming seems to be due to similar errors in solving complex problems, removing complexity can significantly enhance failure independence. Noting Hoare's famous quote from his Turing award lecture (Hoare 1980):

> There are two ways of constructing a software design: One way is to make it so simple that there are *obviously* no deficiencies, and the other way is to make it so complicated that there are no *obvious* deficiencies.

The safety-critical systems industry has followed a somewhat related approach: with the difficulty – or impossibility – of eliminating complexity, asymmetric fault tolerant architectures were developed. A system is built with two channels, where the first one covers the intended functionality and the second, independent and much simpler, is used essentially to monitor the execution and trigger fault recovery mechanisms if it detects some problem. Additionally, it can also provide some minimal functionality. Two essential benefits are attainable: first, the secondary channel can be made much more reliable due to simplicity. Second, given the radical difference in design between the two channels, independence of failure is highly increased.

2.3 Trustworthy (Copernic) Monitors

The work in (Littlewood and Rushby 2012) thoroughly analyses the reliability of two asymmetric configurations: *1-out-of-2 systems* and *monitored architectures*. In a 1-out-of-2 system, both channels can control the system and any of the two can interrupt normal execution and put the system in a safe state. In monitored architectures, the *operational* channel implements the desired functionality and the *monitor* channel signals alarms if it believes the operational channel to have failed or to be causing unsafe behaviour. For the purpose of this paper, we focus on monitored architectures.

For an optimal trust on the monitor channel, three essential features are required: (i) independence from the operational channel, (ii) detection of all unsafe conditions, and (iii) no emission of *false positives* (raising an alarm inappropriately).

Ideally, the monitor channel will have no faults. Eased by its expected simplicity, formal verification technologies could be used to formally prove the monitor implementation against its requirements. Alternatively, a technique as *runtime verification* can be used to synthesize monitors from formally specified requirements (Havelund and Rosu 2004). The main question left then is what should the source of those requirements be.

A first natural answer is the set of requirements for the operational channel. There is, however, an important concern in such a choice: experience in industry has shown that the most significant problems in the software development domain arise from errors in the requirements (National Research Council 2007). This would significantly lessen the likelihood of the monitor channel compensating for conceptual flaws in the system requirements. Additionally, as put in (Littlewood and Rushby 2012), requirements for the operational channel are typically focused on functions to be performed, rather than safety properties to be maintained.

Therefore, we support the argument of the last cited authors that monitoring should concentrate on the properties that are closely related to the safe behaviour of the global system, and that are more likely to be violated when some problem arises. The same authors assert that those very properties can be provided by the claims and assumptions of the system argument-based safety case.

To elaborate some more on the properties to monitor, we start by resorting back to Hoare's above mentioned quote. The immediate next paragraph reads:

> The first method is far more difficult. It demands the same skill, devotion, insight, and even inspiration as the discovery of the simple physical laws which underlie the complex phenomena of nature.

A hazardous state is very likely to reveal physical or logically implausible scenarios. Such scenarios can be defined by a set of properties concerning:

- A given parameter, with:
    - Single point values

- Relative distributions
- A relation between parameters, both:
  - Unquantified, or
  - Knowledge-based bounded
- One-state evaluation;
- Two-state evaluation.

Single point parameter value properties typically concern boundary (minimum and maximum) allowed values for relevant process parameter, like (i) 'the temperature sensed in room R can never exceed value T'. Relative distributions over a given parameter are properties like (ii) 'the difference of pressure between chamber A and chamber B cannot exceed P'.

Properties defining relations between quantities allow a whole new spectrum of possibilities not covered by single parameter properties. Relations encompass much of the way we perceive and develop our knowledge of the world. That has been realized since ancient history, culminating, most notably, in Aristotle's *Metaphysics*. Relations are at the core of knowledge representation formalisms such as ontologies.

Unquantified relations allow the expression of properties like (iii) 'if the brakes are being pressed, speed cannot increase'. By knowledge-based bounded relations we mean properties like (iv) 'at an altitude between A1 and A2, the atmosphere pressure must be between P1 and P2'.

Properties can be evaluated over one single state or over two distinct states. In the example given, (i, ii, and iv) are single state properties; (iii) requires two distinct states to be assess the evolution of the speed value.

Monitoring individual parameters is typically associated with the suspicion of anticipated faults, for which some detection mechanism is put in place. Properties based on relations between parameters provide a substantially higher flexibility. Arguably, these encompass the necessary expressivity to cover all the safe behaviour of the system. We maintain that properties based on relations between parameters can also contribute to the detection of unanticipated unsafe scenarios, for which no individual fault or execution scenario was foreseen.

3 Requirements Specification and Design Stages

The techniques explored in the previous section essentially apply to *fault detection* at execution time. A complementary mechanism for high dependability is *fault prevention*. Fault prevention means to prevent the occurrence or introduction of faults; it comprises the set of methods used for attempting to produce programs that are free of design faults. In this regard, multiple formal techniques for the specification, modelling, development, and verification of computational systems have been devised over the last years. It so happens that a set of such techniques closely relates to the principles explored above for the monitoring channels. This section briefly overviews existing work on the topic.

The idea of monitoring a system model can be found in (Halbwachs et al. 1993), which introduces the concept of a *synchronous observer*: a second program which observes the behaviour of the first one and raises a signal flag some condition is satisfied. A significant novelty in the work was that the same language was used to write the program and its desired properties.

Furthermore, the observer could also be used to express known properties and assumptions about the program environment. In general, not all properties of a reactive system can be expected to hold in an unconstrained environment; a given set of assumptions about the environment must be made explicit. Including them in the observer allowed checking whether the system desired properties would hold provided the environment assumption also did.

The cited work was devised to model check specifications in the language LUSTRE (Halbwachs et al. 1992). More recently, Rushby recovered the topic and illustrated the use of synchronous observers in SAL (Rushby 2014).

Most distinguishingly, SAL[1] supports the specification and verification of system models with *uninterpreted functions* – a function of which we define only its denoting symbol name and arity, and say nothing about its behaviour. In this context, uninterpreted functions can be applied, for example, to discover suitable environment assumptions of the system: starting the model with an "empty assumption" (i.e., an uninterpreted functions), if this is not enough to guarantee the system desired properties, the verifier will generate a *counter-example*, illustrating a scenario that violates the properties. This information can then be used to devise the respective necessary assumption.

Also of notable mention is that properties closely related to those introduced in the previous section as 'relations between parameters, two-state evaluated' can likewise be specified using synchronous observers. The last cited work illustrates an example where similar properties are encoded in synchronous observers using *relational abstractions* for hybrid automata (Sankaranarayanan and Tiwari 2011).

In essence, synchronous observers present a powerful and intuitive way to perfect system specifications. They allow the system requirements, assumptions and desired safety properties to be specified in the same notation. An equally interesting alternative is, for example, TLA$^+$ (Lamport 2002). TLA$^+$ is a high-level specification language especially suited for describing and reasoning about concurrent systems; it is support by a strong set of tools[2] and has been successfully applied in the development of industrial commercial products (Verhulst et al. 2011).

In TLA$^+$ the specifications and their intended properties are expressed in the same language, and refinement is implication: a more detailed model D implements a higher-level specification H if and only if D implies H. Like so, once more, properties akin to the 'two-state relations between parameters' can be de-

---

[1] SAL home page: http://sal.csl.sri.com/. Accessed 30 September 2014

[2] TLA$^+$ Tools homepage: http://research.microsoft.com/en-us/um/people/lamport/tla/tools.htmlt. Accessed 30 September 2014

scribed and checked: the 'high-level specification' serves as the synchronous observer, and the 'detailed model' as the system specification. Also, counterexamples found by the associated model checker (TLC) can similarly be used to reveal necessary environment constraints.

4 Scope, Soundness and Completeness

An ideal monitor channel detects all unsafe scenarios and does not raise false alarms inappropriately; two properties that are very close to what is designated in logic by *completeness* and *soundness*, respectively. The monitor is to be part of a wider system which interacts with an uncontrolled environment, configuring an *open system*. A provable complete description of the environment and of all safety related scenarios is very quickly unattainable for any non-trivial system. So, the question as to whether all possible scenarios and hazards have been identified in a safety analysis is typically left open. Careful judgement, potentiated by accumulated experience, and the combination of multiple analysis techniques are among the best possible answers.

We argued in favour of building the monitor channel from a set of properties relevant to monitor, instead of an approach more directly related to the behavioural description of the operational channel. Admittedly, this is likely to challenge the system engineers' confidence on the completeness of the set of properties monitored. A similar concern is actually one of the rationales behind languages designed to specify system requirements, such as, for example, SCR (Heitmeyer et al. 2005), using state-based behaviour descriptions. Ultimately, adopting a combination of both approaches and adding properties "at will" can be a pragmatic answer in industrial systems.

Not raising false alarms seems an easier task at first sight. The spirit of the properties exemplified appears to be an obvious guarantee: alarms are raised only when safety boundaries are violated, or when readings reveal implausible scenarios. In fact, the actual concern is more related to the way in which some properties can be inferred from the implementation. The monitoring of some quantities or events may be challenging; it may only be achievable through indirect inference or limited sensing technology.

The significance of this aspect is illustrated, for example, by the accident in the landing of Lufthansa Flight 2904, on September 14, 1993, at Warsaw airport: in a landing situation, the thrust reversers and the spoilers are activated to assist bringing the plane to a stop. To prevent the thrust reversers from being inadvertently activated during flight, the control software must be sure that the airplane is on the ground. Two central conditions were defined for this: (i) the weight carried on each of landing gear strut must be higher than 6.3 tons, and (ii) the wheels must be turning faster than 133 km/h. The thrust reversers cannot be activated if the first condition is not true, and the spoilers are only activated if at least one of the two conditions is true.

In this case an unexpected tailwind led the plane to land inclined; only nine seconds after touchdown did the left gear carry enough weight to allow the thrust reversers activation. In addition, heavy rain caused the wheels to hydroplane; only an extra four seconds later did the wheels rotation reached the 133 km/h. Ultimately, the plane went off the end of the runway and two people lost their lives (Wikipedia 2014a).

Like in many other accidents, the control software behaved exactly as expected. A series of unfortunate conditions and the way through which the intended physical phenomena were transformed into input variables to the control software were at the heart of the accident causes.

A notable framework for describing and reasoning about the interaction of a software control system with its environment and for the clear distinction between system requirements and software requirements is the *four-variable model*, introduced by the seminal paper (Parnas and Madey 1995). Multiple related extensions have also been published; e.g., (Lamsweerde 2009, Jackson 2000, Miller and Tribble 2001, Wassyng and Lawford 2010).

In general, all these considerations apply equally to the soundness and completeness of the analysis made at the specification stage. Regarding soundness, in particular, two extra considerations should be considered: not all theories are decidable or can be computed in "reasonable" time. It is therefore normal that automated tools make some compromises, sometimes at the expense of compromising the soundness of the verification. Bounded model checking, for example, is a technique that may return results hindered by soundness issues. Also, in some recent technologies for hybrid systems, some subtle soundness issues have been spotted (Platzer and Clarke 2009). Nonetheless, the maturity and efficiency of these technologies is steadily growing and its benefits are well recognized (National Research Council 2007).

In the next section we revisit the challenge of inappropriately signalling components as faulty when they are actually working as expected.

5 Going Further

In the attempt to protect a system from sensing failures, critical systems often disregard extreme values as invalid data. This is, if a given input variable – corresponding to some physical quantity – reports a value that is too extreme, the controller assumes that the detection mechanism is in failure given the physical implausibility of the data. This would mean to complement a property like 'the temperature sensed in room R can never exceed value T' in the system safety logic, with an extra condition like 'if the sensed value of the temperature exceeds T_{Imp} consider the temperature sensors to be malfunctioning'.

As useful as this extra condition can be, such a design has also perversely contributed to serious safety problems. For example, (Littlewood and Rushby 2012) describes how a related issue was at the core of the near miss of American Airlines Flight 903 of 12 May 1997. Another intriguing example occurred during the final descent of the Air France Flight 447 of 1 June 2009. The plane crashed into the Atlantic Ocean after a nose-high descent on aerodynamic stall. In normal conditions, the standard flight-control mode (called 'Normal Law') automatically intervenes to protect stall. Given the incidents before, at the interval we describe the flight-control system was operating on a reduced regime ('Alternate Law') that does not include stall protection. In Alternate Law, a stall warning sounds on the cockpit whenever the plane's angle of attack is above a given threshold (5 degrees). And the warning sound becomes more intense as the angle increases.

At a critical moment of the flight, the angle of attack was so high (around 40 degrees) that the system rejected the data as invalid. This caused the stall warning to temporarily stop. And it led to a perverse situation that whenever the pilot happened to lower the nose, making the angle of attack decrease just enough to be below the defined extreme plausibility value, the stall warning sounded again, configuring a negative reinforcement that may have contributed to the pilot's repeated tendency of pitching up. Though it is not clear how decisive the stall warning intermittence was to the pilot's behaviour, from this sequence the crash, the angle of attack never dropped below 35 degrees (Wikipedia 2014b).

From this example, it seems natural to speculate that a slight increase on the "intelligence" of the flight-control system could prevent the rejection of the extreme (angle of attack) data values as invalid. Keeping a trace of the registered values and taking into account the slope of the curve appears worth of consideration for the definition of physical plausibility, refining the monitoring mechanisms. The extent to which this is considered in present implementations is unknown to the authors.

6 Conclusions

The first step in correcting a problem is acknowledging it existence. This paper covered a set of techniques for the detection of problems in dependable systems. Detection can occur at runtime, preventing random hardware faults or software implementation faults to turn into a failure; but also at design time, preventing the propagation of design faults into the implementation.

While traditional redundant architectures are quite effective preventing degradation hardware faults, "symmetric" redundancy does not serve software (or complex digital hardware) so well. Introducing asymmetry can both save development costs and increase independence of failure.

Putting the emphasis in the properties to be monitored, instead of possible execution paths and sequences of events, creates a different mind-set in developers that increases diversity from the operational requirements, potentially magnifying the detection of unsafe conditions.

Sources to identify the properties to monitor include the system safety case; provided it exists. We maintain that properties based on relations between parameters not only provide a very high expressive power, and can also detect possibly unforeseen scenarios in the safety analysis.

Good monitoring is no substitute for strong development engineering practices in the first place. Coincidentally, the ideas described for execution monitoring very closely relate to techniques that can be adopted in the specification stage, provided a suitable formalism is adopted. We briefly described their use through synchronous observers, as possible in the SAL suite, or, alternatively, with the high level language TLA$^+$.

For the future we would like to have the opportunity to perform a more in-depth evaluation of the principles here described, applying them on existing applications of monitored architectures and measuring possible improvements. The assessment of the completeness of the monitored properties with respect to the global system safety is also an open topic worthy of further exploration.

Acknowledgments This paper is essentially a review of existing work. We therefore generally thank all the researchers and contributors to the evolution of system safety and dependability, with particular emphasis to the authors more cited here. We also thank the late Samuel Jackson for advising us that '[i]t is insufficiently considered that men more often require to be reminded than informed'.

References

Armoush A (2010) Design patterns for safety-critical embedded systems. PhD Thesis. RWTH Aachen University

Chen L, Avizienis A (1978) N-Version Programming: A Fault-Tolerance Approach to Reliability of Software Operation. FTCS-8: The Eighth Annual International Conference on Fault-Tolerant Computing.

Halbwachs N, Lagnier F, Ratel C (1992) Programming and verifying real-time systems by means of the synchronous data-flow language LUSTRE. IEEE Trans. Softw. Eng. Vol. 18.

Halbwachs N, Lagnier F, Raymond P (1993) Synchronous Observers and the Verification of Reactive Systems. AMAST 1993: 83-96. Springer Verlag

Hampton P (2012) Survey of Safety Architectural Patterns. In: Dale C, Anderson T (eds) Achieving Systems Safety - Proceedings of the Twentieth Safety-Critical Systems Symposium. Bristol, UK. Springer

Havelund K, Rosu G (2004) Efficient monitoring of safety properties. International Journal on Software Tools for Technology Transfer Vol. 6, Issue 2

Heitmeyer C, Archer M, Bharadwaj R, Jeffords R (2005) Tools for constructing requirements specifications: The SCR toolset at the age of ten. Int. J. Comput. Syst. Sci. Eng.

Hoare CAR (1980) The emperor's old clothes. ACM Turing award lecture. In: ACM Turing award lectures. 2007. ACM, New York, NY, USA.

Jackson M (2000) Problem Frames: Analyzing and Structuring Software Development Problems. Addison-Wesley Longman Publishing Co., Inc., Boston, MA, USA.

Knight J (2012) Fundamentals of Dependable Computing for Software Engineers. Chapman & Hall/CRC Innovations in Software Engineering and Software Development Series

Knight J, Leveson N (1986) An Experimental Evaluation of the Assumption of Independence in Multi-version Programming", IEEE Trans. Softw. Eng. Vol. SE-12, No. 1

Knight J, Leveson N (1990) A reply to the criticisms of the Knight & Leveson experiment. SIGSOFT Software Engineering Notes Vol. 15, No. 1

Lamport L (2002) Specifying Systems: The TLA$^+$ Language and Tools for Hardware and Software Engineers. Addison-Wesley Longman Publishing Co., Inc., Boston, MA, USA.

Lamsweerde A (2009) Requirements Engineering: From System Goals to UML Models to Software Specifications. Wiley

Leveson N (2012) Engineering a Safer World: Systems Thinking Applied to Safety. Cambridge, Massachusetts. MIT Press

Littlewood B, Rushby J (2012) Reasoning about the Reliability of Diverse Two-Channel Systems in Which One Channel Is "Possibly Perfect". IEEE Trans. Softw. Eng. Vol. 38, No. 5.

Miller S, Tribble A (2001) Extending the four-variable model to bridge the system-software gap. Digital Avionics Systems, 2001. DASC. 20th Conference, Vol. 1

National Research Council (2007) Software for Dependable Systems: Sufficient Evidence?. Jackson D, Thomas M, Millett L (eds). National Academy Press, Washington, DC, USA.

Parnas D, Madey J (1995) Functional documents for computer systems. Science of Computer Programming. Vol. 25, No. 1

Platzer A, Clarke E (2009) Computing differential invariants of hybrid systems as fixedpoints. Formal Methods in System Design. Vol. 35, N 1. Springer US

Rushby J (2014) The Versatile Synchronous Observer. A Festschrift Symposium in Honor of Kokichi Futatsugi. Proceedings, S. Iida, J. Meseguer, and K. Ogata (Eds.), published as Springer LNCS Vol. 8373.

Sankaranarayanan S, Tiwari A (2011) Relational abstractions for continuous and hybrid systems. In: Computer-Aided Verification, CAV '2011. LNCS Vol. 6806. Springer-Verlag

Verhulst E, Boute R, Faria J, Sputh B (2011) Formal Development of a Network-Centric RTOS: Software Engineering for Reliable Embedded Systems. ISBN 978-1-4419-9735-7. Springer

Wassyng A, Lawford M (2010) Integrated software methodologies - An engineering approach. Transactions of the Royal Society of South Africa Vol. 65(2)

Wikipedia (2014a) Lufthansa Flight 2904. http://en.wikipedia.org/wiki/Lufthansa_Flight_2904. Accessed 5 October 2014

Wikipedia (2014b) Air France Flight 447. http://en.wikipedia.org/wiki/Air_France_Flight_447. Accessed 5 October 2014

The Data Elephant

Paul Hampton and Mike Parsons

CGI UK Ltd
London, UK

Abstract *The contribution software and hardware can make to hazardous failures has long been understood and well covered by standards and guidance. We suggest the role of data in influencing the safe operation of systems is equally important but this has not attracted the same level of attention; there is no standardisation and little guidance on how the risks associated with data should be managed. The issue is becoming more acute as many types of data are now used to deploy, configure, operate, test and justify safety systems; the volume of data in systems is also growing at an unprecedented rate, along with the attendant risks – this is the "Data Elephant", the 'elephant in the room' that can no longer be ignored. This paper presents progress the SCSC Data Safety Initiative Working Group (DSIWG) has made over the last year towards establishing guidance on how the risks associated with data can be appropriately identified and managed.*

Fig.1. The Data Elephant

© Paul Hampton and Mike Parsons 2015.
Published by the Safety-Critical Systems Club. All Rights Reserved

1 Introduction

System safety, as a discipline, has been studying the contributions software and hardware can make to hazardous failures for the best part of half a century. During this period, many industry standards and guidance have been developed and refined and these have arguably had a significant impact on reducing safety risks and therefore, reducing accidents in many sectors such as aviation (Boeing 2013). In those formative years, hardware and then software were likely the 'elephants' of their day, but over time, through standardisation and other activities the associated problems are now better contained; to use an analogy, as in the figure: our ancestral software and hardware elephants have now evolved into the smaller, and more manageable, Rock Hyrax.

However, the role of *data* in influencing the safe operation of systems is equally important. Many types of data are now used to deploy, configure, operate, test and justify safety systems but this has not attracted the same level of attention; there is little standardisation and guidance on how the safety risks associated with data should be managed. As we shall see, the hardware and software aspects of a system can be subject to significant assurance rigour but this can be completely undone by, for example, a simple and innocuous configuration setting. The problem is *here and now* – consider for example, what rigour and assurance has been applied to the growing capture and exchange of medical records? The problem is also growing at an unprecedented rate as organisations seek more and more value, and novel uses of the increasing volumes of data – likened to a data goldrush and a new industrial revolution (EC 2014). For example, consider the emergent use of social media data to inform safety decisions such as the spread of viruses in populations (BBC News 2014).

The SCSC Data Safety Initiative Working Group (DSIWG) was formed in January 2013 with the aim of making data a "first-class citizen" in safety systems development. The working group is therefore trying to raise the profile of data and promote a *data-centric* view of systems.

1.1 Existing Guidance

The DSIWG has already published an initial version of the Data Safety Guidance Note, (DSIWG 2014), and this was made available at SSS'14. The paper summarizing work to date, *"Stopping Data causing Harm: towards Standardisation"* (Parsons and Hampton 2014), was well received with several delegates commending the initiative as being very timely and asking for further information. Many constructive review comments have also been subsequently received. The second version of the Guidance Note is to be made available at this symposium. This version will address review comments that have been practicable to include, as well as new material that has been developed over the past year.

Significant progress has been made since the last publication, and the main results are presented in this paper and will cover areas such as establishing principles for data safety, different user perspectives of data, data HAZOP guidewords, assessing data safety culture, integration with existing standards and guidance and plans for disseminating data safety guidance to various sectors.

2 Progress to Date

The following figure shows the journey the group has taken and the planned roadmap for future work.

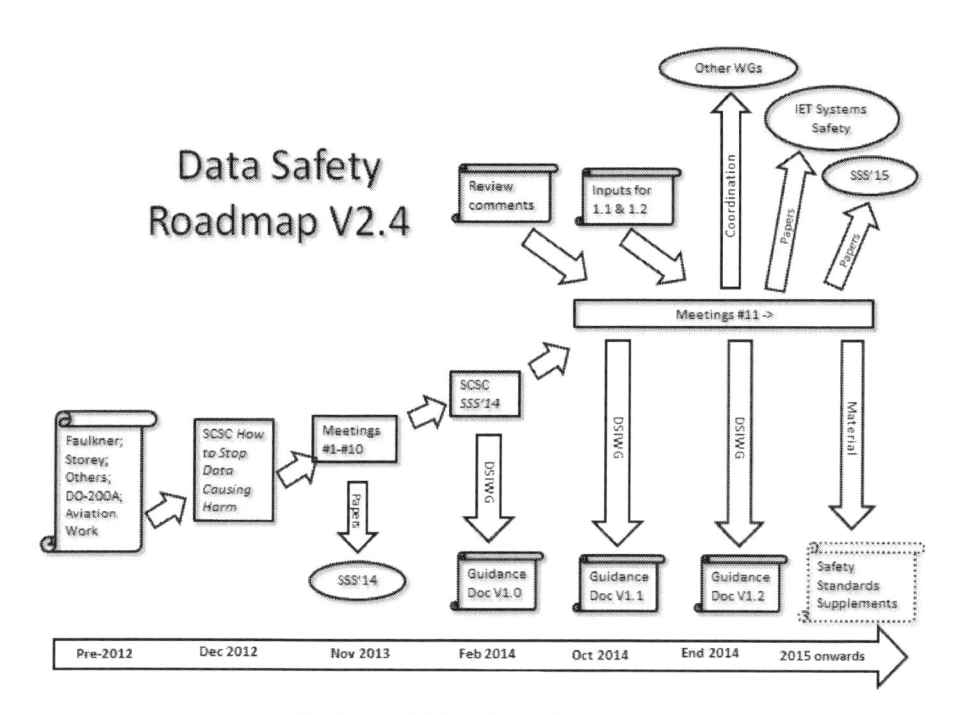

Fig.2. Data Safety Initiative Roadmap

Since its formation in early 2013, the working group has made significant progress and been productive in producing a toolset of components required to manage data safety risks. Progress of the group after its first year of work was report at SSS'14 (Parsons and Hampton 2014) and showed that the group now had:

> ➢ A language, dictionary and ontology for articulating ideas and concepts across different sectors in an unambiguous way;

> Techniques for quickly assessing the organisational risk of data so that an organisation can understand its high level risk exposure and to inform and justify further activities;
> A framework for conducting a more detailed analysis exploring tangible hazards in the use of data in a structure that aids review and comparison between systems;
> A classification scheme for assessing and categorising the criticality of data and its properties;
> Models of different data lifecycles and guidance on when assurance techniques should typically be applied for a given data type;
> Recommendations for assurance techniques to be applied for a given data type and data integrity level.

2.1 Developments

The group has now built on this previous work and the remaining sections describe the following developments:

Data Types: a set of cross sector data type definitions that allow different types of data to be categorised to allow more structured analysis and treatment of the safety risks they can give rise to;
Data Safety Principles: a set of cross sector principles that represent good practice in data safety assurance case development;
Perspectives on Data Safety: introduction to the concept of perspective when analysing risks for an organisation that is part of a wider data supply chain;
Data HAZOP Guidewords: providing guidance on guidewords that could be used in Data HAZOP meetings to help elicit hazards that data could give rise to;
Data Safety and Security: discussing the relationship between safety and security from a data perspective;
Data Safety Culture Questionnaire: a proposed questionnaire to help understand the level of data safety awareness and data safety culture within a project or organisation;
Data Maturity Models for Data Safety: reporting on work to develop a formal framework for assessing the maturity of an organisation in its management of data safety;
Integration with Existing Standards: reporting on progress toward integrating the guidance into existing standards;
Dissemination Plans: plans for how the guidance will be disseminated to various stakeholders to ensure widespread adoption of the content;

As with any document seeking standardisation in approach, it is important that any terms on which the guidance is based are carefully defined. Core to the guidance is the concept of 'data types' and the next section describes further refinements the group has made in categorising and defining these types.

Note that it is appreciated that data has no safety significance in its own right or presents safety risks until it is considered within a particular context and use. When the term *safety-related data* is used, this should be taken to be shorthand for:

'Data that, through a failure to preserve particular critical properties, can contribute to hazardous system states that can give rise to harm'.

Other terms are used in the text for conciseness, but the more precise definition is intended.

3 Data Types

There is a wide spectrum of types of data that contribute to, are used by, produced by, or affected by, safety-related systems. For example, from a system development perspective: data used to specify and support the design of a system, the data used to assure a system (eg. test data) and the user data that flows within the built system during operation are very different and require different treatment in terms of understanding and managing the safety risks presented. There are also some types of data that may not immediately seem to carry safety risks but could potentially have significant safety impact.

The role of data in accidents and incidents is often obscured by the procedural, hardware and software aspects of the system at fault, and is rarely explicitly identified in reports. Where data contributions are mentioned at all, reports may typically state that a 'software problem' led to the accident. It is therefore not always obvious that data has had a part to play in an incident, but a recent incident illustrates the risks data has to play.

Qantas Boeing 737 Loading Incident
Can data on a piece of paper cause a plane to crash?

On 9th May 2014 a Qantas Boeing 737 was preparing for departure from Canberra to Perth. There were 150 passengers, 87 of which were primary school children. These children were all seated together at the rear of the cabin. All had been mistakenly assigned an "adult weight" of 87 kg (ATSB 2014).

During take-off the aircraft appeared nose heavy. Significant back pressure was required to rotate the aircraft and lift off from the runway. The aircraft exceeded the calculated take-off safety speed by about 25 kt. The aircraft rose at a higher initial climb speed than usual, but the crew did not receive any warnings.

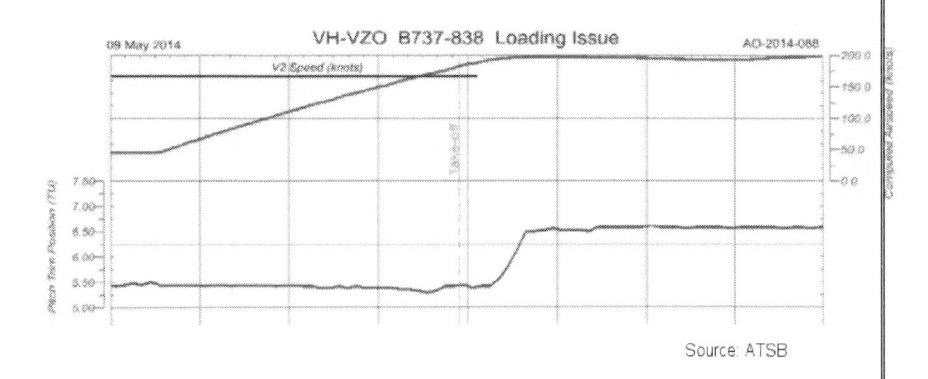

Fig.3. Actual flight data showing safe takeoff speed being exceeded

After investigation, it emerged that a "name template" had been completed by a travel agent on behalf of the school group. This group was travelling from Perth to Canberra and returning back to Perth. Despite being marked as mandatory, the "Gender Description" field in this template was left blank; options for this field were "Adult", "Child" and "Infant".

As per company procedures, two days before the Perth-Canberra leg of their journey this group was "advance accepted" into the booking system. Since the fields recording the number of children and young passengers in the group were blank, the Customer Service Agent assumed all of the group were adults. No loading-related issues were experienced during this flight.

Two days before the return flight the group was again "advance accepted" as all adults. They were checked in at Canberra Airport and assigned seats at the rear of the aircraft. The load discrepancy caused the issues noted above.

Fortunately for the Qantas incident, there were no serious consequences. However, this incident demonstrates the importance of data. This includes checking mandatory fields are completed and default or assumed values are not used.

To allow the guidance to articulate how different assurance techniques can be applied to different data types, it is necessary to be more precise about defining and categorizing the types of data considered. This categorization is challenging; it needs to be sufficiently generic to be applicable to a broad range of sectors and industries but not so abstract that it is difficult to correlate data type definitions to real situations in an obvious and meaningful way.

It is appreciated that the set of definitions will not be perfect; it is difficult to arrive at one set that can achieve completeness of coverage and applicability to all sectors and scenarios. Furthermore, to keep the definitions recognizable and accessible to the readership, vernacular terms have been used. A consequence of this is that they will not form a canonical set and some overlapping in definitions is to be expected.

Nevertheless, there is great value in developing a sector-agnostic common set for the guidance as it:

- Raises awareness of data types that might not have previously been considered as given rise to safety issues;
- Allows a more structured analysis and categorization of data types that may be used in an organization;
- Can be tailored and adapted for more specialized cases where appropriate;
- Allows the guidance to articulate where specific methods and techniques are applicable.

The group has continued to refine the set of data types for use in the guidance and the following table shows the current cross-industry consensus.

Table 1. Safety-related Data Types

Category	Type	Description	Explanation	Typical containers
Context	Predictive	Data used to model or predict behaviours and performance	Data for studies, models, prototypes, initial risk assessments, etc. This is the data produced during the initial concept phase which subsequently flows into further development phases	Prototype results, evaluations, analyses, etc.
	Scope, Assumption & Context	Data used to frame the development, operations or provide context	Restrictions, risk criteria, usage scenarios, etc. explaining how the system will be used and any limitations of use	CONOPS, Safety Case Report part 1
	Requirements	Data used to specify what the system has to do	Data encompassing requirements, specifications, internal interface or control definitions, data formats, etc.	Formal specifications, ICDs, User Requirements documents, Safety Case Report part 1.
	Interface	Data used to enable interfaces between this system and other systems: for operations, initialisation or export from the system	Data that exists to enable exchange between this system and other external systems. Covers start-of-life operations (data import or migration), end-of-life operations and ongoing operational exchange of data between systems.	Protocols, Interface Spec, Schemas, ICDs, Transition Plans, ETL tool specs, Cleansing and Filtering rules
	Reference or Lookup	Data used across multiple systems with generic usage	Data comprising generic reference information sets used by multiple systems (i.e. not produced solely for this system). Typically updated infrequently, and not specific to this system.	Dictionaries, materials information, sector data reference sets, encyclopaedias, etc

Category	Type	Description	Explanation	Typical containers
Implementation	Design & Development	Data produced during development and implementation	This is data encompassing the design & development process artefacts: everything from design models and schemas to document review records. It also includes test documents (specification and results) but not the test data itself.	Design documents, Review records, Hardware, Software and design, Test scripts, Code inspection reports, etc. Safety Case report part 2.
Implementation	Verification	Data used to test and analyse the system	This is data comprising the test values and test data sets used to verify the system. It may include real data, modified real data or synthetic data. It includes data used to drive stubs, and any data files used by simulators or emulators.	Test data sets, Stub data, Emulator and Simulator files
Configuration	Infrastructure	Data used to configure, tailor or instantiate the system itself	Data used to set up and configure the system for a particular installation, product configuration, or network environment	Network configuration files, Initialisation files, Hardware pin settings, Network addresses, Passwords, etc.
Configuration	Behavioural	Data to change the functionality of the system	Data to enable / disable or configure functions or behaviour of the system	XML config files, CSV, schemas, etc
Configuration	Adaptation	Data to configure to a particular site	Data used to tailor or calibrate a system to a particular physical site or environment, incorporating physical or environmental conditions	Configuration files
Capability	Staffing & Training	Data related to staff training, competency, certification and permits	Data which allows staff to perform a function within the wider context of the safety-related system. This may include training records, competency assessments, permits to work, etc.	HR records, training certificates, card systems
The Built System	Asset	Data about the installed or deployed system and its parts, including maintenance data	Data related to location, condition and maintenance requirements of the system under consideration. This may cover hardware, software and data.	Inventory, asset and maintenance database systems
The Built System	Performance	Data collected or produced about the system during trials, pre-operational phases and live operations	Data produced by and about the system during introduction to service and live service itself. Includes fault data and diagnostic data. This may be the results of various phases of introduction and may include trend analysis to look for long-term problems.	Field data, Support calls, Bug reports, NCRs, DRACAS data

The Data Elephant 343

Category	Type	Description	Explanation	Typical containers
Application	Release	Data used to ensure safe operations per release instance	Explanation of particular features or limitations of a release or instance. May include specific time-limited workarounds and caveats for a release.	Release notes, Certificates of Design, Transfer documents, Safety case part 2 or part 3
Application	Instructional	Data used to warn, train or instruct users about the system	This is data that explains to users the risks of the systems and gives any mitigations that may be required to be implemented by users, e.g. by process, procedure, workarounds, limitations of use	Manuals, SOPs, On-line help, Training courses, etc. Safety case part 3
Application	Evolution	Data about changes after deployment	This is data that covers enhancements, formal changes, workarounds, and maintenance issues. It also covers data produced by configuration management activities, such as baselines or branch data	Change Requests, Modification Requests, Issue and version data, CM system outputs
Application	End of Life	Data about how to stop, remove, replace or dispose of the system	This is data covering all activities related to taking the system out of service or mothballing / storage / dormant phases	Transition, Disposal and decommissioning plans
Application	Application	Data manipulated by the system during operations	This is the data processed or produced by the system which has end-user meaning. It may be displayed and used within the system or may be for transfer or distribution to other systems or downstream users. It is data that has some real domain meaning, i.e. is not to do with the system internals.	May be stored internally within the system (e.g. in databases or text files), or transferred into or out of the system through interfaces (e.g. Ethernet).
Compliance & Liability	Standards and Regulatory	Data that governs the approaches, processes and procedures used to develop safety systems.	This is data predominantly in the form of documents that describe and dictate the activities, processes, competencies etc. to be used for a particular development in a particular sector.	Standards documents, guidelines, legal directives and laws
Compliance & Liability	Justification	Data used to justify the safety position of the system	Data used to justify, explain and make the case for starting or continuing live operations and why they are safe enough. Often passed to external bodies (regulators, HSE, ISAs) for their review.	Safety Case report, Certification case, Regulatory documents, COTS Justification file, Design Justification file
Compliance & Liability	Investigation	Data to support accident or incident investigations (i.e. potential evidence)	This is data collected or produced during an incident or accident investigation which may be used in investigation reports, lessons learnt or prosecutions. This can be process data, trace data, site data (e.g. photographs of crash site) or may be derived (accident simulations, analyses, etc)	Incident/accident Investigation reports and supporting documents

3.1 Data Types in Scope for the Guidance

Note that given the wide range of types identified it would be too ambitious to provide guidance on all these types initially; focus in the first issues of the guidance will therefore be on: **Verification, Behavioural, Application and Justification**.

Cedars Sinai Medical Centre - CT Scanner
The default configuration values – what harm can they do?

A data misconfiguration in a CT (Computed Tomography) scanner used for brain perfusion scanning at Cedar Sinai Medical Center in Los Angeles, California, resulted in 206 patients receiving radiation doses approximately 8 times higher than intended during an 18 month period (HealthImaging 2009). Some patients reported temporary hair loss and erythema.

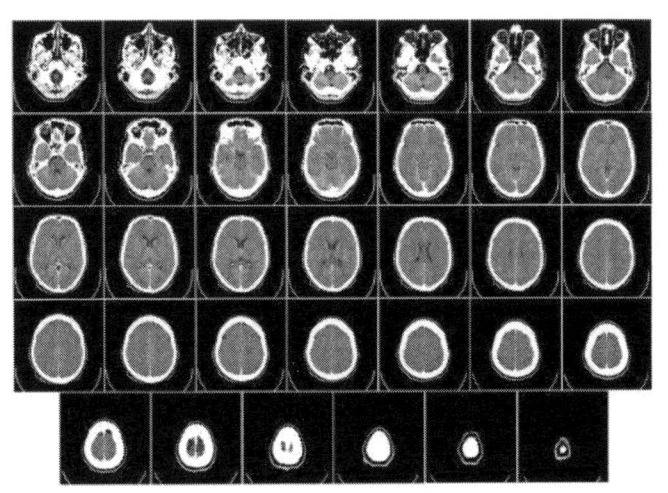

Fig. 4. Example of a CT Scan

The problem reportedly resulted from an error made by the hospital in resetting the CT machine after it began using a new protocol for the procedure in February 2008, but it wasn't detected until one of the patients reported patchy hair loss in August 2009.

"There was a misunderstanding about an embedded default setting applied by the machine," according to a statement from Cedars-Sinai. *"As a result, the use of this protocol resulted in a higher than expected amount of radiation."*

This incident shows that simple errors in configuration data can undermine safety of a medical device where hardware and software is operating correctly. While the hardware and software of the scanner is subject to regulation, there is no such equivalent rigour for the data.

Having established that data can give rise to hazards, the group wanted to establish a set of good practice principles that could apply to managing risks associated with data. As shall be seen in the next section, the group explored what could be learned and derived from similar principles established to reflect good practice in software assurance.

4 Data Safety Principles

There are many standards either directly or indirectly addressing software safety assurance. Although the detail and approaches in these standards can vary, underlying these variances is a set of fundamental principles that can be observed in most of them and that represent good practice in software assurance case development (Kelly 2014). There are 4 principles that have been developed with 1 overarching principle that applies to all and these are referred to as the *4+1 Principles*.

As these principles represent cross-industry good practice in managing software risks, the group investigated how these could be mapped to a set of equivalent principles that would apply to the management of data safety risks.

The following shows the resulting data safety assurance principles that the group has developed:

Data Safety Principle 1: Data Safety requirements shall be defined to address the data contribution to system hazards

Data pervades active system operation, as well as the system's specification, realisation, verification, validation, certification, maintenance, and retirement. Moreover, data may be passed from one system to another; sometimes with a significant passage of time. It may be assimilated, and converted from prior uses into new uses, or simply used as is by many systems. It is stored in media whose storage integrity decays.

A system can enter unsafe states because of the data that it directly consumes, or, indirectly from data that was used to shape it. Examples of this are: flawed test data that leads to an incorrect conclusion of the system's suitability. Flaws in data tables that define, perhaps, control laws or data transforms that result in subtle unexpected system misbehaviours e.g. Mars Climate Orbiter (NASA 1999). Missing data requirements that determine system interface compatibility e.g. Ariane 5 (ESA 1996).

This principle therefore asserts that as well as considering mechanical, human, environmental factors, software etc. consideration should be given to the contribution data has to make to the system hazard analysis.

Data Safety Principle 2: The intent of the data safety requirements shall be maintained throughout requirements realisation.

Data safety requirements establish the system's safety properties for data, for the system's use of data, for the management of data, and for the engineering lifecycle of both the system and its associated data. The system's requirements' hierarchy must preserve the intent of the data safety requirements (and hence the system's data safety properties). Moreover, the applied engineering process for both the system's realisation and subsequent lifecycle stages shall demonstrate that the data safety properties are preserved.

Data Safety Principle 3: Data safety requirements shall be satisfied.

Evidence is required that the system satisfies all of the data safety requirements imposed on it for all anticipated operating conditions. Moreover, the data safety requirements that pertain to the data's lifecycle outside of the system shall be evidentially demonstrated prior to the system acting on such data, or else that the system is able to adequately defend against broken data safety requirements. In other words, either the data can be shown to conform to its safety properties prior to being used, or the system can implement adequate defences and mitigations against data that does not conform to the required safety properties.

Data Safety Principle 4: Hazardous system behaviour arising from the system's use of data shall be identified and mitigated.

This is an intentionally broad statement because data is conceptual and not physical; it is the contextualised use of data that could result in a system hazard. Data Principle 1 deals with system level hazards arising from data, whereas Data Principle 4 is concerned with hazards that arise from the way the system uses its data; that is, whether the system's design and implementation introduce further hazards. An example is a ship navigation system's display of hydrographic map data, where a wide field display results in small shallow underwater features disappearing due to image scale when it is critical that situational awareness of such hazards is maintained.

Data Safety Principle 4+1: The confidence established in addressing the data safety principles shall be commensurate to the contribution of the data to system risk.

The confidence in the evidence that demonstrates establishment of the first four Data safety Assurance Principles shall be proportionate to the contribution data has with the system hazards.

Having established principles for how data safety risks can be managed, the group now turned to the challenges of conducting data-centric risk analysis and as discussed in the next section, it emerges that the analysis is more effective when conducted from particular perspectives.

5 Perspectives on data safety

One reason why data has become the 'elephant in the room' is arguably its fluidity. Hardware and software can undergo significant amounts of product assurance and once assured do not change frequently. Where change is required to hardware or software, it can be carefully managed and the impact on the safety case appraised. This is not always the case for data, which can be much more fluid: data change can be much more frequent and may be harder to control when there are multiple parties and systems in a data supply chain. It is not always obvious who owns or has governance over the data throughout its lifecycle, if it is copied, modified or retransmitted to other parties then provenance and ownership can be blurred or even lost.

Dallas Hospital turns Ebola patient away
We ignore the data elephant at our peril...

On 26th September 2014, a Dallas hospital mistakenly sent home a man who had the Ebola virus having missed what would have appeared to be an obvious potential case: a Liberian citizen with fever and abdominal pain who said he had recently travelled from Liberia (NBC News 2014). He returned to the hospital, was eventually diagnosed with the illness, but subsequently died. Two nurses that had treated the man also contracted the virus but later recovered.

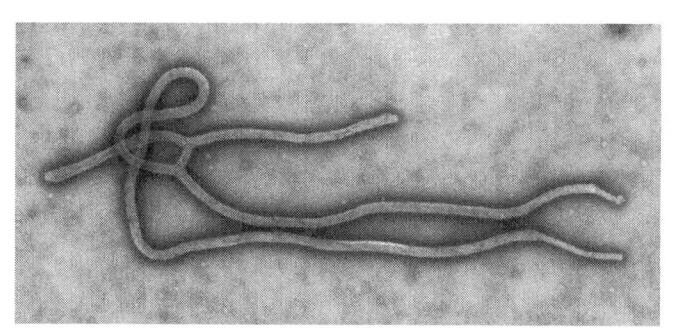

Fig. 5. The Ebola Virus virion

There have been mixed reports on the cause of the problem, but what is clear is that external social phenomena such as the Ebola outbreak, which are outside the hospital's electronic health record (EHR) system and processes, can change the *safety significance* of data held in the EHR. If the importance of the data is not recognised and elevated appropriately in the support tools and processes, then the risk of unintended harm can increase.

This conclusion is reinforced by system vendors who are now updating their systems to reflect the Ebola crisis in light of the Dallas incident.

In the Dallas Hospital incident, the system was behaving as designed, there were no software errors, and the staff followed the defined protocols exactly. The issue arguably arose as a data-centric view of the data in the EHR system was not being considered to keep abreast with a social phenomenon occurring external to the system. Lack of a data-centric view of the significance of data had very serious consequences.

This problem has been recognized in the Aeronautical Information domain where the quality of certain classes of data exchanged between national air navigation services is important from a safety perspective and there is standardisation in this domain (DO-200A 1998); however, this is largely the exception and the issue is poorly covered in most other domains.

The drive for interconnectivity of systems is developing increasingly more complex data supply chains; this is increasing the dependency on data and therefore increasing the risks it may pose; the safety risks are therefore increasing and the problem is becoming more complex in an ever changing world. There is a therefore a pressing need to develop approaches and strategies to understand and manage these risks.

One important concept that has emerged from the group's work is that of *Perspective*. It is becoming increasingly rare that an organization has complete responsibility for all of the data it uses throughout its lifecycle. More common is the situation where an organization is part of a larger data supply chain and will be a consumer and/or producer of data. Some aspects of the organisation's processes or systems may also be subcontracted to other parties such as IT systems providers, system integrators and outsourcing companies and each of these relationships will be governed by potentially complex contractual arrangements, service level agreements, terms and conditions, etc.

To tackle the data safety problem across the entire supply chain is problematic – who will own, fund and provide governance for the overall exercise? How will cooperation from all parties be secured and maintained? How will work that lies outside of contractual agreements be specified and funded?

As with the Aeronautical Information example, such cooperation is possible but is far from the norm and for many organisations who are new to data safety, the prospect of changes to working practices and systems may be not be palatable or indeed viable.

While addressing the problem in a coordinated manner across the entire supply chain may be too ambitious in some cases, the problem can however be tackled from a different angle by looking at the problem from an individual organization's perspective. A perspective based analysis will likely be conducted along contractual boundaries and will consider the following:

> ➢ What commercial exposure does the organization have to safety risks that arise from data?
> ➢ What is the organization's attitude to the risks that data poses and how much is it prepared to invest in managing the risk?

> As a consumer of data, what properties of the data being received must be preserved to avoid possible safety issues?
> What assurances does the data producer give that these data properties will be preserved when it is 'handling' the data (e.g. acquiring, generating, processing, transforming, transmitting the data, etc.).
> As a producer of data, what assurances do consumers require for the data the organization is providing to them?
> Knowing the data properties that need to be preserved, how will the organization ensure these are maintained within its own processes as the data is exchanged and used amongst its internal systems?

An important point here is that if the organization is part of a larger supply chain, they may be several steps removed from where end accidents can actually occur. Consider a healthcare system provided by a 3rd party that stores a national summary of clinical data held in local General Practice and Hospital systems. These local systems can query the national repository but the actual presentation, use and dependency on, say, correctness and availability of the data, is not known to the repository manufacturer.

While the end accident may be remote in this case it does not prevent the manufacturer from seeking data assurance requirements from the consumers of its data. If those consumers are also part of a supply chain then they too can then seek data assurance requirement from their consumers and so on. If all parties approached the analysis in a consistent manner then a complete dependency chain could be established and verified across the entire end to end process.

Verification is important to ensure the data properties can be maintained in a consistent manner. For example, if a data producer is contracted to produce a data set at DIL2 (data integrity level 2, see section 3 of DSIWG 2013 for a description of DILs) but the consumer of that data set is expected to pass the data on to another consumer who is expecting the data to be assured to DIL3, then there is clearly an assurance shortfall.

A framework has been developed (Faulkner and Nicholson 2014) that could be used by organizations across the supply chain to provide structure and consistency to the data-centric analysis of safety risks. This framework uses a layered model to express the hierarchy of systems within an organization and particular focus is given to the horizontal and vertical interfaces between components in the hierarchy. It is through consideration of these interfaces with external systems that assurance requirements can be defined in a consistent manner and an 'interface contract' agreed to ensure requirements will be met.

A report template (Faulkner 2013) has been developed along with some worked examples of using the assessment framework available on the SCSC website.

The concept of perspective is therefore very powerful and with a standardized approach to assessing data safety risks within an organization, an overall picture of data safety risks at all stages across the entire supply chain can be established and verified.

When analysing safety risk from a data-centric perspective, it would seem natural to undertake a structured hazard analysis process such as a hazard and operability study. However, in this case it is behaviours resulting from variations in the properties of data that should be considered rather than the unintended functional behaviour of software. This difference necessitates a change to the guidewords that would be used in a data-based HAZOP.

6 Data HAZOP guidewords

As discussed earlier, common principles of most software standards involve the consideration of the contribution software can make to hazardous system failures. A common approach for identifying hazards is a hazard and operability study (HAZOP) where a multidisciplinary team collaborate to identify potential hazards and operability problems. Structure and completeness are supported through the use of *guideword* prompts; for example, considering the implications if software components perform functions early, late, not at all, etc. These are intended to stimulate imaginative thinking, to focus the study and elicit ideas and discussion.

The work of the group has shown a mapping of software safety principles to data safety principles, and so, as with software, a data HAZOP emerges as a method of identifying changes in the properties of data that can give rise to hazards. However, as the consideration is now on data rather than software, the guidelines will necessarily be different. The group has therefore developed a set of guidewords that could be used in a data HAZOP as shown in the following table.

Table 2. HAZOP Data Guidewords

Property	Description	HAZOP Data Properties	HAZOP Data Guidewords
Integrity	the data is correct, true and unaltered	Loss, partial loss, incorrect, multiple	Correctness, truth, original, trustworthy, coherency, stability, perfect, unquestionable, faithful, certain, ordered, unadulterated, unmodified, unchanged, clean, uncontaminated, untainted, proper, flawless, organized, exact, undistorted, faultless, guided, connected, linked, traced, unbiased.
Completeness	the data has nothing missing or lost	Loss, partial loss, incorrect, multiple	Whole, complete, entire, finished, done, stable, qualified, certified.
Consistency	the data adheres to a common world view, e.g. units	Loss, partial loss, incorrect, multiple, too early, too late, loss of sequence	Coherent, compatible, congruent, congruous, harmonious, deconflicted, consistent, appropriate, suitable, sound, cleansed.

Property	Description	HAZOP Data Properties	HAZOP Data Guidewords
Format	the data is represented in a way which is readable by those that need to use it	Loss, partial loss, incorrect, multiple	Conformant, suitable, valid, configured, well-formed, setup, composed, well structured, arranged, compliant, organised, exact, unaliased, migrated, transformed.
Accuracy	the data has sufficient detail for its intended use	Loss, partial loss, incorrect, multiple	Accurate, true, correct, undistorted, unbiased, faultless.
Resolution	the smallest difference between two adjacent values that can be represented in a data storage, display or transfer system	Loss, partial loss, incorrect, multiple	Exact, untruncated, retention of detail, clarity, determination, distinguishable, clear, within range, distinct, separated, discernible, discriminatable, unconfused, divisible, unaliased, granularity, precision.
Traceability	the data can be linked back to its source or derivation	Loss, partial loss, incorrect, multiple, too early, too late, loss of sequence	Traceable, verifiable, indexed, linked, connected, justified, proven, evidenced, substantiated, continuous, unfragmented, complete, networked.
Timeliness	the data is as up to date as required	Loss, partial loss	Timely, early, ready, expected, unique, appropriate, opportune, ordered, organised, anticipated, seasonable, converging, settling, on-time, latency, lag, lead time, time slots, real-time, determinism, predictable.
Verifiability	the data can be checked and its properties demonstrated to be correct	Loss, incorrect, partial loss, multiple, too early, too late, loss of sequence	Verifiable, provable, checkable, supportable, demonstrable, sustainable, certifiable, defensible, excusable, justifiable, undisputable, irrefutable, validated
Availability	the data is accessible and usable when an authorized entity demands access	Loss, partial loss, multiple, too early, too late	Ready, available, obtainable, reachable, accessible, serviceable, operable, functional, usable, capable, released, issued, disseminated, distributed
Fidelity / Representation	how well the data maps to the real world entity it is trying to model	Loss, incorrect, partial loss, multiple, too early, too late	Representative, accurate, faithful, trustworthy, characteristic, normal, standard, real, expected, natural, typical, regular, fit for purpose, validated, separable, associated, correct units/dimensions, stable, unbiased

Property	Description	HAZOP Data Properties	HAZOP Data Guidewords
Priority	the data (items) are presented / transmitted / made available in the order required	Loss, incorrect, partial loss, multiple, too early, too late	Current, ordered, included, precedence, hierarchy, pre-eminence, retained, ahead, readiness
Sequencing	the data (items) are preserved in the order required	Loss, incorrect, partial loss, multiple	Ordered, contiguous, unique, ordered, clear, continuous, successive, uninterrupted, sequential.
Intended Destination/Usage	the data items are only sent to those that should have them	Loss, incorrect, partial loss, multiple, too early, too late, loss of sequence	Directed, delivered, copied, sent, transmitted, correct recipient, unintercepted, unseen, integral, received, acknowledged, forwarded, filtered.
Accessibility	the data items are visible only to those that should see them	Loss, incorrect, partial loss, multiple, too early, too late	Secure, open, visible, reachable, seen, usable, accessible, obtainable, uncompromised, secure, encrypted, preserved.
Suppression	the data items are intended never to be used again	Loss, incorrect, partial, too early, too late, too much, too little	Hidden, encrypted, private, confidential, erased, unlinked, unavailable, unaccessible, redacted
History	the data has an audit trail of changes	Loss, incorrect, partial loss, multiple	Justifiable, traceable, provable, supportable, demonstrable, sustainable, certifiable, defensible, excusable, justifiable, undisputable, irrefutable.
Lifetime	when does the safety-related data expire	Loss, too early, too late, incorrect, multiple, loss of sequence	Expiry date, age, validity, currency, applicability, durability, duration, lifespan, stretch, tenure, half-life, longevity, span, in-date, best-before, window, established.
Disposability / Deletability	the data can be permanently removed when required	Loss, incorrect, partial, too early, too late	Unavailable, unaccessible, redacted, hidden, filtered, lost, deleted, destroyed, backup, archive, locked, secured, unlinked.

The above terms touch on aspects such as confidentiality (Intended Destination/Usage), integrity and availability; these have strong resonance with aspects typically considered to be in the security domain and so it is important to understand the relationship between safety and security from a data perspective.

7 Data safety and security

The relationship between safety and security, as engineering concepts, can be summarised by their relationships to cultural, developmental and aspirational properties of systems development. Culturally, embedding both safety and security into an organisation is seen as a key strategic goal for creating systems that are both safe and secure. Developmentally, safety and security are quality factors, generating transverse requirements that impact the entire system. Most importantly, at the aspirational level, both safety and security have the common goal of preventing harm from accidental and malicious interventions respectively.

For an organisation aiming to create systems that are both safe and secure, these connections can be both a benefit and a burden. The shared goal of preventing harm means that both quality factors seek to identify routes to harm through analysis of the system being developed. This can result in shared processes and tools, which in turn can save time and money during systems development. However, safety and security interact in a more volatile way at the functional level. Security failings can undermine the safety case for a system and conversely, safety requirements can prevent the implementation of standard security solutions. In addition, "Fail-Safe" states can often leave a system with exposed security vulnerabilities.

These links between safety and security infer that there are connections between the sub-categories of data safety and information security: both attempt to take a data-centric view of the system of interest in order to improve the associated quality factor; and both attempt to prevent harm through the preservation of the properties of data within that system.

In the security domain, the three key properties of data considered are confidentiality, integrity, and availability. Confidentiality, (the failure of which is termed "Information Disclosure" in the Microsoft security model) is typically not a safety concern, as without malicious intent, information sharing is not inherently unsafe. However, when considering systems where confidentiality is an important property, the interaction between data safety and security cannot be trivially resolved. For example, accidental disclosure of information can form part of a causal chain, which leads to harm from a malicious actor.

Data integrity is a critical property for both domains. The Microsoft security model describes malicious removal of the property of integrity as "tampering". Whether by accident or through malicious intent, the potential harm from loss of data integrity can be disastrous to a safety critical system, from the values of drug dosages to control system parameters. Data availability is also important to both domains. Loss of availability, or "denial of service" in the Microsoft security model, is another property that can be lost accidentally or through malicious intervention. Loss of availability prevents systems from functioning properly and can result in undefined behaviour if not mitigated by design.

8 Data safety culture questionnaire

The group has previously developed a technique to assess, at a high level, the risks data poses to an organization (Parsons and Hampton 2014). As well as considering the level of risk data may pose to the organization's system, operation, or service, it also considered wider aspects such as the legal and regulatory environment, organizational maturity, and levels of responsibilities for data system. The key device for performing the assessment was a questionnaire called the Organizational Data Risk (ODR) Assessment Form. The form presented 8 questions with 5 multiple choice answers. Each question's response carries a score; once completed, all the scores are summed to give a final total that is compared against specific ranges to give a ODR level running from 0 (no risk) to 4 (highest risk).

The questionnaire format has proved very effective in establishing an initial high-level assessment of risk; it is relatively quick to complete and is readily accessible to top management who are responsible for allocating funding and resources to managing the risk.

One question in the ODR relates to assessing the organization's maturity in managing data safety risks and responses are aimed at establishing the depth of awareness of data safety and the associated management processes within the organization. However, how is the level of awareness of processes and concepts in an organization measured or evaluated? There may be sufficient high-level knowledge of this for the purposes of the ODR but this seems an area that warranted further exploration; for example, what good is having detailed processes for managing risks if staff have poor awareness or understanding of them?

The group has therefore developed another questionnaire to explore the specific area of measuring the data safety *culture* for a particular activity; whether this be the organization as a whole or for a particular project, service or activity. However, here the focus is on a personal view rather than a project or company's view so the questionnaire would be completed by all or a significant subset of staff working on, say, a particular project. Responses could then be aggregated to give an overall data safety culture value. A key aspect of this approach is that it can be periodically repeated to determine trends – if overall scores are declining, this may suggest that further training and briefings will be required.

The following figure shows an extract from the questionnaire; the full version is available on the SCSC website.

Data Safety Culture Questionnaire Form

This form is used to assess the safety culture related to data for a particular programme (the DSC value).

You play a key role in protecting the organisation from data safety risks and your views are important.
This self-assessment survey is designed to assess our current level of data safety culture within the programme.
The output can help us to improve our safety position.

Please tick the box which reflects your view and answer as honestly as possible. Space is provided for explanatory comments. Your response will only be of value if it reflects what you actually believe is the case, rather than what you believe should happen.

If you would like to remain anonymous please print and send this form by post.

The survey should take no longer than 10 minutes. It is anticipated that this form will be used on a regular basis (e.g. annually).

Programme Name:
Completed By: Date:

Answer each question as you see it – there is no right answer!

QUESTION 1 – MY VIEW OF OUR SUPPLY						
		Don't Know	Strongly Disagree	Disagree	Maybe	Agree	Strongly Agree
1a	I see data as an important factor in the safety of my programme.	☐	☐	☐	☐	☐	☐
1b	I am familiar with the safety aspects of our data.	☐	☐	☐	☐	☐	☐
1c	I think that data in our solution could contribute to an accident.	☐	☐	☐	☐	☐	☐
1d	I think we could be blamed if there were an accident due to our data.	☐	☐	☐	☐	☐	☐
Comments:							

QUESTION 2 – WHAT WE'RE DOING						
		Don't Know	Strongly Disagree	Disagree	Maybe	Agree	Strongly Agree
2a	I think that the programme is aware of data safety risks.	☐	☐	☐	☐	☐	☐
2b	I believe we need to implement measures to manage data safety risks.	☐	☐	☐	☐	☐	☐
2c	I think that the programme meets its obligations (e.g. has a Data Management Plan in place and a role with specific responsibilities in this area)	☐	☐	☐	☐	☐	☐

Fig. 6. Data Safety Culture Questionnaire (extract)

Questionnaires to assess awareness of data safety management within an organisation will give insight into the maturity of the organisations management of data safety issues. The questionnaire is appealing: it is quick to carry out the assessment and gives a view of what is actually taking place not just what is intended to happen. However, it does lack structure and rigour of a more formal maturity assessment model. The group therefore considered what could be learned from existing process maturity models in the context of data safety.

9 Data Maturity Models for Data Safety

More formalised capability maturity models for common processes have been established for some time; these typically look at process areas such as Development and Services. The Capability Maturity Model Integration (CMMI) developed by the Software Engineering Institute (SEI) is a typical example of a process improvement training and appraisal programme. These models formalise the processes for assessing and categorising the maturity of an organisation's processes.

Adaptation of these types of model to allow assessment of an organisation's Safety management Maturity have already been suggested such as the 'The Safety Maturity Model – SaMM' similar to the SEI CMMI (Nagarajan and Davuluri 2014).

However, our concern here is not just on safety management in general, but on management of *data safety* – is there a way to formally measure and assess the maturity of an organization's management of data safety risks? As with safety maturity, the group looked at what models might exist for data to see if these could be adapted to address data safety. One such model has recently been developed by the CMMI Institute called the Data Management Maturity (DMM). (DMM 2014). This has been developed to address the realisation that organisations can now live or die by the quality of their data, and the challenges of building, optimizing and controlling an organisation's data assets are becoming overwhelming with the unprecedented increase in the volume of data to be managed.

The model comprises 20 data management process areas as well as 5 supporting process areas based on CMMI process areas. Level of capability can then be evaluated for each of these to determine an overall maturity level for the organisation.

An analysis of the goals and questions in DMM showed that these can very readily be mapped to equivalent goals and questions to assess data safety maturity. For example, The DMM goal of 'The data management function is aligned with data governance on data management priorities and decision' would simply become: 'The data safety management function is aligned with data safety governance on data safety management priorities and decision'.

Questions similarly map in a straightforward way, so the DMM question 'Is the data management function defined such that it is clear to all relevant stakeholders?' would become 'Is the data safety management function defined such that it is clear to all relevant stakeholders?'

This is a promising area to explore but it should be noted that the DMM is not the only model – there are other variants (DMBOK from DAMA, MIKE2.0, IBM Data Governance Council Maturity Model etc.) that are all very different and an analysis has shown that there are gaps between them. Also the models are focussed on business risks not safety risks so may need tailoring to refocus on systems safety engineering aspects.

There has been significant progress on the guidance and the work has reach sufficient maturity that the group feels it is able to start engagement with external standardisation bodies and to develop plans for raising awareness of the work within the relevant communities. The following sections describe the group's activities in this respect.

10 Integration with existing standards and guidance

While the guidance could be developed into a standalone document that applies to all industries, the group's preference is to incorporate the material into existing standards and guidance as part of their normal review cycles. This is because it may be difficult to reach consensus on a normative set of requirements that apply equally to all sectors and there is therefore risk that some sectors may not feel it is applicable.

Furthermore, if the material is incorporated into the existing standards, then this is likely to be an easier path to realising the guidance as a separate formal standard. The work can also, if required, be tailored more appropriately to the particular industry or sector the existing standard applies to.

The group has therefore approached a number of standards committees and the response has been positive. DEF STAN 00-55 interim issue 3 is currently undergoing review and the intention is that guidance on data safety will be incorporated into that standard as an appendix.

The relatively recently formed Standardisation Committee for Care Information (SCCI) responsible for healthcare standards for NHS England, have also been approached and a process has been established for incorporating data safety considerations as guidance into the current standards: ISB-0129 (Clinical Risk Management: its Application in the Manufacture of Health IT Systems) and ISB-0160 (Clinical Risk Management: its Application in the Deployment and Use of Health IT Systems).

These bodies have been selected initially as their review cycle happens to be timely but there are many more standards bodies that will be engaged in the fullness of time. The material is therefore intended to provide normative guidance in the interim, until superseded by content formally embodied in those standards.

11 Dissemination plans

As the guidance is intended to apply across a large number of sectors and industries, the group acknowledged that there would be particular challenges in disseminating the material to ensure it gained widespread adoption. Clearly, adoption into existing standards and guidance is a key activity but there are a large number of stakeholders that need to be made aware of the activity, how it may affect them and most importantly, for them to understand why this is necessary now.

The group have therefore developed an engagement approach and sector specific engagement plans. The plans are sector specific as, for example, sectors such as Defence and Healthcare have widely different stakeholders and levels of maturity in safety management and so require more tailored dissemination approaches.

The dissemination plans for each sector follow a similar format. Each contains a list of entities types or communication channels as follows.

1. Conferences (present /attend)
2. Papers
3. Publications / articles
4. Presentations to influencers (industry associations, companies, organisations)
5. Professional bodies (BCS, IET, ...)
6. Standards Committees / Working Groups
7. Regulators
8. Government / other sponsors
9. Existing initiatives
10. Academia

The plan would then address for each of these:

- What specific entity (e.g. the sector regulator) will be engaged or what communication channel (e.g. publication or conference) will be used;
- How the particular entity or communication channel would be engaged;
- Who would be tasked with leading the engagement;
- When the activities would be carried out.

The group has currently drafted plans for Civil Nuclear, Defence and Healthcare and further plans will be developed in due course.

12 Conclusions

The work of the Data Safety Initiative Working Group continues to progress with a sustained and high degree of motivation and conviction. The work has been very positively received in the community and is no longer an isolated activity with a number of standards committees already engaged with work the group is doing to see how best to incorporate the guidance in future standards such as DEF STAN 00-55 and ISB-0129/ISB-0160 healthcare standards.

The DSIWG is pleased to issue the next version of the guidance at this symposium.

Acknowledgments The authors would like to thank the SCSC DSIWG for their continued support in this initiative and for all the contributions they have made over the last year to make this paper possible. Special thanks go to members of the working group for providing the sections on data safety & security and data safety principles.

References

ATSB (2014), Loading issue involving a Boeing 737, VH-VZO at Canberra Airport, ACT. Australian Transport Safety Bureau. http://www.atsb.gov.au/publications/investigation_reports/2014/aair/ao-2014-088.aspx. Accessed 24 Oct 2014.

BBC News (2014) Ebola: Can big data analytics help contain its spread? http://www.bbc.co.uk/news/business-29617831. Accessed Nov 2014.

Boeing (2013) Statistical Summary of Commercial Jet Airplane Accidents, Worldwide Operations, http://www.boeing.com/news/techissues/pdf/statsum.pdf. Accessed Nov 2014.

DMM (2014) Data Management Maturity (DMM) http://whatis.cmmiinstitute.com/data-management-maturity. Access 25 Oct 2014.

DO-200A (1998) DO-200A Stadards for Processing Aeronautical Data, RTCA.

DSIWG (2014) Data Safety, SCSC. http://scsc.org.uk/paper_127/Data Safety (Version 1.0).pdf?pap=954 , Accessed 24 Oct 2014

EC (2014) The data gold rush http://europa.eu/rapid/press-release_SPEECH-14-229_en.pdf. Accessed Nov 2014.

ESA (1996), Ariane 501 – Presentation of Inquiry Board Report, http://www.esa.int/For_Media/Press_Releases/Ariane_501_-_Presentation_of_Inquiry_Board_report Accessed. Nov 2014.

Faulkner A (2013) http://scsc.org.uk/file/gd/158-002 Dataware Framework Report v0-01-21.doc, Accessed Nov 2014.

Faulkner A, Nicholson M (2014) An Assessment Framework for Data-Centric Systems. In Dale C, Anderson T (eds) Addressing System Safety Challenges. SCSC

HealthImaging (2009) Update: Cedars-Sinai explains CT perfusion radiation overexposure. http://www.healthimaging.com/topics/diagnostic-imaging/update-cedars-sinai-explains-ct-perfusion-radiation-overexposure. Accessed 24 Oct 2014.

Kelly T (2014) Software Certification: where is Confidence Won and Lost? In Dale C, Anderson T (eds) Addressing System Safety Challenges. SCSC

Nagarajan K, Davuluri AK (2014) Safety Maturity Model. In Dale C, Anderson T (eds) Addressing System Safety Challenges. SCSC

NASA (1999) Mars Climate Orbiter Team Finds Likely Cause Of Loss, http://mars.jpl.nasa.gov/msp98/news/mco990930.html. Accessed Nov 2014.

NBC News (2014), http://www.nbcnews.com/storyline/ebola-virus-outbreak/texas-hospital-makes-changes-after-ebola-patient-turned-away-n217296, Accessed 8 Nov 2013.

Parsons M, Hampton P (2014) Stopping Data causing Harm: towards Standardisation. In Dale C, Anderson T (eds) Addressing System Safety Challenges. SCSC

Approximate verification of swarm-based systems: a vision and preliminary results

Benjamin Herd, Simon Miles, Peter McBurney, Michael Luck

Department of Informatics, King's College London

London, UK

Abstract *Swarm-based systems, i.e. systems comprising multiple simple, autonomous and interacting components, have become increasingly important. With their decentralised architecture, their ability to self-organise and to exhibit complex emergent behaviour, good scalability and support for inherent fault tolerance due to a high level of redundancy, they offer characteristics which are particularly interesting for the construction of safety-critical systems. At the same time, swarms are notoriously difficult to engineer, to understand and to control. Emergent phenomena are, by definition, irreducible to the properties of the constituents which severely constrains predictability. Especially in safety-critical areas, however, a clear understanding of the future dynamics of the system is indispensable. In this paper we show how agent-based simulation in combination with statistical verification can help to understand and quantify the likelihood of emergent swarm behaviours on different observational levels. We illustrate the idea with a simple case study from the area of swarm robotics.*

1 Introduction

Swarm-based systems — systems comprising a possibly large number of interacting and autonomous components which collaborate in order to achieve a common goal — have become increasingly important. Inspired by nature, the principle of swarm intelligence provides a powerful paradigm for the construction of fully decentralised systems. As opposed to other multiagent systems, swarms are typically composed of very simple individual components. Instead of being built into the system explicitly, the complex behaviour that can be observed at the macro level *emerges* from the actions and interactions of the constituents. Due to their lack of a central coordination mechanism, swarm-based systems typically offer a high level of scalability and fault tolerance. A high level of redundancy, the capability to self-organise and robustness are further characteristics which make

© B. Herd, S. Miles, P. McBurney, M. Luck. 2015. Department of Informatics, King's College London. . Published by the Safety-Critical Systems Club. All Rights Reserved.

swarm-based systems highly attractive for application in safety-critical areas (Winfield et al. 2006).

The capability of a set of simple components producing emergent behaviour is a powerful advantage of swarms since it allows for the construction of complex systems at relatively low cost. On the other hand, it is precisely the emergent nature of swarms which makes them notoriously difficult to engineer. For example, due to the irreducibility of emergent phenomena to the behaviour of the constituents, it is exceptionally hard to construct a system with a particular global-level behaviour in mind. A clear understanding of the future dynamics of the system is indispensable, especially in safety-critical areas.

It is thus not surprising that questions of veracity play an important role in the construction of swarm-based systems. Safety analysis is usually carried out through *experimentation, computational simulation* or *formal analysis*. In the case of *experimentation*, the agents (e.g. robots) are placed in a real-world environment and their behaviour is observed over time. It is obvious that experimentation, albeit realistic and trustworthy, may be subject to hard resource constraints. *Simulation* is typically carried out using a dedicated simulation tool such as, for example, the Player/Stage system (Gerkey et al. 2003). Thanks to its lightweight nature, simulation enjoys several advantages over real experimentation. However, without proper statistical analysis, it is unclear how insights obtained from individual simulation runs can be generalised. *Formal analysis* represents the most rigorous way of analysing the behaviour of a swarm. It involves the construction of a mathematical model, e.g. a (probabilistic) finite state automaton (Liu et al. 2007) and its subsequent analysis, e.g. through temporal logic model checking (Konur et al. 2010). State space explosion is a common problem in this context and an efficient way to circumvent it is to restrict the focus to the macro level of the system. Despite its power, macro-level modelling limits the analysis to the *mean-field behaviour* of the system which may, in some cases, be too restrictive.

In this paper, we describe our ongoing research efforts on the verification of swarm-based systems using a combination of *agent-based simulation* and *statistical runtime verification*. Swarm-based systems can be seen as a special type of multiagent system and it is thus natural to consider agent-based simulation as a means to study their temporal dynamics. We show how a combination of probabilistic simulation and statistical model checking can help to combine the advantages of both informal and formal verification and estimate the likelihood of emergent events. As opposed to pure simulation, it allows a modeller to (i) describe the correctness criteria in a formal way using temporal logic, and (ii) quantify the precision of the results (obtain a confidence interval for the likelihood of an observed phenomenon). As opposed to formal macro-level analysis, it preserves the individual richness of the simulation approach and allows for the exploration of complex scenarios without giving up much of its formal rigour. It allows a modeller to analyse the emergent behaviour of swarm-based systems in a quantifiably accurate way.

The paper is structured as follows. We give a brief overview of related work on swarm analysis in Section 2. Theoretical preliminaries on agent-based simulation, linear temporal logic and statistical model checking are given in Section 3. In Section 4, we describe the architecture of MC²MABS, our statistical runtime verification framework. The application of the framework to a simple scenario in the area of swarm robotics is described in Section 5. The paper concludes with a summary and an overview of limitations and opportunities for future work in Section 6.

2 Related work

Given the increasing interest in swarm-based systems, formal approaches to analysis and verification become more and more critical. A good overview of macroscopic modelling approaches to swarm analysis is given in (Lerman et al. 2005). Here, the system under study is represented as a finite state automaton and formulated as a set of difference equations which are mathematically tractable. Macroscopic approaches are based on the assumption that, despite stochastic variance in the agent population, the overall collective behaviour is statistically predictable. This works well if agents are homogeneous, spatial characteristics do not need to be taken into account and agents are fairly independent. As soon as those requirements are weakened, however, macroscopic modelling can become extraordinarily complex (Liu et al. 2007).

The applicability of formal techniques to the individual-based verification of swarm-based systems in the context of the NASA Autonomous Nano Technology Swarm (ANTS) mission has been shown in (Hinchey et al. 2005, Rouff et al. 2004). The authors use an integration of four established techniques – Communicating Sequential Processes (CSP), Weighted Synchronous Calculus of Communicating Systems (WSCCS), X-Machines and Unity Logic – to model and verify the behaviour of ANTS spacecraft swarms with a particular focus on the occurrence of race conditions. The authors conclude that a blend of the aforementioned techniques is promising but more work needs to be done on their integration.

A proof-based approach to the verification of a foraging robot swarm has been presented in (Behdenna et al. 2009). The authors first describe a propositional approach which allows for fairly detailed individual representation, yet remains limited to a small number of agents due to exponential growth of the underlying transition system. A solution for the verification of arbitrarily-sized swarms using First-Order Temporal Logic is also presented. It avoids some of the complexity problems of the propositional approach, yet it requires a simplified representation of the overall system (similar to the macroscopic modelling approaches).

The application of temporal logic model checking to the verification of robot swarm behaviour has been shown in (Dixon et al. 2012). Again, in order to cope with complexity, the system is represented in a highly simplified way (small grid-based environment and small number of robots). The authors are planning to work on abstraction and reduction techniques to tackle the state space explosion problem.

An interesting approach to verify the emergent behaviour of robot swarms using probabilistic model checking has been presented in (Konur et al. 2010). In order to tackle the combinatorial explosion of the state space, the authors exploit the high level of symmetry in the model and use a *counter abstraction*. Instead of creating the parallel composition of the single agents' state machines, they represent the system with a single, system-level state machine. This global representation is similar to the individual state machines but contains an additional counter variable which stores the number of individual agents being in the respective state. In doing so, the authors manage to transform the originally exponential into a polynomial problem[1]. This is a significant improvement, however, since the resulting problem is still exponential in the number of agent states, the approach remains limited to relatively small-scale systems.

A work related to ours (albeit not focussed on swarm-based systems) is that of (Yasmeen et al. 2012). The authors present a methodology which is based on the analysis of individual simulation traces. In order to reduce complexity, *trace reduction* represents a central step in the methodology. As opposed to ours, the approach is solely focussed on individual traces and does not make any attempt to generalise the results to the overall state space.

3 Background

Agent-based simulation: Agent-based modelling (ABM) or agent-based simulation (ABS) is rapidly emerging as a popular paradigm for the simulation of complex systems that exhibit a significant amount of non-linear and emergent behaviour (Macal and North 2007). It uses populations of interacting, autonomous and often intelligent agents to model and simulate various phenomena that arise from the dynamics of the underlying complex systems. Influenced by (distributed) artificial intelligence, complexity science and computer simulation, agent-based simulation is applied successfully to an ever-increasing number of real-world problems and could in many areas show advantages over traditional numerical and analytical approaches; as a consequence, it is often being employed as a decision support tool for policy making and analysis. Although social science has been its traditional domain, agent-based simulation is also increasingly being used for the analysis of complex technical, in many cases also safety-critical, systems in areas such as avionics (Bosse and Mogles 2013), airport performance modelling (Bouarfa et al. 2013) or the design and analysis of robot and UAV swarms (McCune and Madey 2013, Wei et al. 2013).

---

[1]To be precise: The resulting problem is polynomial in the number of agents and exponential in the number of agent states

Linear temporal logic: Correctness properties about real-world systems typically have a temporal flavour. For example, one wants to state that certain undesirable things will *never* happen in the system under consideration, certain good things will *eventually* happen, or certain invariants are guaranteed to *always* hold. This raises the question how those properties should be formulated in order to be automatically answerable. Temporal logic is a convenient formalism to represent statements that involve tense in a formal and unambiguous way. The treatment of time in temporal logic can be roughly subdivided into *branching time* (CTL, CTL*) and *linear temporal logic* (LTL) (Baier and Katoen 2008). Branching time logics assume that there is a choice between different successor states at each time step and thus views time as an exponentially growing tree of 'possible worlds'. Linear time logic views time as a linear sequence of states. The approach described in this paper is based upon the analysis of individual finite paths representing simulation output; we thus focus on LTL here.

LTL formulae are evaluated over traces — sequences of system states. They comprise as their constituents (i) *atomic propositions* that represent statements which are either true or false in a certain state, (ii) logical combinators ('and', 'or', and 'not'), and (iii) temporal operators such as, for example, 'until' (U), 'globally/always' (G), or 'finally/eventually' (F). Examples for typical types of correctness questions and their formulation in LTL are given below:

- p will always be true (in each state): Gp ('globally p')
- p will never be true: $\neg Fp$ ('not finally p')
- Whenever a is true, then b will also eventually be true: $G(a \Rightarrow Fb)$
- p will eventually hold forever: FGp

Due to its recursive nature, LTL allows for the formulation of complex and arbitrarily nested temporal properties in a convenient and unambiguous way. A full description of LTL is beyond the scope of this paper; a good formal introduction is, for example, given in (Baier and Katoen, 2008).

Statistical model checking: Given a system whose correctness is to be ascertained and a correctness property, verification refers to the process of checking whether the system satisfies the property. In the context of temporal logic model checking, the description of the system is assumed to be given as some kind of finite-state automaton M (an abstract, formal representation of the system), and the correctness property p is assumed to be given in temporal logic. The verification problem then is to determine whether M satisfies p, denoted $M \vDash p$. Conventional model checking aims to find an *accurate* solution to a given property by *exhaustively* searching the state space of M which is only possible if the space is of manageable size (Baier and Katoen 2008). One solution to this problem that works for probabilistic systems is to use a *sampling approach* and employ statistical techniques in order to generalise the so obtained results to the overall state space. In this case, n paths or *traces* are sampled from the underlying state space and the property is checked on each trace; techniques for statistical inference, e.g. *hypothesis testing*, can then be used to determine the significance of the results. Approaches of this kind are summarised under the umbrella of *statistical model checking*; a good overview is given in (Legay and Delahaye 2010). Due to its independence from the underlying state space, statistical model checking allows for the verification of large-scale (or even infinite) systems in a timely, yet approximate manner.

One particular approach described by Hérault and Lassaigne, *Approximate Probabilistic Model Checking*, provides a *probabilistic guarantee* on the accuracy of the approximate value generated by using *Hoeffding bounds* on the tail of the underlying distribution (Hérault and Lassaigne 2004). According to this idea, $\ln \frac{2}{\delta} / 2\epsilon^2$ samples need to be obtained in order to achieve a result Y that deviates from the real probability X by at most ϵ with probability $1 - \delta$, i.e. $\Pr(|X - Y| \leq \epsilon) \geq 1 - \delta$. The number of samples grows exponentially with increasing confidence and accuracy. However, it is important to note that a high level of confidence can be achieved much more cheaply than a high level of accuracy. A particularly attractive feature of this approach is that the number of traces necessary for achieving a certain level of confidence and accuracy is independent from the size of the underlying system. This makes it particularly suitable for the verification of agent-based simulations (Herd et al. 2014).

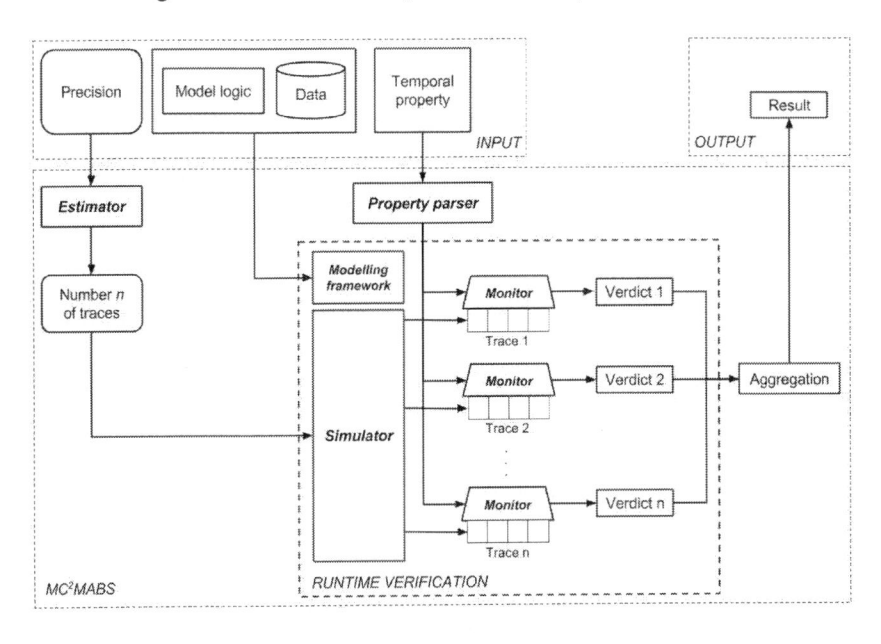

Fig. 1: The overall architecture of MC^{2}MABS

4 MC²MABS – Monte Carlo Model Checker for Multiagent-based Simulations

Our current research on the approximate verification of large-scale multiagent scenarios resulted in the development of MC²MABS, a statistical runtime verification framework for agent-based simulations. MC²MABS uses a Monte Carlo approach to estimate the probability of a given temporal property. A high-level overview of the tool is shown in Figure 1. The framework comprises as its central components (i) an estimator, (ii) a modelling framework, (iii) a property parser, (iv) a simulator, and (iv) a runtime monitor. The typical sequence of actions in a verification experiment can be described as follows:

1. The user provides (i) the logic of the underlying multiagent system by utilising the modelling framework, (ii) an associated correctness property, and (iii) the desired precision of the verification results as inputs to the framework.
2. The correctness property is translated into a runtime monitor by the property parser.
3. The estimator determines the number of simulation traces necessary to achieve the desired level of precision.
4. The simulator uses the model together with additional configuration information to produce a set of simulation traces.
5. Each simulation trace is observed by a runtime monitor which assesses the correctness of the trace using a given correctness property; due to the online nature of the monitor, a verdict is produced as soon as possible.
6. The individual results are aggregated into an overall verification result and presented to the user.

Due to the decoupling of simulation and verification, MC²MABS supports both *ad-hoc* and *a-posteriori* verification. Ad-hoc verification is synonymous to runtime verification and assesses the correctness of a system during its execution. A-posteriori verification assumes the existence of traces prior to the actual verification. The latter mode can be useful, for example, if the traces have been obtained with a different simulation tool, e.g. a dedicated swarm robotics simulator. In that case, the simulator of MC²MABS is merely used to 'replay' the pre-existing output for the purpose of verification.

The individual components of MC²MABS are briefly described below.

```
Input: Accuracy ε, confidence δ
Output: sample size
    cdf(n) = ($\binom{n}{(i \cdot p) - (i \cdot \epsilon)}$) $p^i (1-p)^{n-i}$
    L := [1..ln($\frac{2}{\delta}$) · $\frac{1}{2\epsilon^2}$]
    for i := 1 to |L| do
        P := cdf(L[i])
        If P < δ return L[i]
    end
```

Algorithm 1: The sample size calculation procedure

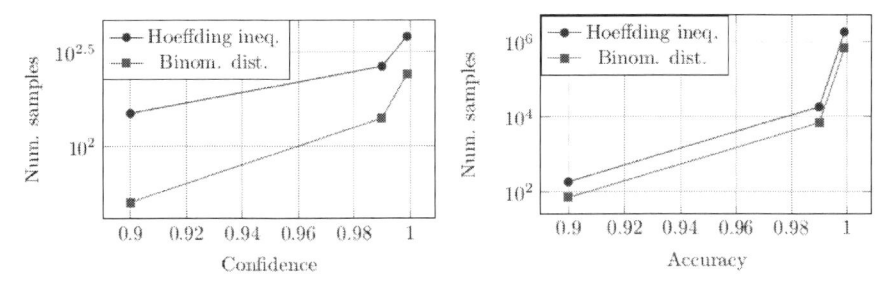

Fig. 2: Relation between sample size and confidence (left) and sample size and accuracy (right) for Hoeffding bounds and the accurate calculation using the Binomial distribution

Estimator: The main purpose of the estimator is to determine the number of traces necessary to achieve a certain level of precision (provided by the user) with respect to the verification results. Sample size determination is a common problem in statistics; in the current version of MC²MABS, we use a simple algorithmic procedure to calculate the necessary number of traces. As opposed to a probabilistic *bound* (such as the Hoeffding bound briefly mentioned in Section 3) that is represented by a nice mathematical formula but often overestimates the actually necessary sample size by a significant degree (as exemplified in Figure 2), the procedure currently used by MC²MABS is algorithmic but accurate since it operates directly on the Binomial distribution. The procedure is shown in Algorithm 1. For example, let the desired accuracy be 1% and the desired confidence be 95%; in this case, the algorithm would suggest a sample size of 6,900. In contrast, the Hoeffding bound would suggest a much larger sample size of 18,445.

It is important to note that the algorithm does not avoid the exponential growth of the sample size necessary for detecting rare events and thus suffers from the same limitations as, for example, the Hoeffding bound. It does, however, return a lower *total sample size* which reduces the number of simulation traces that need to be analysed. In the presence of resource constraints, this can represent a critical practical advantage. Furthermore, the calculation of the cumulative distribution function can become expensive as n grows. In a real implementation, it is thus advisable to avoid a naïve iteration from 1 to N and perform, for example, a binary search instead.

Modelling framework: Instead of providing a dedicated model description language – a path which is taken by most existing verification tools – we decided to allow for the formulation of the underlying model in a high-level programming language. This is motivated by the observation that agent-based simulations often contain a significant level of functional complexity, e.g. probability evaluations, loading and manipulation of external data, and location-based search algorithms, to name but a few. Any simple modelling language would thus significantly limit the range of models which it is capable of describing. As a consequence, we decided to take a different path and realise the interface between the model description and the monitor by means of a *service provider interface* (SPI) which provides a basic architectural skeleton for the underlying model and limits the prescriptive part of the framework to a set of callback functions. In order to maintain a high level of performance (which is crucial for the generation of large batches of traces), we use C++ as the modelling language. As a compiled multi-paradigm language, we believe that C++ offers a good balance between usability and performance.

Property parser: The property parser is responsible for translating a user-provided text string into a *monitor* which is then used to observe the temporal dynamics of a simulation trace. The parser uses a formal grammar which defines the space of valid correctness properties. The language understood by the parser is *simLTL,* a specification language that is based on LTL and tailored to the particular characteristics of agent-based simulations (Herd et al. 2014b). As opposed to conventional LTL, simLTL allows for a much more fine-grained formulation of properties about *individual agents* as well as about *subgroups* within existing groups of agents. This is achieved by a subdivision of the language into two layers: an *agent layer* and a *group layer*. Furthermore, the language is augmented with *quantification* and *selection* operators. These features make it possible to formulate properties such as, for example, the following:

It is true for **every** agent that the energy level will never fall below 0
No more than 20% of the agents will eventually run out of energy
All agents of group x will eventually run out of energy

Furthermore, as mentioned above, the formulation of properties is closely linked with the way the simulator performs the sampling from the probability space underlying the simulation model.

Simulator: The simulator is responsible for executing the simulation model repeatedly in order to obtain a set of traces used for subsequent verification by the monitor. Technically, by repeatedly executing the simulation model, the simulator performs a sampling from the underlying probability space. By interpreting the probability space in different ways, different levels of granularity with respect to property formulation can be achieved. So, for example, by interpreting a trace of length *k* produced by the simulation model not as a single sample from the distribution of traces of length *k* but instead as a set of *k* samples from the distribution of states, properties about *individual states* and their likelihood become expressible; by interpreting the trace as a set of *k/2* samples from the distribution of subsequent states, properties about *transitions* and their likelihood become expressible, etc. In general, a single trace of length *k* can be interpreted as a set of samples of trace *fragments* of length $1 \leq i \leq k$. Furthermore, by relating probabilities of individual properties, statements about *correlations of events* as well as *conditional* and *causal relationships between events* can be made. This allows for a high level of granularity and expressivity with respect to property formulation (see below) and verification.

Technically, the simulator is tightly interwoven with the modelling SPI. At the current stage, all simulation replications are executed sequentially. Parallelisation is a central task for future work, especially since the individual replications are entirely independent and therefore efficiently parallelisable.

Monitor: The runtime monitor is the central component of the verification framework. Its main purpose is to observe the execution of a single trace as generated by the simulator and check its correctness against the background of a given property on-the-fly — while the trace is being produced. In the case of thousands of traces that need to be assessed, online verification represents a critical advantage: as soon as a property can be satisfied or violated, the monitor is able to produce a verdict and move on to the next trace. For properties which are satisfiable or refutable at some point along the trace, this leads to significant speedup over an exhaustive approach. As indicated in the previous paragraph, a monitor is a direct product of a temporal property. Monitor creation strongly relies upon the notion of *expansion laws*. Informally, expansion laws allow for the decomposition of an LTL formula into two parts: the fragment of the formula that needs to hold in the *current* state and the fragment that needs to hold in the *next* state in order for the whole formula to be true. It is useful to view both fragments as *obligations* — aspects of the formula that the trace under consideration needs to satisfy immediately and aspects that it promises to satisfy in the next step. For example, in order for a statement such as Gp ('it is always the case that *p* holds') to be true in a given state, two requirements need to be satisfied: (i) *p* needs to be true in the *current* state (*immediate obligation*), and (ii) Gp needs to be true in the *next* state (*future obligation*).

Expansion laws play an important role for the idea of runtime verification, since they form the basis for a decision procedure that a model checker can use to decide in a certain state whether a given property has already been satisfied or violated. By decomposing a formula into an immediate and a future obligation, optimality can be achieved: as soon as the immediate obligation is satisfied and no future obligation has been created, the entire formula is satisfied and the evaluation finishes.

5 Case study: A swarm of foraging robots

In this section, we describe the results of a set of preliminary experiments in which we applied to the verification of a simple swarm robotic scenario. We focus here on *foraging*, a problem which has been widely discussed in the literature on *cooperative robotics*. In a nutshell, foraging describes the process of a group of robots searching for food items, each of which delivers energy; individual robots strive to minimise their energy consumption whilst searching in order to maximise the overall energy intake. The study of foraging is important because it represents a general metaphor to describe a broad range of (often critical) collaborative tasks such as *waste retrieval*, *harvesting* or *search-and-rescue*. A detailed overview of multirobot foraging is beyond the scope of this work; a good overview has been given in (Cao et al. 1997).

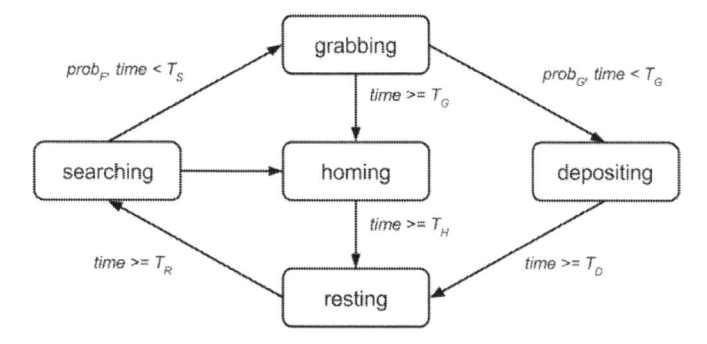

Fig. 3: Transition diagram for an individual agent in the foraging scenario (Liu et al. 2007)

Model description

The model described in this chapter is based on the work of (Liu et al. 2007). In the model, a certain number of food items are scattered across a two-dimensional space. Robots move through the space and search for food items. Once an item has been found, it is brought back to the nest and deposited which delivers a certain amount of energy to the robot. Each action that the robot performs also consumes a certain amount of energy. The model is kept deliberately simple. Each robot can be in one of five states: *searching* for food in the space, *grabbing* a food item that has been found, *homing* in order to bring a food item back to the nest, *depositing* a food item in the nest, and *resting* in order to save energy. Transitions between states are probabilistic and dependent upon time as well as upon the state of the other agents. A state transition system for an individual agent is shown in Figure 3 (for better readability, self-loops have been omitted). $prob_F$ describes the probability of finding food, $prob_G$ describes the probability of grabbing food, and

T_S, T_R, T_H, T_G, and T_d represent the time spent searching, resting, homing, grabbing and depositing, respectively. *time* represents the time the agent has already spent in the respective state. The probabilities $prob_F$ and $prob_G$ are functions of the number of other agents currently foraging (searching, grabbing, or depositing) and grabbing. This creates a high level of interdependence between the agents which makes the dynamics more interesting but also harder to analyse analytically. The overall swarm energy is the sum of the individual energy levels.

Investigating the overall performance of the algorithm is beyond the scope of this paper. We keep verification deliberately simple and focus on the conceptual ideas rather than on the authenticity of the verification experiments. The simulation itself was implemented in C++ using the modelling framework provided by MC²MABS. It is realised in a bottom-up, agent-based way: each robot is represented explicitly including its individual behaviour which is adjusted over time according to the state of the other agents. For space limitation, we cannot give the full C++ code here; it does, however, closely follow the description in (Liu et al. 2007).

All experiments have been conducted using MC²MABS described in Section 4 The simulation for each experiment comprises a population of 100 agents and runs for 1,000 ticks. All experiments were conducted on a Viglen Genie Desktop PC with four Intel® Core™ i5 CPUs (3.2 GHz each), 3.7 GB of memory and Gentoo Linux (kernel version 3.10.25) as operating system.

Verification

Despite its conceptual simplicity, the model already exhibits a significant level of complexity. If we just take into account the basic states that an agent can be in, the state space of a model of *n* agents already amounts to $O(5^n)$ states. The fact that transition probabilities of the agents are dependent upon the number of other agents being in a certain state further complicates an independent assessment of individual agents' temporal behaviours. This represents a significant problem for purely formal approaches to verification.

Instead of providing a full analysis of the foraging scenario, we focus on a very small and simple fragment: the influence of each agent's initial level on the system's energy level over time. The overall energy level is a simple sum of the individual energy levels, i.e. each agent makes a direct contribution to it. The dependence between the individual agents' transition probabilities (and thus also their energy level) and the activity of the other agents (which are, in turn, influenced by their own energy levels) creates a feedback loop which is typical for complex adaptive systems and which significantly exacerbates predictability. In order to assess the energy level over time, we verified two different temporal properties:

1. **Property I**: "The swarm must never run out of energy":

   $G\ (SWARM\_ENERGY > 0)$

2. **Property II**: "No single robot must ever run out of energy":

   $G(\forall(ENERGY > 0))$

It is important to ensure that the swarm *as a whole* will never run out of energy. This is described by the first property. However, despite the overall swarm energy always being positive, there may well be robots within the swarm which run out of energy. This is the focus of the second property which makes a universal statement about the *individuals* within the swarm. Both properties are examples of *safety properties* which state that "something bad will never happen". It is important to stress that these two criteria only provide a very superficial view on the correctness of the swarm. One problem of safety properties is that they are always satisfied by a system which does not do anything. In order to provide a coherent picture of the dynamics of a system, they therefore need to be augmented with *liveness properties* (properties which state that "something good will always happen"). As mentioned above, however, our goal here is to show the conceptual workings of MC^{2}MABS rather than to provide a deep analysis of a particular foraging algorithm. We thus restrict our focus to the two properties mentioned above.

We checked the properties above with three different levels of precision which are given below (the number of traces has been calculated using the estimation procedure described in Section 4).

1. confidence $\delta = 0.01$, accuracy $\varepsilon = 0.01 \Rightarrow$ 13,700 traces
2. confidence $\delta = 0.001$, accuracy $\varepsilon = 0.01 \Rightarrow$ 24,000 traces
3. confidence $\delta = 0.001$, accuracy $\varepsilon = 0.001 \Rightarrow$ 2,389,000 traces (partial estimates)

In the first set of experiments, we obtained estimated probabilities which deviate from the actual probabilities by at most 1% in at least 99% of all cases; in the second experiment, we obtained estimated probabilities which deviate from the actual probabilities by at most 1% in at least 99.9% of all cases; in the third experiment, we obtained estimated probabilities which deviate from the actual probabilities by at most 0.1% in at least 99.9% of all cases. For time constraints, the third experiment has only been run partially. Some of the scenarios would have taken several days to finish and the times have thus been estimated.

Property	Init. energy	Probability	Total time (1 core)	Total time (8 cores)
G(SWARM_ENERGY > 0)	25	0%	00:00:14	00:00:02
	26	23.1%	00:08:25	00:01:03
	27	91.9%	00:33:02	00:04:08
	28	100%	00:35:52	00:04:29
G(∀(ENERGY > 0))	30	0%	00:00:16	00:00:02
	40	7.3%	00:08:37	00:01:05
	50	46.1%	00:50:29	00:06:19
	60	100%	01:49:00	00:13:37

Table 1: Verification results for 13,700 traces

The results for the first experiment are shown in Table 1. It shows the initial amount of energy per agent, the resulting probability estimate of the respective property, the total simulation and verification time on a single processor core and an estimate of the total time for an eight-core machine. We can see that some verification results can be gained very cheaply, even for large groups of agents. The numbers clearly illustrate that the total time for verification and simulation strongly depends upon the satisfiability/refutability of the given property. For example, in the case of an initial energy level of 25, the swarm will quickly run out of energy. As a consequence, the monitor can report a result very quickly and move on to the next trace. In the case where all traces need to be examined exhaustively (e.g. initial energy level = 28), the total time required for simulation and verification is about 150 times higher than in the immediately refutable case. The numbers also show that the verification of the individual-based Property II is significantly higher than that of the group-based Property I.

Property	Init. energy	Probability	Total time (1 core)	Total time (8 cores)
G(SWARM_ENERGY > 0)	25	0.13%	00:00:23	00:00:03
	26	23.9%	00:15:36	00:01:57
	27	91.9%	00:57:52	00:07:14
	28	100%	01:02:50	00:07:51
G(∀(ENERGY > 0))	30	0%	00:00:28	00:00:04
	40	6.6%	00:14:34	00:01:49
	50	46.1%	01:28:26	00:11:03
	60	100%	03:10:57	00:23:52

Table 2: Verification results for 24,000 traces

The results for the second experiment are shown in Table 2. As expected, due to the higher level of confidence, verification takes longer. Scenarios in which the property is satisfiable/refutable early along the trace (initial energy level = 25) are only marginally slower to verify than in the previous case, but for scenarios which require exhaustive monitoring of the traces, the difference becomes more apparent (e.g. initial energy level = 28). However, the numbers also show that a higher level of confidence can be achieved relatively cheaply. As opposed to the previous experiment, we increased the level of confidence from 99% to 99.9%, yet the overall verification time is only about twice as long.

Property	Init. energy	Probability	Total time (1 core)	Total time (8 cores)
G(SWARM_ENERGY > 0)	25	0%	00:39:18	00:04:55
	26	23.1%	≈ 1d	≈ 4hrs
	27	91.9%	≈ 4d	≈ 12hrs
	28	100%	≈ 13d	≈ 1.5d
G(∀(ENERGY > 0))	30	0%	00:46:35	00:05:49
	40	7.3%	≈ 1d	≈ 4hrs
	50	46.1%	≈ 6d	≈ 18hrs
	60	100%	≈ 13d	≈ 1.6d

Table 3: Verification results for 2,389,000 traces

The third result increases the accuracy by reducing the measurement error from 1% to 0.1%. As described briefly in Section 3, achieving higher accuracy is very expensive as the number of traces required increases exponentially with the level of accuracy. This is also reflected in the results shown in Table 3 most of which, for time constraints, have only been estimated based on the results of the previous two experiments[1]. The numbers show that, apart from immediately satisfiable/refutable scenarios, the desired precision is too high to be achieved in a realistic time frame — at least on the hardware platform which was used for the experiments.

However, it is important to note that a high level of accuracy is not always necessary. Especially in the case of frequent events — events whose probability is sufficiently high (e.g. 50%) — it is often irrelevant whether the actual measurement error is 1%, 2%, or, in some cases, even 5% or more. In general, the accuracy will often be made a function of the variance of the underlying distributions. For frequent events whose variance is large, a low level of accuracy is sufficient. For rare events with a small level of variance, a high level of accuracy is necessary. Especially if the variance is small, however, a lower number of samples is necessary to quantify the likelihood of the respective event (Mount 2013b) . We plan to further investigate this idea since it may allow for the detection of rare events using a significantly smaller number of traces than recommended by our current estimation procedure.

---

[1] One nice aspect of the idea of statistical runtime verification is that estimates can be easily made by extrapolating results obtained with a lower precision.

6. Conclusions and future work

Safety analysis represents a central task in the development of critical systems. In the presence of emergent behaviour, however, determining the likelihood of an event can be a difficult problem. Formal verification, particularly model checking, has proven hugely successful for the analysis of a wide range of systems. Due to combinatorial explosion, however, the verification of large-scale systems remains a critical problem. Informal verification, e.g. simulation-based testing, on the other hand, is able to circumvent most of those complexity problems, yet it suffers from a lack of verification strength.

In this paper, we described our ongoing research which aims to combine simulation with statistical model checking in order to allow for the verification of large-scale swarm-based systems in an approximate, yet accurately quantifiable way. With properties formulated in temporal logic, high scalability due to the focus on individual simulation traces and its ability to provide confidence intervals for the results, the approach is able to combine some of the strengths of both formal and informal verification techniques into a convenient and easy-to-use framework. The major strengths are briefly summarised below:

- **Rigour:** Properties can be formulated in a formal way by means of linear temporal logic.
- **Flexibility:** Due to the reliance on traces rather than on a formal model, arbitrary multiagent scenarios are monitorable.
- **Automation:** The verification of properties can be automated efficiently.
- **Predictability:** The number of traces required to achieve a certain level of precision is independent of the size of the underlying system.
- **Optimality:** Due to the 'online' nature of the verification, verification results are guaranteed to be reported as eary as possible
- **Anytime:** The functionality of the approach is independent of the number of traces being monitored. A verification result can always be produced, even if only a single trace is being examined.

As the experiments above illustrated, it is currently possible to verify properties about fairly large-scale systems (100s of agents) with a decent level of confidence (99%) and accuracy (1%) in a reasonable time, even on general purpose hardware. However, we are well aware that this level of precision is not sufficient for highly safety-critical areas. It is also important to stress that safety analysis involves much more than the verification of simple properties. We hope, however, that the proposed approach may eventually serve as a useful building block in the engineering process which helps to gain a better understanding of the behaviour of emergent systems.

Before MC^2MABS can be used efficiently in a real-world scenario, there are, of course, still plenty of limitations and open problems which need to be overcome. Some of the issues are briefly mentioned below.

Accuracy: To achieve a high level of accuracy with respect to verification, an exponentially increasing number of samples is required which represents a critical limitation. One way to remedy this problem is to exploit the fact that high precision is mostly important for *rare events* which are located in the tail of the underlying distribution and thus do not have the same level of variance as more common events. The calculation procedure described in this paper does not make any difference between rare and frequent events which could further reduce sample sizes.

Efficiency: One critical advantage of trace-based verification is that the approach is easily parallelisable. At the current stage, MC^{2}MABS performs verification purely sequentially and does not exploit the capabilities of modern parallel hardware. A second important starting point for performance improvements is the architecture of the simulation itself. The model used for the case study was implemented in a fairly straightforward, procedural and largely non-object-oriented way. It did not make use of some of C++'s advanced features such as template metaprogramming which allow for the execution of code at compile time. By structuring the program in the right way and exploiting compile time constants, runtime may reduced significantly. And finally, MC^{2}MABS itself is technically subdivided into a simulation (written in C++) and a verification part (written in Haskell). *Marshalling* (translating and transferring the data structures between the two languages) represents a significant bottleneck which also negatively influences the capability of the tool to analyse large batches of simulation traces.

References

Baier, C and Katoen, J.-P. (2008) Principles of Model Checking. The MIT Press

Bauer, A and Leucker, M and Schallhart, C (2010) Comparing LTL semantics for runtime verification. Journal of Logic and Computation, 20(3):651-674

Behdenna, A and Dixon, C and Fisher, M (2009) Deductive verification of simple foraging robotic behaviours. International Journal of Intelligent Computing and Cybernetics, 2(4):604-643

Bosse, T and Mogles, N (2013) Comparing modelling approaches in aviation safety. In Curran R (ed) Proc. 4th International Air Transport and Operations Symposium (ATOS2013), Toulouse, France

Bouarfa, S and Blom, H and Curran, R et al (2013) Agent-based modeling and simulation of emergent behavior in air transportation. Complex Adaptive Systems Modeling, 1(1):1-26

Cao, Y U and Fukunaga, A S and Kahng, A (1997) Cooperative mobile robotics: Antecedents and directions. Autonomous Robots, 4(1):7-27

Dixon, C and Winfield, A and Fisher, M et al (2011) Towards temporal verification of swarm robotic systems. Robotics and Autonomous Systems, 60(11):1429-1441, 2012. Towards Autonomous Robotic Systems.

Donaldson, R and Gilbert, N (2008) A Monte Carlo model checker for probabilistic LTL with numerical constraints. Technical Report 282, Department of Computing Science, University of Glasgow.

Gerkey, B and Vaughan, R.T. and Howard, A (2003) The player/stage project: Tools for multirobot and distributed sensor systems. In Proc. 11th International Conference on Advanced Robotics, pages 317-323.

Hérault, T and Lassaigne, R and Magniette, F et al (2004) Approximate probabilistic model checking. In Proc. 5th International Conference on Verification, Model Checking and Abstract Interpretation (VMCAI'04), volume 2937 of Lecture Notes in Computer Science, pages 307-329. Springer.

Herd, B and Miles, S and McBurney, P et al (2014) Verification and validation of agent-based simulations using approximate model checking. In Alam, S J and Parunak, H V D (eds) Multi-Agent-Based Simulation XIV, Lecture Notes in Computer Science, pages 53-70. Springer.

Herd, B and Miles, S and McBurney, P et al (2014b) An LTL-based property specification language for agent-based simulation traces, Technical Report TR-14-02, Department of Informatics, King's College London

Hinchey, M.G. and Rou, C.A. and Rash, J.L. et al (2005) Requirements of an integrated formal method for intelligent swarms. In Proc. 10th International Workshop on Formal Methods for Industrial Critical Systems (FMICS '05), pages 125-133, New York, NY, USA, ACM.

Konur, S and Dixon, C and Fisher, M (2010) Formal verification of probabilistic swarm behaviours. In Dorigo, M and Birattari, M and Di Caro, G et al (eds) Swarm Intelligence, volume 6234 of Lecture Notes in Computer Science, pages 440-447. Springer.

Legay, A and Delahaye, B and Bensalem, S (2010) Statistical model checking: an overview. In Barringer, H and Falcone, Y and Rosu, G et al (eds) Proc 1st International Conference on Runtime verification (RV'10), pages 122-135. Springer.

Lerman, K and Martinoli, A and Galstyan, A (2005) A review of probabilistic macroscopic models for swarm-robotic systems. In Swarm robotics, pages 143-152. Springer.

Liu, W and Winfield, A and Sa, J (2007) Modelling swarm robotic systems: A case study in collective foraging. In Wilson, M S and Labrosse, F and Nehmzow, U and Melhuish, C et al (eds) Towards Autonomous Robotic Systems, pages 25-32.

Liu, W and Winfield, A and Sa, J et al (2007) Strategies for energy optimisation in a swarm of foraging robots. In Şahin, E and Spears, W and Winfield, A (eds) Swarm Robotics, volume 4433 of Lecture Notes in Computer Science, pages 14-26. Springer.

Macal, C M and North, M J (2007) Agent-based modeling and simulation: desktop ABMs. In Proc. 39th Winter Simulation Conference (WSC '07), pages 95-106.

McCune, R and Madey, G (2013) Agent-based simulation of cooperative hunting with UAVs. In Proc. Agent-Directed Simulation Symposium. Society for Computer Simulation International.

Mount, J (2013) A bit more on sample size. http://www.win-vector.com/blog/2013/03/a-bit-more-on-samplesize/, Last accessed: 08/2014.

Mount, J (2013) Sample size and power for rare events. http://www.win-vector.com/blog/2013/12/sample-size-and-power-for-rare-events/, Last accessed: 09/2014

Rou, C and Vanderbilt, A and Hinchey, M et al (2004) Verification of emergent behaviors in swarm-based systems. In Proceedings of the 11th IEEE International Conference and Workshop on the Engineering of Computer-Based Systems, pages 443-448.

Wei, Y and Madey, G and Blake, M (2013) Agent-based simulation for UAV swarm mission planning and execution. In Proc. Agent-Directed Simulation Symposium, page 2. Society for Computer Simulation International.

Winfield, A and Harper, C J and Nembrini, J (2006) Towards the application of swarm intelligence in safety critical systems. In The 1st Institution of Engineering and Technology International Conference on System Safety, pages 89-95.

Yasmeen, A and Feigh, K M and Gelman, G and Gunter, E L. Formal analysis of safety-critical system simulations. In Proc. 2nd International Conference on Application and Theory of Automation in Command and Control Systems, ATACCS '12, pages 71-81, Toulouse, France, IRIT Press. 14

Demonstrating Compliance in the Arctic

Nick Golledge

Aker Solutions

Aberdeen, UK

Abstract *The drive to realise new hydrocarbon assets deeper into the Arctic Circle, would require the development of an improved environmentally-centric technology assurance and acceptance criteria. These criteria demand that all installed systems are of high integrity. The high consequence of failure on the Arctic Circle environment demands that all systems are to be installed to safety standards as a minimum, for all aspects of the petroleum infrastructure. Offshore Oil and Gas duty holders will claim, and be required to demonstrate that the 'Environmental Integrity Level' (EIL) has been achieved in order to demonstrate conformance to the stringent Arctic Circle legislation.*

1 Introduction

In recent years a greater attention (via governments and the media at large), has been focussed upon increased activity in the Arctic Circle. This increased activity is in part due to the influence of climate change reducing sea ice and leading to increased maritime traffic. The Sub Arctic has been connected with petroleum activities for many years, Norman Wells, located in the Canadian Northwest Territories was discovered in 1920 and then continuously operated since 1932, and Sakhalin has been in operation since 2005. In the mid 1960's to the late 1980's numerous explorations and subsequent production facilities were established within the Arctic Circle. However these hydrocarbon assets were either located on land or relatively close to land (Prudhoe Bay and Beaufort Sea are typical examples). This year however; *Geological prospecting work in the Laptev Sea is being carried out ahead of license commitments subject to all requirements of environmental protection legislation of the Russian Federation* (Offshoreenergytoday.com, 2014).

The Norwegian Continental Shelf (NCS) and the associated petroleum industry's contribution to the Norwegian government budget is forecasted to fall by

NOK[1] 58 billion (Approximately £5.8 billion) during the period 2010 to 2015. To recover from the forecasted decline, new commercially attractive exploration needs to be found and sanctioned. This concern expressed by the 'Norwegian Oil and Gas Association' is due to the long lead times to realise First Oil (10-15 years)[2] and the subsequent production of hydrocarbon revenue streams, (Norsk Olje Og Gass, 2012). The NCS will continue to bear its fruits for a long time to come as stated by the 'The Norwegian Petroleum Directorate' which estimates Norway's total resources to be 84 billion barrels (13.4 billion standard cubic metres) of oil equivalent. Apparently one third of this estimate has been recovered, with another one third thought to have been identified but not produced, and the final one third remains to be discovered, (Oljedirektoratet, 2014). Nevertheless, the NCS current trend in the fall of hydrocarbon output is expected to continue with exploration activity expected to be low for at least the next 18 months. This is further compounded by the new discoveries which are small in comparison with the previous Elephant fields of past successes. In contrast with the above the United Kingdom Continental Shelf (UKCS), has according the Department of Energy and Climate Change (DECC), an average estimate of 20 billion barrels of oil equivalent, still remains to be discovered, (DECC, 2013).

2 The Arctic Circle Opportunity

The Arctic Circle encompasses about 6 percent of the Earth's surface, an area of more than 21 million km2 (8.2 million mi2), of which almost 8 million km2 (3.1 million mi2) is onshore and more than 7 million km2 (2.7 million mi2) is on continental shelves under less than 500 m of water and ocean waters are typically deeper than 500 m. The extensive Arctic continental shelves may constitute the geographically largest unexplored prospective area for petroleum remaining on Earth. (U.S. Geological Survey, 2008).

The Arctic region contains jurisdictional elements of eight countries — Canada, Denmark / Greenland, Finland, Iceland, Norway, Russia, Sweden and the United States. Finland and Sweden do not border on the Arctic Ocean and are the only Arctic countries without jurisdictional claims in the Arctic Ocean and adjacent seas. Sixty one large oil and natural gas fields have been discovered so far within the Arctic — 43 are in Russia, 11 in Canada, 6 in Alaska and 1 in Norway. It has been estimated that the Arctic could contain over 20% of the world's undiscovered oil and natural gas reserves, (Budzik, 2009).

---

[1] Norwegian Krone

[2] "The Snøhvit gas field in the Barents Sea was discovered in 1984, but more than two decades passed before its resources could be produced and exported as liquefied natural gas (LNG) from the Melkøya terminal in northern Norway. The time lag since the licence award was even longer." (Norsk Olje Og Gass, 2012)

International Energy Agency (IEA) forecasts via World Energy Outlook 2010 (WEO), that by 2035 global energy consumption will increase by 36 per cent above 2008 consumption, (IEA, 2010). This forecast is based upon policy commitments in which governments stated their future needs at the time. Oil consumption is projected to increase at a rate of one per cent per year, from 85 million barrels a day in 2008 to 99 million barrels a day by 2035. Which is further supported according to OPEC's World Oil Outlook (WOO) for 2013 which states, *With world energy demand set to grow by 52 per cent up to 2035*[1], (OPEC, 2014).

Deep-water and Arctic oceanic reservoirs are often referred to as 'Unconventional oil resources' as they are in difficult the reach locations, and in environmentally sensitive areas. WEO suggests that from the year 2000, more than half of the oil discovered was in deep-water oilfields, and further the production of oil from these reservoirs is expected to grow. A significant number of onshore areas in the territories of Canada, Russia, and United States of America (Alaska) have already been successful, resulting in the discovery of more than 400 oil and gas fields north of the Arctic. These fields account for approximately 240 billion barrels of oil equivalent and natural gas, which is almost 10 per cent of the world's known conventional petroleum resources (cumulative production and remaining proved reserves). In the case of offshore reserves located north of the Arctic Circle, undiscovered resources are estimated to be 90 billion barrels of oil, of which 84 per cent occurs offshore. (U.S. Geological Survey, 2008).

3 The Risks

The risks associated with any petroleum exploration are multifarious and complex; when considered against the backcloth of a 'New frontier' as others describe the Arctic Circle, the Rio Declaration on Environment and Development is a suitable reminder of mankind's responsibilities:

Principle 15, In order to protect the environment, the precautionary approach shall be widely applied by States according to their capabilities. Where there are threats of serious or irreversible damage, lack of full scientific certainty shall not be used as a reason for postponing cost effective measures to prevent environmental degradation (United Nations, June 1992).

This precept has formed the basis of subsequent multinational legislation[2], which seeks a long term approach based upon scientific knowledge. The Meta principle here; the polluter pays for and is wholly responsible for damages and other derived compensation from one country to another etc.

---

[1] The author notes the publication suggests 88% of the forecast increased is expected to be derived from Asia.

[2] Coastal states may also establish exclusive economic zones (EEZs), to 200 nautical miles from the coast lines (UNCLOS art.57), the 'Law of the Sea Convention'.
http://www.un.org/depts/los/convention_agreements/texts/unclos/part5.htm

For the critical infrastructure protection community to implement a risk-based prioritization of resources — whether at the facility, community, or other level — it requires information about the threats, vulnerabilities, and consequences of a variety of potential scenarios (French, 2007).

3.1 The Hazards

Currently the Arctic Circle represents enormous infrastructure and knowledge gaps, constraining development and increasing the risks of future so called 'Frontier Projects'. Let us look at the causality of potential hazardous events, which can be foreseen, applicable to this problem space, with reference to figure 1.

Activities before commissioning covers a multitude of technology issues and technical disciplines; some topical research and development risk reduction activities are:

- Environmental Risk Assessment for Marginal Ice Zones (MIRA)
- Ice Loading Designs - Structures
- Material Selection for Extreme Cold Temperatures
- Iceberg Trenching Counter Measures
- Long Distance Fibre Optic Networks
- Lithium Sea Water Batteries

Some of these activities, designed to improve scientific knowledge and increase technology maturity are performed via Joint Industry Projects (JIP) such as Arctic Oil Spill Response Technology,[1] and the Arctic Council's Arctic Monitoring Program (AMAP) includes extensive work related to oil and gas activities in the Arctic.

All technologies which are considered for Subsea deployment are required to undergo 'Marinisation'. Such that the equipment in general becomes, smaller with a greater degree of robustness, improved maintainability and a well understood reliability model. The process described above is generally referred to as a Technology Qualification Programme (TQP), see figure 2.

---

[1] The JIP is supported by nine international oil & gas companies – BP, Chevron, ConocoPhillips, Eni, ExxonMobil, North Caspian Operating Company (NCOC), Shell, Statoil, and Total – making it the
largest pan-industry programme dedicated to this area of research and development.

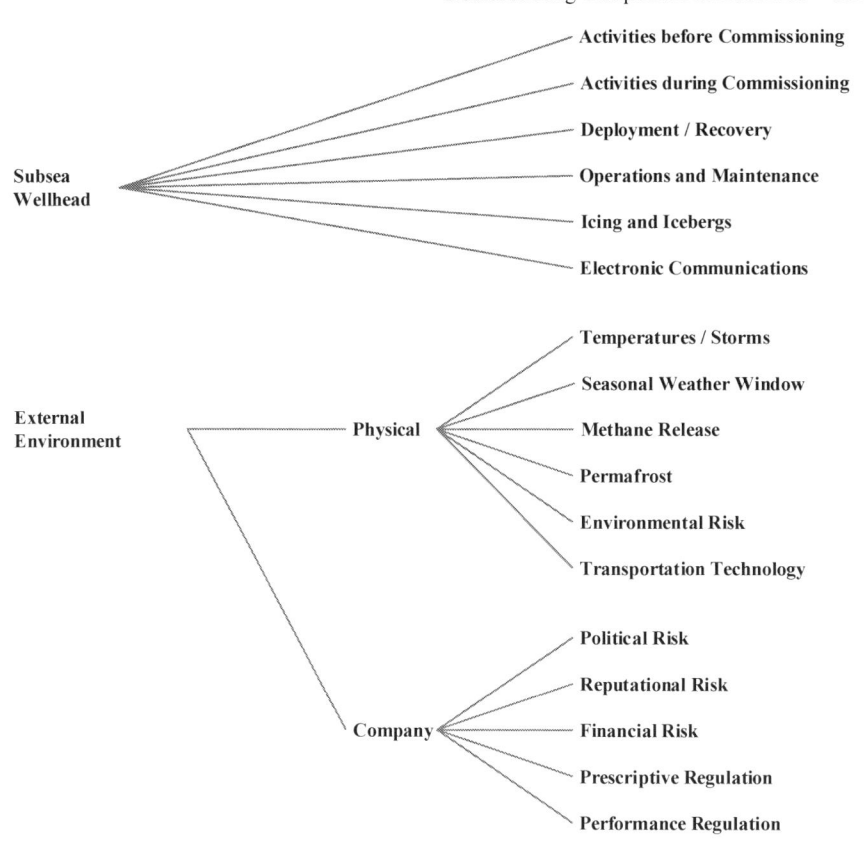

```
Subsea Wellhead
 ├── Activities before Commissioning
 ├── Activities during Commissioning
 ├── Deployment / Recovery
 ├── Operations and Maintenance
 ├── Icing and Icebergs
 └── Electronic Communications

External Environment
 ├── Physical
 │    ├── Temperatures / Storms
 │    ├── Seasonal Weather Window
 │    ├── Methane Release
 │    ├── Permafrost
 │    ├── Environmental Risk
 │    └── Transportation Technology
 └── Company
      ├── Political Risk
      ├── Reputational Risk
      ├── Financial Risk
      ├── Prescriptive Regulation
      └── Performance Regulation
```

Fig. 1: Causes of Threats - adapted from EN 50159:2010, (BSI, 2010)

This is a petroleum industry approach which seeks to determine the 'Fit, Form and Function for Subsea deployment', in order to systematically exclude uncertainties. This provides a technical body of evidence for the specific instance, of that technology package. The TQP process follows a prescribed DNV-RP-A203 guideline. For safety related technology the petroleum industry follows IEC 61508 and IEC 61511. The TQP guidelines have tended over time, since its introduction in 2001, to focus upon the demonstration of risk reduction processes and the verification and validation of requirements.

Therefore in many ways the TQP process is not dissimilar in nature, to that of the processes described in IEC 61508 and other associated safety critical frameworks. Allied with the TQP process is the concept of 'Technology Readiness Levels' (TRL) these are adapted from United States Department of Defence 'Technology Readiness Assessment' (TRA) Deskbook (July 2009), and further referenced in API 17N. How can these technology maturity scales, with any scientific certainty, suddenly become applicable for use in The Arctic?

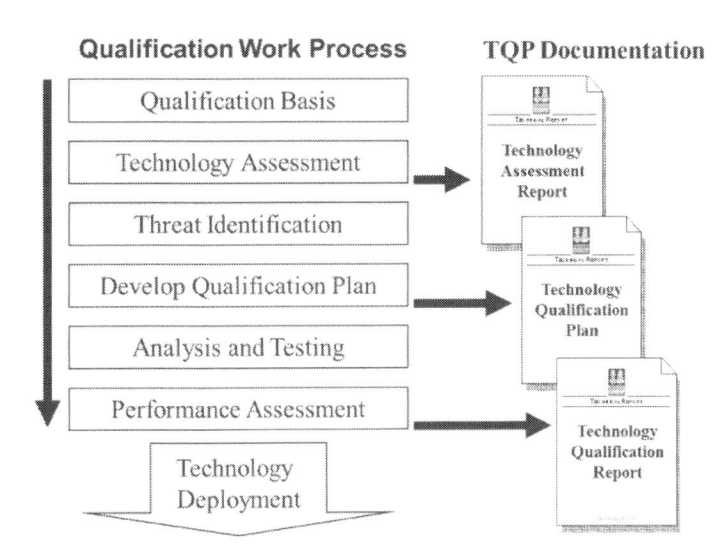

Fig. 2: Technology Qualification Process – DNV-RP-A203

Activities during commissioning are typically facilitated by the use of a dive support vessels, equipped with Dynamic Positioning (DP) for the control of the ships position and heading. These ships typically require some twenty or more Global Positioning Sensors (GPS_{en}) to keep the ship in position above the Subsea well. Additionally the vessel requires to have Arctic winterisation and icebreaking capabilities, in order to deploy the Remotely Operated Vehicles (ROV). The challenges become further complicated by the potential for rapid changes in temperature, high winds - storm associated conditions aggravating the installation process, with increased wave heights and swell. Whilst the Subsea equipment resides upon loading deck of the vessel, where the icing conditions that may cause machinery to seize up.

The Arctic remains an operations risk environment. These operational risks continue to be potent threats for many months of an Arctic year. Complicated supply logistics enhance the potential realisation of the risk events. The Russian company Gazprom stated in 2012 its intention to *announce an open tender for offers for building a sea-based helicopter platform for the Shtokman gas condensate field in the Barents Sea*, in order to mitigate risks with logistics and 'Search and Rescue' (SAR) . The guide price suggested by Gazprom was approximately $550 million, (Interfax Group, 2012).

Operations and maintenance activities will require improved technology maturity, greater than exists to date. It can be conceived that autonomous underwater vehicles (AUVs) will eventually become mature enough to be place on station, to perform gainful work, equipped with environmental monitoring sensor arrays amongst other suitable robotic appendages. The sensor arrays are the basis for executing knowledge based 'Sensor Fusion' algorithms in order to make better decisions prior to notifying the host control centre. Multiple AUVs positioned and organised in a swarm organisational hierarchy can then communicate via a mesh networks to assist one and other to in performing complex computations, such as in parallel computing applications. The AUVs will require low energy processors units (Similar in topology to that of current Quad Core mobile phones etc.), essentially dormant in order to conserve the battery life. The AUVs thus, are located to the seabed via locking onto 'Lithium Sea Water Battery' charging stations, which essentially are trickle charging whilst in a low energy observation mode. Improved vision image processing allows in-conjunction with the sensor array for 'Plant Tours' to be performed and when operated in a hive mentality the AUVs assist each other in the required navigation duties. It is difficult to understand without the use of such drones, how the petroleum asset can be viable assuming the location of the field is such that a permanent human presence is not feasible or conducive, all year round. As a comparison, the current state of the art is described by (Tristan Crees, 2010).

Icing and icebergs, Icing is a particularly serious hazard for all Arctic vessels, which can cause essential machinery or petroleum infrastructure to seize up and the maritime ship becomes top heavy. Additionally this can also be a threat for coastal infrastructure, in places exposed to sea-spray. Statoil's Melkøya Liquid natural gas (LNG) plant, which is located outside Hammerfest in Norway, noted to be the only LNG plant above the Arctic Circle, has reported numerous operational technical difficulties, some of which are pertinent to LNG plant's location, the threats are derived from rapid changes to temperature and thus icing. The Norwegian press has suggested that the technical issues cost Statoil $34–$51m a week in lost revenue, (Platts, 2011).

Icebergs are formed from the sea facing end of a freshwater glacier or ice shelf. Arctic icebergs are found to be smaller and more random in shape than Antarctic variants, however Arctic Icebergs have a greater tendency to repeatedly capsize. Icebergs commonly run aground, with a significant threat to any Subsea infrastructure as the scour can leave a furrow many metres deep into the seabed. The scour is sometimes seen as a circular depression emanating from a small iceberg foot in which the iceberg twists and turns creating a significant depression. *Grounded bergs have a deleterious effect on the ecosystem of the seabed, often scraping it clear of all life* (Encyclopædia Britannica, 2014). This threat has led Drover and others to postulate the requirement and even claim patents, for the submergence of the Subsea assets into the seabed. Drover suggests the opposing forces from Subsea structures interacting with free floating icebergs are not great enough to cause the iceberg to heave or rotate around or pass over the Subsea structure, (Drover, 2012). The techniques for performing Subsea trenching are well understood and practiced. What is not well understood is how deep, and how to manage the Subsea asset subsequently when effectively buried, it has never been done before.

Electronic communications challenges are sourced from magnetic and solar phenomena, interference and geostationary satellite geometry (satellites in orbit over the equator do not cover the Arctic because of the curvature of the earth), which affects high frequency radio communications via geostationary satellites, which have poor or no coverage north of 78°- 79° degrees northern latitude and Global Positioning Systems (GPS) are degraded above 74.5° north latitude. 'New exploration blocks on the Russian side, such as those in the Perseyevsky Field, are in a satellite shadow', (Lunde, 2015).

The Arctic and Antarctic both exhibit aurora[1], where streams of charged solar particles enter the earth's upper atmosphere. These natural anomalies vary in strength with solar activity. The effects of these charged solar particles on large scale, predominantly ferrous, petroleum infrastructures should also be addressed.

Like all modern control system communications, the network infrastructure is required to span beyond the local area network, this is especially a Subsea requirement for deployment into the Arctic Circle due to the inhospitable nature of the dominion and the need for real time data links. The Iridium constellation of communications satellites provides communication services that operate in the Arctic environment, albeit with limited bandwidth. Whilst the Iridium[2] system has a worldwide coverage, it is designed primarily to support a bit rate 2.4-2.8 Kbps uncompressed, 24 Kbps compressed, which is used primarily for voice communications. Iridium has responded with a proposed new system called Iridium NEXT, which is expected to be launched in 2015. Whilst and improvement, *Iridium NEXT will deliver a high-speed user experience with sustained data speeds up to 512 Kbps and bursts up to 1.5 Mbps over an end-to-end packet data solution,* (Iridium Satellite LLC, 2014). Two fibre optic broadband projects (Arctic Fibre[3] & Arctic Link) are currently laying fibre infrastructure. This involves a distance of 9,320 miles from Japan to Europe via the southerly Northwest Passage, which will spur to adjacent land nodes. This then is a potential improvement, but a threat

---

[1] The aurora borealis that glows and flicker across the northern sky are potentially destructive. "Space weather" can affect GPS navigation and even knock out power grids. In 1989, the Hydro-Québec power grid was taken down completely by a space weather event, (Athabascau University, 2014), (The Engineer, 2012).

[2] Iridium's 66 low-Earth orbiting (LEO) cross-linked satellites – the world's largest commercial constellation – operate as a fully meshed network that is supported by multiple in-orbit spares, (Iridium Communications Inc, n.d.).

[3] Arctic Fibre Subsea cable project is estimated to cost up to US $1.5 billion.

which must be resolved sooner rather than later in order to derive a suitable technology maturity level.

3.2 Mitigation

Risk has to be mitigated throughout the organisation from the drilling ship deck through to the boardroom making the corporate decisions. Companies cannot be allowed to set the risk appetite, despite any assurance of prudent corporate risk management. Deepwater Horizon - amongst other lessons tells us that simple truth. This point is further elaborated upon by (Hauge, 2011), in his report which considers Environmental Risk Acceptance Criteria (ERAC), a topic which is discussed further in section 4.0. Local communities and other interest parties influence government regulatory frameworks; governments decide risk appetite not the duty holders.

Cleaning up of any contamination in the Arctic, particularly in ice-covered areas, would present multiple obstacles which together constitute a unique and hard-to-manage activity. Natural biodegradation of oil in the Arctic could be expected to be lower than in more temperate environments such as the Gulf of Mexico, although there is currently insufficient understanding of how oil will degrade over the long term in the Arctic. System failures that lead to environmental incidents are likely to incur prohibitive costs, how can restorative actions on 'pristine' environments be quantified in terms of cost? Some argue for the removal of a liability cap for investors and more stringent regulations.

In Greenland (Denmark) as part of the licensing award (Baffin Bay) a $ 2 Billion bond has to be transferred at the time of award for small companies towards any potential clean-up costs. Companies and investors need to be aware of the severe obligations from binding agreements with the licensing government(s). If a major accident occurred the financial exposure from future international legislation – such as that proposed to the EU Offshore Directive to override national jurisdictions will prevail and hence drive a cultural attitude of a 'Chronic Unease'. In this type of regulation, reduction to the bonds or equity holding required prior to a licensing award, could be offset by the independent verification of rigorous engineering governance. Such offsets could focus upon engineering practices not just toward the protection of human safety, or asset protection, but additionally; a greater weighting towards environmental safety. In such a regulatory framework the cost to implement triple redundancies as a de facto standard; more layers of protection than previously thought necessary would become funded: Established safety management treatments of risk to reduce the frequency, probability of occurrence or mitigations leading to reduced severity require re-evaluation for systems deployed in these pristine environments.

4 Discussion

The brief outline provided in this paper has established that many scientific knowledge gaps exist in the technology used for Arctic Oil and Gas installations within numerous technology domains. This outline highlights a requirement to establish criteria to protect the environment. This leads directly to questions such as - how do we associate risk reduction activities normally attributable to the protection of the individual, or society and activities that introduce additional risks to the environment? We need to seek and develop a consensus for risk reduction and acceptance criteria for environment. Do we have the scientific and technological maturity to proceed further into the Arctic Circle as required by, Principle 15 of the Rio Declaration?

Damage to the environment can be expressed at different levels such as organism level, population level, habitat level or complete ecosystem level, and several environmental components can be damaged. Environmental risk assessment is about making estimates of harm to plant and animal life and to the ecosystem integrity that can later be compared to previously agreed risk acceptance criteria. However, due to practicality, environmental risk analysis for a complete ecosystem is normally not performed and the risk is assessed for vulnerable singular components within the environment, e.g. specific species or habitats. These serve as risk indicators and has previously been considered to be sufficient in order to estimate the environmental risk within temperate regions.

The Norwegian Norsok Z-013 standard was first published in 1998. Through subsequent revisions the text has helped to establish a methodology in order to determine ERAC. Although in the latest edition (2010), the method is no longer published. The perspective taken by the standard was that when a hazardous event occurred; the duration of environmental pollution becomes insignificant in relation to the expected time interval between hazardous events. Implying that different categories of environmental pollution may be associated with an acceptable frequency of occurrence, according to the duration of the hazard. Five categories for environmental damage are described. These categories are; Insignificant damage, minor, moderate, considerable and serious. The categorisation of environmental pollution is based on the recovery time as shown below, i.e. the time required before the harm to; a specific species or habitat, has recovered to the condition prior to the accidental spillage.

1. Insignificant damage: recovery time less than 1 month
2. Minor damage: recovery time 1 month to 1 year
3. Moderate damage: recovery time 1 year to 3 years
4. Considerable damage: recovery time 3 years to 10 years
5. Serious damage: recovery time more than 10 years

The recovery time interval following an environmental pollution event is deemed insignificant compared to the period between such hazards recurring. Different values for what can be regarded as "insignificant" will result in different acceptance criteria. For example, if say 5% is regarded as insignificant, then this interpretation of insignificant is a minor environmental hazard with an average recovery time of 0.5 years may not occur more frequently than once every ten years.

The MIRA guideline which is issued by 'Norwegian Oil and Gas Association' (formerly OLF) continues with the above methodology.

The establishment of risk acceptance criteria for pollution is based on the guiding principle that the frequency of harm shall be "insignificant" compared to the consequence of the harm, measured in terms of the restitution time of affected marine and coastal resources. The restitution time is the time needed for a resource to return to its original state after being affected by pollution. With resource we understand any valued faunal or floral population or habitat, including shoreline, seabed or even bodies of water, (Hauge, 2011).

The responsibility to determine what constitutes as insignificant has been left to the offshore 'Duty Holders'. According to (Hauge, 2011) the duty holders have typically selected 5% as the level of tolerable harm. Where are the risk reduction 'countermeasures' with this methodology? Is it an industry practice that simply deals with a release response plan?

It is noted that these environmental recovery times are developed based on experience in temperate regions – these recovery times will be significantly extended within Arctic (and Antarctic) based upon lower temperature and very low flushing rates. Environmental damage will persist for extended periods and may eradicate whole ecosystems.

5 The Importance of Environmental Integrity Level

Current methods for determining environmental risk acceptance criteria are directed towards to consequences of a hydrocarbon release or spillage, which then concentrates upon the restitution time and the frequency of the next environmental pollution event. The license duty holder has the sole discretion to choose the tolerable risk sensitivity criteria. Despite espoused values from government agencies concerning the attitudes towards the environment, why are these agencies not actively driving the tolerable risk acceptance criteria in conjunction with the scientific community?

The current ERAC methodology does not address any form of prevention, additional barriers or risk reduction criteria. The current ERAC in accordance with the MIRA guidelines, suggests due consideration in relation to concentration or proximity of potential hazards. But this is compounded by, and therefore diluted with emergency preparedness policies. The greater the concentration of hazardous sources i.e. shipping, drilling rigs etc., the higher the environmental risk acceptance criteria, but in converse the more remote, i.e. the greater the distance between the potential hazards the lesser the criteria. ERAC maturity in the current guidelines cannot be presented as 'Fit, form and function' for deployment into the Arctic Circle.

Risk management is the systematic process of understanding, evaluating and addressing these risks to maximise the chances of objectives being achieved and ensuring organisations, individuals and communities are sustainable. Risk management also exploits the opportunities uncertainty brings, allowing organisations to be aware of new possibilities. Essentially, effective risk management requires an informed understanding of relevant risks, an assessment of their relative priority and a rigorous approach to monitoring and controlling them, (The Institute of Risk Management, 2014).

Despite leading efforts by (Hauge, 2011) which are to be applauded and in a facsimile approach by (Zhou, 2013) to determine a linkage between Safety Integrity Level (SIL) and EIL. These embryonic efforts seem to be approached from a purely mathematical bias by employing a calibrated risk graph, without the scientific data to support a coherent risk assessment. Engineers cannot make judgements about the consequences or restitution from a potential cocktail of pollutant releases. Scientists must be funded to study the phenomena and then advise Engineers how best to treat the aftermath.

In consideration of the pristine 'Arctic Circle Frontier', a method is required which links the SIL and EIL with Asset Integrity Levels (AIL). Environmental risks are interwoven in complexity with geographical attributes and the associated financial implications for avoidance, mitigation and any subsequent rectification (Clean-up costs).

This paper has presented a strong case for action; damage to the Arctic environment will persist for extended periods and may eradicate whole ecosystems. Once these pristine regions are damaged the cost of restitution may not be measured in financial cost, and the effects of this environmental damage may be felt globally for generations to come.

Acknowledgements I would like to thank the following, for their valuable and constructive suggestions during the planning and development of this paper. In addition to their willingness to give their time so generously which has been greatly appreciated: Dr Alastair Faulkner (Abbeymeade Ltd), Dr Alan Barclay and Dr Phil Bagley (Aker Solutions)

References

Athabascau University, 2014. Beware the aurora: How AU is heading off the dangerous effects of aurora borealis. [Online]
Available at: http://news.athabascau.ca/news/aurora-research-autumn-magnetometers/
[Accessed 29 Novemeber 2014].

BSI, 2010. Railway applications — Communication, signalling and processing systems — Safety-related communication, s.l.: British Standards Policy and Strategy Committee.

Budzik, P., 2009. Arctic Oil and Natural Gas Potential. [Online]
Available at: http://www.eia.gov/oiaf/analysispaper/arctic/pdf/arctic_oil.pdf
[Accessed 14 Nov 2014].

DECC, 2013. Oil and gas: Review of UK offshore oil and gas recovery. [Online]
Available at: https://www.gov.uk/oil-and-gas-review-of-uk-offshore-oil-and-gas-recovery
[Accessed 12 November 2014].

Drover, E. A. e. a., 2012. Bounds on Iceberg Motions During Contact With Subsea Equipment and the Seabed. Rhodes, Greece, International Society of Offshore and Polar Engineers (ISOPE).

Encyclopædia Britannica, 2014. "iceberg". [Online]
Available at: http://www.britannica.com/EBchecked/topic/281212/iceberg
[Accessed 19 November 2014].

French, G. S., 2007. Intelligence Analysis for Strategic Risk Assessments. In: S. A. S. P. Liz Jackson, ed. Critical Infrastructure Protection: Elements of Risk. Washington DC, USA: George Mason University School of, p. 111.

Hauge, S. e. a., 2011. Barriers to prevent and limit acute releases to sea, s.l.: Sintef.

IEA, 2010. World energy Outlook 2010 edition, Paris: The International Energy Agency Publications;.

Interfax Group, 2012. Gazprom ready to pay over $500 mln for sea-based helicopter platform at Shtokman. [Online]
Available at: http://www.interfax.com/newsinf.asp?id=311225
[Accessed 8 November 2014].

Iridium Communications Inc, n.d. Company Profile. [Online]
Available at: https://www.iridium.com/About/CompanyProfile.aspx
[Accessed 19 Novemeber 2014].

Iridium Satellite LLC, 2014. Notice of Inquiry: Department of Commerce National Telecommunications And Information Administration, Washington, D.C.: US Gov..

Lunde, H., 2015. Satellite Communications in the Arctic User requirements and possible technical solutions. Oslo, Norway, Telenor Satellite Broadcasting AS.

Norsk Olje Og Gass, 2012. Need To Open Additional Areas. [Online]
Available at: http://www.norskoljeoggass.no/en/Facts/New-areas/
[Accessed 12 November 2014].

Offshoreenergytoday.com, 2014. Rosneft and ExxonMobil have started 2D seismic exploration. [Online]
Available at: http://www.offshoreenergytoday.com/rosneft-exxonmobil-start-2d-seismic-studies-in-laptev-sea/
[Accessed 12 November 2014].

Oljedirektoratet, 2014. NCS in numbers, maps and figures. [Online]
Available at: http://npd.no/en/
[Accessed 12 November 2014].

OPEC, 2014. OPEC Bulletin April, Vienna, Austria: OPEC Secretariat.

Platts, 2011. Statoil says Snohvit LNG. [Online]
Available at:
http://www.platts.com/RSSFeedDetailedNews/RSSFeed/Oil/8362448
[Accessed 7 November 2014].

The Engineer, 2012. Flare path: protecting infrastructure from space weather. [Online]
Available at: http://www.theengineer.co.uk/in-depth/the-big-story/flare-path-protecting-infrastructure-from-space-weather/1007598.article
[Accessed 29 November 2014].

The Institute of Risk Management, 2014. Risk management. [Online]
Available at: http://www.theirm.org/about/risk-management/
[Accessed 29 November 2014].

Tristan Crees, e. a., 2010. Preparing for UNCLOS – An Historic AUV Deployment in the Canadian High Arctic , Port Coquitlam, British Columbia , Canada: ISE .

U.S. Geological Survey, 2008. Circum-Arctic Resource Appraisal: Estimates of Undiscovered Oil and Gas North of the Arctic Circle. [Online]
Available at: http://pubs.usgs.gov/fs/2008/3049/
[Accessed 12 November 2014].

United Nations, June 1992. Rio Declaration on Environment and Development. [Online]
Available at:
http://www.unep.org/Documents.Multilingual/Default.asp?DocumentID=78&ArticleID=1163
[Accessed 10 Nov 2014].

Zhou, J., 2013. Determination of Safety/Environmental Integrity Level for Subsea Safety Instrumented Systems. s.l.:Norwegian University of Science and Technology.

The Ethics of Acceptable Safety

Ibrahim Habli[1], Tim Kelly[1], Kevin Macnish[2], Christopher Megone[2], Mark Nicholson[1], Andrew Rae[3]

[1] Department of Computer Science, University of York, UK
[2] Inter-Disciplinary Ethics Applied Centre, University of Leeds, UK
[3] Safety Science Innovation Lab, Griffith University, Australia

Abstract *Engineers of safety-critical systems have a duty to address ethical issues that may arise in the development, assessment, operation and maintenance of these systems. Dealing with ethical dilemmas during safety risk assessment is particularly challenging, especially when making and justifying decisions concerning risk acceptability. This is complicated by organisational issues and contractual limits that do not necessarily align with the boundaries of ethical responsibility. In this paper, we explore some of these dilemmas and discuss the duties of engineers to identify, analyse and respond effectively to ethical concerns about safety risk decisions. We illustrate these through short case studies that highlight particular issues relating to the ethics of safety advice, safety and cost tradeoffs, novel technologies and institutional support.*

1 Introduction

Any discussion of engineering ethics risks belabouring the obvious and ignoring the true challenges. No engineer (we hope) sets out to be evil. All have, at some stage of their career, agreed to follow a carefully worded code of conduct. Why do we, as professional educators of engineers, feel that it is necessary to write a paper about the ethics of safety?

The engineering of safety-critical systems is a constant process of ethical decision making. Some of these decisions are subtle, such as balancing cost, performance and reliability in the selection of a component, or choosing the right wording for a customer memo. Other decisions are explicit and life-changing, such as deciding whether to persevere in a dysfunctional organisation in the hope of eventually making a positive difference.

Throughout this paper we aim to provide some insight into what makes safety ethics difficult. We first set out principles on which ethical decision making may be founded by introducing the Royal Academy of Engineering's Statement of Eth-

© Habli et al. 2015. Published by the Safety-Critical Systems Club. All Rights Reserved ical Principles. We then show how these principles are stretched, challenged and sometimes directly threatened by contractual obligations, organisational capability, competing ethical concerns, and the uncertain nature of risk acceptability. We explore and illustrate these complications through a series of short case studies. Finally, we conclude with some practical suggestions for ways forward.

This paper does not provide simple answers. As we highlight in some of the examples, we are fellow-travellers on the safety journey, not beacons or signposts. Safety ethics is just as problematic when educating and advice-giving as it is in the design, maintenance and operation of safety-critical systems. We hope that by explicitly acknowledging the challenges and complications our work will help engineers reflect on their own practice and be more confident in recognising and resolving ethical problems.

2 Royal Academy of Engineering's Statement of Ethical Principles

An engineer applies scientific and technical knowledge to address problems within their domain of expertise. As members of a professional discipline, they are required to take account of ethical considerations in their work. What exactly does it mean to 'take account of' ethical considerations, and what can be done to enhance and develop engineering practice in this regard?

In the UK, various engineering institutions have developed slightly differing codes of conduct or codes of ethics. In 2005, the Royal Academy of Engineering, working in conjunction with the Engineering Council and a number of leading engineering institutions, developed a Statement of Ethical Principles (RAEng 2011a). This was partly to address issues of uniformity, but more importantly to raise the profile of ethical considerations amongst engineers. The aim was to have a succinct statement to which all practising engineers could subscribe. The Statement presented a set of four principles designed to cover all of the fundamental ethical considerations that are most prominent in an engineer's day to day work[1]. The principles are:

- Accuracy and Rigour;
- Honesty and Integrity;
- Respect for Life, Law and the Public Good; and,
- Responsible Leadership, Listening and Informing.

Engineers are a practical group of people. Faced with such brief and rather abstract principles, they are going to ask how exactly these should be used to guide

---

[1] As noted many professional engineering institutions also have codes of ethics, but this statement has been endorsed by both the Engineering Council and leading institutions, so constitutes a good starting point for those in all areas of safety engineering.

behaviour. In the short launch document for the Statement there is a brief elaboration of each of the principles with several short bullet points. Even this elaboration is still quite abstract, and so the question remains as to how exactly these general principles can help an engineer meet high ethical standards.

One of the factors that makes it difficult to apply the principles is that it will often be the case in real-life situations that the principles intersect. It will not simply be a case of recognising one principle and working out how to address its requirements in the particular circumstance faced, but of considering what to do given that more than one principle is in play. Furthermore the principles may sometimes conflict, making a decision even more difficult.

Ethicists will often talk of weighing or balancing different considerations in order to arrive at good judgements. For those in technical disciplines it is easy to assume that this will involve some kind of algorithm or a precise mathematical process. However, as Aristotle pointed out a long time ago, although there is truth in ethics as there is in mathematics, the kind of precision that is expected in mathematics is not appropriate in ethics. Thus, although it is tempting to seek a grid or a formula which reduces the complexity of ethical choice, the apparent precision achieved is gained at the cost of over-simplifying the decision process. An algorithmic approach to ethics leads to significant factors being ignored or not properly taken account of.

So, the intersection of ethical principles, including possible conflict between them, is one challenge to their ready use in decision making. Another is the fact that the principles are not self-interpreting rules. Reflection is required in order to work out what any given principle might mean in context. Some ethical principles may sometimes in practice require absolute prohibitions on certain types of behaviour, whilst on other occasions they may point to ideals of behaviour to which we might aspire but realistically must inevitably fall short of.

Consider the first principle which highlights the need for accuracy and rigour. How much accuracy and rigour is it reasonable to expect on any given occasion? Accuracy and rigour are 'limit' concepts; they admit of ideals to which our ordinary practice can approach, but never fully instantiate. It will always be open to question how much time, effort and cost should be expended on achieving ever greater accuracy. A sub-principle here advises that engineers "should identify and evaluate and, where possible, quantify risk" (RAEng 2011) but the identification of risks can be an endless task if one includes every single risk which an engineering intervention might introduce. We touch on this point further below.

A final issue to mention here is that the Royal Academy's ethical principles must be applied in context. This means they will need to be acted on at times when the practitioner will be facing commercial considerations of cost and contract, personal issues such as their own needs and the demands of family and friends, and matters of organisational culture. In applying the principles all these factors complicate matters. We take these ideas further in the next section, but one thing to note immediately is that, given the importance of organisational culture, effective implementation of the principles requires that the organisations in which engineers work find suitable ways to embed these principles within that culture.

Despite these difficulties, the Statement of Ethical Principles provides a solid starting point for thinking about the ethics of safety. The genuine ethical concern that we feel when the principles cannot be simultaneously met, or when their ideals cannot be instantiated, indicates that they speak to what it means to practise engineering ethically. Recognising the principles, and identifying when they are challenged by situations and actions is the first step towards ethical behaviour.

3 Day to Day Application of Principles for Safety Management

Having introduced both the four Ethical Principles and some of the challenges that accompany them, in this section we expand upon these challenges by applying these principles to safety management within real world organisations. These challenges include recognising the fact that one is presented with an ethical decision; deciding what is reasonable; dealing with uncertainty; reconciling competing duties; making money; everyday considerations; and choosing the greater good.

Recognising the Ethical Nature of Decisions

Engineers are seldom presented with a stark choice between ethical and unethical options. Problems involving pulling a lever to kill one person instead of five people appear in philosophy tracts, not requirements documents (Thomson 1976). Often even the fact that a choice is being made is not evident. When an engineer records the appropriate Safety Integrity Level (SIL) or Design Assurance Level (DAL) for a component, this may appear to be a deterministic calculation. In fact since SILs and DALs determine assurance effort, this is a cost/safety tradeoff. As the engineer makes assumptions and chooses parameters under uncertainty, they are exercising the "Respect for Life, Law and the Public Good" principal. As they decide whether it is worthwhile pursuing more information before selecting these values, they are exercising the "Accuracy and Rigour" principle.

Framing engineering or management decisions as ethical choices is often a rhetorical device used to justify a more conservative or expensive option. At other times the decisions are seen as "simply" engineering judgements or management choices, with no explicit ethical content. For example, setting a discount rate or internal rate of return (IRR) is a company strategic decision for making project and investment choices. The IRR can have a significant impact on which mitigations appear feasible or impracticable in a trade-off calculation, as it changes the present value of future costs. Applying the IRR to safety improvements can also result in inter-generational inequity, by valuing future victims less than persons currently alive.

Choosing suppliers is another decision fraught with safety implications. System safety may depend on the reliability of components where there is no direct visibility of their manufacture or quality assurance processes. Selecting a cheap supplier may be a direct cost/safety tradeoff, although it is usually expressed instead as a straight "value-for-money" proposition.

Deciding what is Reasonable

As we foreshadowed when we introduced the four principles, no ethical position is absolute. Yes, public safety is an important value, but that doesn't mean that all enterprise presenting any risk should be abandoned. Instead we look to make reasonableness trade-offs. A product is considered "safe" if the resultant risk is considered acceptable (Lowrance 1976). In the UK "As Low as Reasonably Practicable" (ALARP) recognises that public's right to safety is limited by the mitigations that businesses can be reasonably expected to put in place (HSE 2001). Arguably, there is an ethical *obligation* to treat risks proportionately, to avoid over-emphasis on some hazards at the expense of others presenting greater risk.

Words such as "tolerable", "acceptable", "reasonable" and "practicable" are common currency in safety discussions. Ethical behaviour is not about stepping over a line, but drawing the line appropriately (Dekker 2009). No one would suggest presenting the public with intolerable risk, but we do make frequent decisions about which risks are tolerable and which intolerable.

Reasonableness does not just apply to direct safety, but to related issues such as risk communication. An obligation to share information about risk normally does not extend to revealing commercial secrets. What if a company has invented a safer way of performing an activity? Is it right to treat that as a commercial advantage? What if they plan to introduce a new technology with commercial advantage but new safety challenges? Transparency in safety analysis allows more thorough review and better stakeholder participation, but also gives competitors insight into near-term market strategy.

Dealing with Uncertainty

Related to reasonableness is the problem of limited knowledge. This may be scientific uncertainty related to new technology, or assurance deficits which could, if further resources were allocated, be reduced. The problem is that there will always be assurance deficits, i.e. knowledge gaps that prohibit complete understanding and perfect confidence (Hawkins et al. 2011]. No testing regime proves the absence of bugs, and no safety analysis demonstrates that all hazards have been identified and adequately treated. Organisational paralysis in the absence of perfect information would not be reasonable, but some level of doubt should lead to a pause and rethink of current operations. Indeed, Turner (Turner 1976) describes how many disasters arise not from willfully ignoring problems, but from a mistaken focus on the wrong areas of uncertainty.

Often it is not doubt itself, but the persistence of doubt which causes an ethical problem. When the first cases of blow-by occurred on the space shuttle solid rocket boosters, it may indeed have been appropriate to continue flying the shuttle whilst the problem was assessed and treated. By the time of the final Challenger launch, the *continued* uncertainty was a clear ethical failing. The lack of evidence that the O-rings would fail was a poor argument, since that evidence had not been properly sought. However, an engineer suggesting that "enough-is-enough" will almost always be challenged as to why they have not been more forceful earlier.

Their very willingness to be pragmatic and reasonable becomes ammunition against them when they finally take a stand.

Reconciling Competing Duties

We have already indicated that the four principles may intersect and conflict. Fulfilling one duty can mean compromising another. Even where the principles align, they may clash with other obligations which the engineer holds as a company employee and member of the wider community. For example, engineers not only owe a duty to support the interests of their employer, but any safety influence they have arises from meeting this duty. If a safety practitioner is not trusted to act in their employer's interest, and to act reasonably in balancing that interest with other concerns, they will not be able to meet their duty to the public safety.

A common escape from this apparent contradiction is to suggest that "safety is good business", but this is simply not true. Major accidents are fortunately rare, and in the short term even an organisation which neglects its own safety responsibility is still unlikely to have an accident. Organisations frequently gain short and medium term competitive advantage by ignoring rare, high consequence events (Taleb 2007).

Making Money

Safety management is an inter-organisational endeavour, shaped by contracts and informal business relationships. From a business perspective, the party responsible for performing safety work is the party being paid to perform the work. This viewpoint can be at odds with a "purely ethical" or legal determination of duties.

The Nimrod XV230 accident and subsequent Haddon-Cave Inquiry shows how ethical duties and contractual responsibilities can intersect in complex and counter-productive ways (Haddon-Cave 2009). The amount of work performed by the contractor was limited not by the amount of work *needed*, but by the size of the contract. The effectiveness of the independent assessor was limited by the ambiguous nature of their contractual responsibility, and by perceptions of their business motivations in raising safety concerns.

The business nature of safety presents obstacles to clear risk communication. Interaction between the parties is formalised, and practitioners often discharge their responsibility to accurately communicate risk through multiple layers of management and contracts. For example, it is clearly unethical to keep secret a potential hazard from those who will be exposed to it. An engineer who identifies the hazard must often rely on others to pass on their warnings. If management dilutes the message, or if the customer fails to respond appropriately, the engineer may not even know. If they *suspect* miscommunication, an attempt to investigate or to reinforce the message may breach duties of confidentiality and loyalty, particularly if the suspicion turns out to be unwarranted.

Ethics in Everyday Conduct

Ethical considerations can be present in some of the smallest everyday actions and decisions taken by engineers as they carry out their work. Firstly, there are many issues around human communication and interaction in the conduct of safety engineering activity. For example, subtle choices can be made regarding the language used in statements in a safety case report that may dissuade, rather than encourage, challenge. Safety-related decisions can be heavily influenced by the personality types, and the interpersonal dynamics, of those involved on a project, as highlighted in (Haddon-Cave 2009). For example, the manner in which safety meetings are chaired can have a major influence on how safety issues are explored and sentenced. There are also practical considerations regarding the context and timing of safety engineering activities. For example, is it ethical for an engineer to review a safety report in the evening of an already busy day when they are tired? Should a safety manager schedule a safety review meeting when a known 'difficult' colleague is unavailable? Likewise, is it ethical to ask the lecturer on a safety course to provide endorsement of a complex safety decision within the space of a fifteen minute coffee break between lectures (a situation regularly faced by a number of the authors)? Equally, is it ethical for the lecturer to respond to such questions? Everyday choices such as these often have an ethical dimension and can have a significant effect on safety outcomes.

Choosing the Greater Good
The final challenge emerges from the nature of ethics itself. Different ethical systems can suggest different behaviour in the same situation. A set of ethical principles might suggest a deontological system - an engineer is ethical if they abide by the principles, trusting in the universal goodness of those principles[1]. However, what if the engineer could achieve a better outcome by breaking the principles? A consequentialist approach would suggest that the engineer should be concerned with the overall public good ahead of blind allegiance to principles[2]. The clash of these two approaches is not just a philosophical debate, but a grim reality for senior safety managers. "Choosing one's battles wisely" is the mantra of a consequentialist approach. How does a safety manager judge whether keeping their job is truly serving the greater good rather than self-interest? How does a safety engineer leaving their job reassure themselves that they really did all they could before forcing a final conflict?

4 Case Studies

In this section we discuss a number of case studies to bring the four principles and the challenges surrounding their day-to-day application into context. Producing a definitive set of answers to such scenarios is not feasible here. Instead, each scenario contains a short discussion highlighting the ethical concerns and relating these to the four principles.

---

[1] Although rule-consequentialists adhere to rules.

[2] But again it is not just consequentialists who give weight to consequences.

4.1 Case Study 1: Kudochem

The "Kudochem" study is adapted from the Royal Academy of Engineering's Engineering Ethics in Practice, specifically to illustrate the application of the Academy's third ethical principle 'Respect for Life, Law and Public Good' (RAEng 2011b). In this case the concept of respect for life involves "holding paramount the health and safety of others". The version of the case study below is an edited and reduced version to serve our present purposes.

Scenario

Kudochem is a multinational company producing chemicals for the agricultural industry. Responsibility for engineering issues at the eleven Kudochem plants in Europe lies with Kudochem's European Regional Engineering Director, Sally Proctor.

In the early hours of one morning, Sally receives a telephone call informing her that there has been a serious explosion at one of the plants. There have been some injuries, and damage has been done to property several hundred metres from the plant, but there have been no fatalities. The scale of the damage is huge, and the main site of the chemical plant is almost completely destroyed. In accordance with company policy an inquiry team is set up, involving company employees as well as independent consultants.

After several weeks, the team discovers two possible causes, both relating to a new ammonia production technique for fertiliser. This technique has recently been introduced in all of Kudochem's plants. The team is unable to determine which of the two possible causes are responsible. Given the presence of the production technique in all of Kudochem's plants, it is imperative that the ultimate cause of the explosion is identified, so that urgent steps can be taken to safeguard against similar accidents at other sites.

The inquiry team is very concerned at their inability to determine the precise cause of the accident. Without this knowledge, it will be impossible to satisfactorily modify the plants in order to prevent future explosions of this kind. They make a radical recommendation: to call a meeting with several competitor companies who are also using the new procedure in their fertilizer plants, in order to share experiences and research findings.

This would be a significant departure from standard practice, and some senior colleagues with commercial responsibilities have reservations. To call the meeting would entail releasing information about the safety lapse, as well as discussing sensitive commercial information with business rivals.

However, it may be the case that other engineers in other companies have encountered problems with the new method for producing ammonia, and could offer help in isolating the problem. Whilst such a course of action may be unusual in this case there are industries where safety critical information is routinely shared amongst competitors.

Discussion

In this scenario, the situation could be seen as one in which there is a conflict of interests and duties, such that Sally is required to balance these conflicting concerns. On the one hand she needs to ensure the safety of employees and local residents, and on the other hand she needs to maintain the security of commercially sensitive material. In addition, she needs to balance the risks with the financial costs of possible remedies, and she needs to judge what is appropriate in an abnormal situation.

The Statement of Ethical Principles states that an engineer must "hold paramount the health and safety of others." At the same time, though, she needs to take into consideration any other obligations she may have – including the duty to keep sensitive material secure, and to protect people's jobs by protecting the commercial interests of the company. Of course, if a company is acting illegally or irresponsibly, there may be a duty to 'blow the whistle', and this may defeat any obligation to keep sensitive information secret. However, in this case, there is no indication that the company was acting irresponsibly. As such, Sally could reasonably consider the commercial risks of sharing information with her competitors to be too significant.

Even if this was not her first response, she could be persuaded by commercial managers of the company that this is true. However, it is not clear that these considerations can outweigh the safety concerns. The principle states that she should *hold paramount* the health and safety of others. The same procedures are being used in all of Kudochem's plants and, given that the cause hasn't been identified, she needs to take seriously the possibility that there could be another explosion.

In summary, there does seem to be good reason to share safety information. Of course, where possible this should be done in a way that gives appropriate weight to one's other duties, regarding sensitive information, for example. Ultimately, however, it should be recognised that holding health and safety paramount doesn't just mean ensuring that you are not directly responsible for harms to the public, but that you also have some responsibility to help others improve their health and safety, for example by warning them of dangers they may not be aware of.

4.2 Case Study 2: Cargo Bay Doors

The Turkish Airlines Flight 981 crash was a real event. Knowing the future outcome of the decisions made in this case study gives the ethical considerations a prominence they would not necessarily have had at the time the decisions were made. This case study touches on all four principles, but particular emphasis can be placed on Principle 2 (honesty and integrity) and Principle 4 (responsible leadership). The main practical issues relate to the intersection of the principles with other duties.

Situation

Convair was responsible for the design and construction of the fuselage of the McDonnell Douglas DC-10. As a subcontractor to McDonnell Douglas, they undertook detailed design work, but requirements and major design choices were determined by McDonnell Douglas.

Most doors on an aircraft are of a "plug" design. They open inwards, and are held in place in flight by the pressure difference between the inside and outside of the aircraft. However, the cargo bay door on the DC-10 opened outwards; the pressure difference in flight pushed the door open, so it was important to have a reliable locking mechanism.

Convair conducted a hazard analysis of the door, which postulated several scenarios where failure of the door could lead to loss of the aircraft. They also had good reasons to question the reliability of the locking mechanism as failures had occurred in both ground flight and trials.

Convair were limited in their ability to directly control the safety risk of an open cargo bay door for several reasons:

They supplied safety analysis to McDonnell Douglas and were contractually prohibited from speaking directly to the regulator. Safety issues that Convair raised to McDonnell Douglas were not always included in documentation that McDonnell Douglas passed on to the regulator, including the scenarios involving cargo bay doors opening in flight.

The Federal Aviation Administration (FAA) were aware of the in-flight incident, and elected not to issue an Airworthiness Directive. They negotiated with McDonnell Douglas to issue a less enforceable Service Bulletin requiring minor changes to the door design.

It was unclear who would bear the cost of changes to the door design, particularly if those changes were made at the request of Convair rather than their customer, McDonnell Douglas.

The Director of Product Engineering at Convair issued a memo to his immediate supervisor. In unambiguous terms he challenged the safety of the cargo bay door, and the adequacy of the changes made in response to the incidents. He predicted that at least one aircraft would be lost in-flight during the life of the DC-10. This prediction was fulfilled when Turkish Airlines Flight 981 crashed in France.

Discussion

The supervisor of the Director of Product Engineering at Convair had to weigh up the parties to whom he owed ethical responsibility. He was a subcontractor and under the regulation of the FAA he had to judge how these responsibilities were altered by the fact that his organisation had already made strenuous representations to McDonnell Douglas, and that the FAA had rules on the type and level of action in response to an earlier incident.

If the Convair director was not confident that his concerns had been clearly expressed to McDonnell Douglas, he would need to judge what further action, if any, he should take. He may have had uncertainty about his own interpretation of what had already happened. Even if he was confident that his concerns had been passed on, he still had ethical issues to address depending on the feedback, or lack of feedback, he received.

Finally, this scenario also highlights the impact of contractual and financial arrangements on ethical decisions. Is the ethical concern about reasonableness affected by such arrangements?

4.3 Case Study 3: Independent Safety Advice

This case study concerns the ethical issues that a safety assessor may have to address when providing independent safety advice. It introduces issues relating to principles 2 (honesty and integrity) and 3 (respect for life, law and the public good). It also addresses issues of the importance of decisions and reasonableness.

Scenario

Swift, a Four Wheel Drive (4WD) passenger car manufacturer, has been approached by a large health organisation, Save, that is interested in expanding their rapid response vehicle fleet. Rapid response typically involves a multidisciplinary team attending to a patient often in very inhospitable environments. Quick delivery time of the fleet is crucial.

Save request that Swift involve an Independent Safety Assessor (ISA) in the safety approval process of the new fleet. Swift recruits Trust Limited, a safety assessment consultancy with a long established relationship with Swift. Trust has recently highlighted major safety concerns about the Electronic Stability Control systems of some Swift models. These concerns proved to be incorrect, resulting in expensive and unnecessary tests. Afterwards, Trust managers went to extraordinary lengths to keep Swift happy, leading to occasions where Trust ISAs were prepared to give undue credit during safety audit sessions to assurances made by Swift engineers.

Swift senior managers have a warm feeling that they can meet the safety requirements for the new fleet without making any significant changes to the existing vehicle models. A meeting was organised at the Swift Head Office and involved: Swift (senior business manager and chief engineer), Trust: (experienced ISA), Power (engine supplier powertrain specialist) and Save (director of regional operations).

At the start of the meeting, the senior business manager from Swift praised the safety track record of their vehicle product lines and supported his claims by the excellent feedback they received from their customers. The chief engineer highlighted that Swift complied with best practice and were annually audited by Trust. The powertrain specialist from Power agreed with Swift about their existing safety record but noted his concerns about the impact that the changes in the operational profile might have on the reliability of the engines.

The senior business manager from Swift dismissed these concerns, questioning the motivation behind them (i.e. "Power test people as usual are touting for business"). He asserted that it was always possible to carry out further tests but that this might exceed the budget allocated by Save, asking the powertrain specialist to be reasonable when making any such judgments. The chief engineer from Swift added that any claims about failures to meet the targets set by Save could only be made by the Swift engineers, who were ultimately the designers of the vehicle, and the engine was one of many vehicle components. The director of regional operations from Save explained that he was not an expert in vehicle safety and turned to the ISA for advice on the best course of action. How should the ISA respond?

Discussion

The Trust ISA is obliged to provide frank and honest advice within their domain of expertise. They are also ethically obliged to indicate if the advice that is being sought relates to knowledge or skills outside their domain of expertise. This is complicated by the fact that what appears to be an engineering judgement has an obvious direct impact on the negotiations between Swift, Power, and Save. The ISA is being employed by Swift, and has an implied duty to support them at the meeting. There are also a long term business relationships to consider.

This situation does not directly call into play principle 1 (accuracy and rigour) because the ISA is being asked to provide advice rather than information. Ethically, the ISA could act with honesty and integrity (principle 2) by declining to give an opinion, but would this show respect for life, law, and the public good (principle 3)? The Trust ISA has an opportunity to influence the safety of the vehicle, but their role is complicated by the contractual limits and financial perspectives that do not necessarily align with the boundaries of ethical responsibility.

If there are clear safety concerns, the ISA has a duty to highlight them. However, the ISA is not directly involved in the safety analysis work and as such their safety and domain knowledge is limited. Furthermore, because of the changed environmental and operational characteristics those involved cannot point to direct evidence, or counter-evidence, as to the safety characteristics the power unit and the vehicle will exhibit in operation. This is also complicated by the conflicting judgments made by the different stakeholders that are directly involved in the safety work. There is a trade off between the warm feeling of the designers and the cost of buying more information about the situation. This cost will be both financial and time. It is not clear who should bear the cost of such an information gathering exercise.

Uncertainty means that false alarms are bound to occur when addressing safety issues. The past consequences of such alarms will potentially have an impact on the confidence of the ISA to raise issues. It will also have an impact on the receiving organization trust in the advice given.

Finally, all parties should be conscious of behaviours that can compromise the integrity and independence of the ISA. For example, ambiguous communication and refutation of safety concerns, mixed with talk about business motivation, will limit the effectiveness of the ISA. Prior agreement about the role of the ISA in this meeting could have made the ethical situation clearer for everyone.

Case Study 4: Safety and Novel Technologies

This case study concerns common ethical issues involved in the deployment of a novel technology. In particular, it focuses on principles 1 (accuracy and rigour) and 4 (responsible leadership), although there are aspects of the other principles.

Scenario
Robots R Us (RRU) specialises in developing safety mechanisms for automated cars. A notable success has been the development of a laser tracking system that is able to reliably model a 360 degree environment of the car with a visible radius of up to 150m. This will enable the vehicle to recognize actual and potential hazards as fast as a human driver, or possibly faster. In most cases the vehicle will either be able to avoid the hazard or to alert the driver in such a way that he or she will be able to take control in time to prevent an accident. The working name for the system is HATTAR (Hazard Avoidance Through Technologically Advanced Recognition).

RRU has a good relationship with a car manufacturer, Ovlo, with whom they have worked closely in developing HATTAR. While there is no contractual obligation to offer HATTAR to Ovlo, many at the company think that this would be 'the right thing to do'. Ovlo have also said that they would like to be kept up to date with developments on the HATTAR programme. However, there is also a concern that if the technology is sold to Ovlo, they will use it to gain a competitive advantage. Many feel that *all* automated cars should be as safe as possible and that no one company should benefit from a development that could ultimately save lives. Ovlo are working on their own safety system, which is not as effective as HATTAR. It is possible that they may take their own system to market first to recoup costs and then release HATTAR in five years' time in a new generation of vehicles. Every day in which this technology is not deployed people will die in accidents on the roads that could have been avoided by HATTAR.

A second concern is that the HATTAR technology has not been trialled on a large scale. There are foreseeable problems of HATTAR systems interfering with one another, leading to false positives in the recognition system. However, it is not realistically possible to carry out a trial on the scale necessary to fully test the technology. Ultimately, any hazards of HATTAR being used by a large number of cars simultaneously can only be determined through the actual use of the system on the roads. It will take a number of years until the HATTAR system is being used across the country in a manner such that its ultimate efficacy can be understood.

Discussion
This situation asks RRU to balance public good and private profit under high uncertainty. RRU could sell HATTAR to Ovlo and thereby pass the responsibility of its implementation and any inherent risks to the car manufacturer. This would

likely mean restricting a powerful safety mechanism to one manufacturer. The HATTAR technology might remain in the hands of Ovlo and off the road for five years. On the positive side, the five years could be employed by Ovlo in further testing. Alternatively, RRU could refuse to sell the technology but continue to test it. With this option, RRU take the responsibility themselves for the resilience of the system. However, this will mean a long period, possibly several years, in which the system is not being used to improve road safety and a similar timeframe in which RRU will not benefit financially from this technology.

However, RRU could demonstrate industry leadership by making available the development to the entire car manufacturing industry. This would put the technology 'out there' to anyone who wanted to develop and test it further. This would have the greatest likely impact in terms of making automated vehicles safer. RRU would not stand to benefit financially from the revelation, though. To address this, RRA could patent and sell the technology to all manufacturers willing to pay for it. This has the advantage of making HATTAR widely available and benefitting RRU financially. However, RRU would lose control of testing the technology and the relationship with Olvo is likely to sour through RRU not having offered the HATTAR system to them first.

Whilst the principles offer little guidance to RRU in this situation, posing it as an ethical problem rather than a purely business decision allows RRU to properly evaluate their options with the public good as an explicit consideration.

5 Ways Forward and Practical Suggestions

In Section 2 we noted that the Royal Academy of Engineering has developed a set of ethical principles upon which engineers of safety critical systems can draw in order to develop good (ethical) judgement in their decision making. We then set out some of the challenges to implementing those principles in practice. Amongst other things, the principles intersect; they must be applied in complex circumstances with a range of other pressures bearing down; and they require a high level of interpretation. As shown by the case studies in Section 4, there is no algorithm for ethical decision making.

So how much use are the principles? Looking at matters more positively, we note that a very significant challenge for all of us is identifying an ethical problem *as* an ethical problem. The ability to perform that identification is a complex skill. Successful execution of the identification task depends on how we conceive or frame a situation. We can look at the same factual situation in very different ways.

Consider, by way of analogy, three people are on the same walk through the countryside. The first person might conceive of the situation as one in which they are going bird-watching. If so, they will tend to notice the birds in their visual field. The second might see the same walk as field research into wild-flower conservation, and so notice the flowers. The third might conceive of it simply as physical exercise, and be oblivious to both birds and flowers. What the agent sees is affected by how they frame the situation. The Royal Academy's ethical principles provide guidance on framing engineering challenges so that the framing includes the relevant ethical considerations, not simply the technical issues.

How does one develop the skill of framing situations with the concepts set out in these principles? Simple awareness of the principles is a first step, but they also need to be made meaningful; as we noted above they are not self-interpreting. This is to some extent a matter of practice and experience – so working through the case studies above and others like them is a helpful technique. This quasi-experiential approach is related to the Aristotelian notion of habituation (Megone and Robinson 2001). Organisational culture can also help by finding ways to draw attention to the principles for staff.

Here are two related suggestions. One is to introduce an ethics template into the project planning process for all projects undertaken by an organisation[1]. The template is designed to help highlight the kinds of factors that engineers need to consider so that attention to ethical factors is built into the process of deliberation about project design.

Another process involves building up a database of case studies in which the Academy's key ethical principles are applied, so that in future deliberation the organisation can draw on past decisions. This allows them to consider how principles were weighed in different circumstances and to reflect on what that suggests for new situations. This process could start with a period of consideration on ways in which the principles are likely to apply to a particular organisation's safety critical engineering decisions. This may involve drawing up a provisional list of what seem to be the most significant concrete applications for that organisation. Then, on every subsequent major decision, it will be helpful to record how the principles were applied, as an audit trail of the approach to ethical decision-making. A third stage will then be to set a small amount of time aside, perhaps on an annual basis and possibly with an independent reviewer, to consider the application of the principles in the previous year, with a view to considering any particularly difficult decisions which might suggest that the organisation's guiding principles need refinement or revision.

This is a sensible practical process for embedding ethical principles in an organisation's practice, but it needs to be carefully organised (and perhaps involve some external review) in order to avoid the danger that it may just become a 'tick-box' exercise.

6 Conclusion

Every professional engineer is bound by a code of ethical standards. However, such codes do not provide a blueprint for all subsequent ethical action. In the engineering of safety-critical systems, there are a number of ethical challenges which

---

[1] Ethics Template, 2013, IDEA CETL, University of Leeds.

arise in day-to-day work. These challenges include pragmatic considerations such as recognising when one is faced with the application of an ethical principle, and determining what is reasonable in communicating information about hazards. They extend to the philosophical, such as deciding what is ethical when the principles conflict, or even deciding whether a principles-based approach is applicable.

In this paper we have raised these issues and illustrated some of them through the use of four case studies. We argue that such reflection is itself a practical way forward for ethical practice of engineering. The principles can help engineers and engineering organisations recognise the ethical dimension of decisions, and by doing so, be more confident that they are resolving problems ethically.

We encourage engineers to reflect on their own practice, and organisations to provide safe spaces for discussion of ethical problems.

References

Dekker S (2009) Just Culture: Who Gets to Draw the Line?. Cognition, Technology & Work 11, no. 3

Haddon-Cave C (2009) The Nimrod review. The Stationary Office. London

Hawkins R, Kelly T, Knight J, Graydon P (2011) A New Approach to Creating Clear Safety Arguments. SSS '11, Southampton, 2011

HSE (2001) Reducing Risks, Protecting People, Health and Safety Executive Books

Lowrance W (1976) Of Acceptable Risk: Science and the Determination of Safety. Los Altos, William Kaufmann

Megone C, Robinson S (2001) Case Histories in Business Ethics. Routledge, London

RAEng (2011a) Statement of Ethical Principles. Royal Academy of Engineering

RAEng (2011b), Engineering Ethics in Practice: A Guide for Engineers. Royal Academy of Engineering

Taleb N (2007) The Black Swan: The Impact of the Highly Improbable. Random House

Thomson J (1976) Killing, Letting Die, and the Trolley Problem. 59 The Monist 204-17

Turner B (1976) The Organizational and Interorganizational Development of Disasters. Administrative Science Quarterly 21, no. 3

Combining Organisational and Safety Culture Models

Elizabeth Jacob[1]

Atkins

Bristol, U.K.

Abstract *A good safety culture is widely characterised as communications founded on mutual trust, shared perceptions about the importance of safety, and confidence in harm prevention and protection measures. It provides any organisation both safety and financial benefits. Safety Culture first appears in literature in the mid-1960s but did not take hold as concept until the mid-1980s and different models have been developed to assess the Safety Culture of an organisation over the past 30 years. Many of these safety culture models measure a snap shot of characteristics but do not address the aspects of organisational change. Organisational models have been developed since the early 1960s, normally with the aim to model and understand the organisation's performance and potential effects of change. I have conducted research on combining both organisational models and safety culture models over the past few years, leading to a list of factors/characteristics that could aid development of a combined Organisational and Safety Culture Model. In turn, this improved understanding could help organisations predict the effect of enforcing a particular change and detecting any weaknesses in the organisation structure. In addition, a combined model would aid risk management techniques such as Cost Benefit Analysis and in general, the overall process of measuring the health and safety performance of the organisation.*

1 Introduction

"It could be argued that the most dangerous element in any process plant is not the plant, or the chemicals in it, but the people operating it."

This statement by Flavell-While in the August 2012 issue of The Chemical Engineer sums up the importance of Human Factors in industries. In the world of

---

[1] The Hub, 500 Park Avenue, Aztec West, Almondsbury, Bristol, U.K., BS32 4RZ

© Elizabeth Jacob 2015. Published by the Safety-Critical Systems Club. All Rights Reserved

Human Factors, Safety Culture plays a key role in determining the level of risk that personnel are willing to take in order to complete the task at hand, despite time constraints.

Every organisation experiences some form of change during its lifetime. It can be difficult to control "the number of affected factors changing at the same time" (Burke and Litwin, 1992) such as the magnitude and range of changes to organisation structure, management practises, policy and procedure, work climate etc. The definition of a factor adhered to in this paper relates to a feature that contributes to positive Safety Culture. However, organisational events and literature research show that certain relationships amongst factors have been found. These 'relationships' and other factors can be combined together to form organisational models. These models can then help investigate how the proposed change could impact the behaviour of individuals, groups and structures within the organisation.

Different tools have been created to try and assess the Safety Culture of an organisation and different organisation models have to been developed to aid organisational change. Here lies the objective of this study, which is to conduct a logical and qualitative comparison between Organisational Models and Safety Culture Models within different industries (e.g.: defence, oil & gas, railway, healthcare etc.) to determine if a combined tool already exists or if one could be developed and used across a range of industries.

2 Organisation Models

Organisational studies relate to "the examination of how individuals construct organisational structures, processes and practises and how these shape social interactions within the organisation and create institutions that ultimately influence people" (Bailey and Clegg, 2008). In the last few years, a wide variety of specialists have remodelled the hierarchical approach with the addition of factors such as strategy, mission, leadership in order to create a model for organisation structure that represents the most effective 'routes' to communicate and hence establish good safety culture. A selection of these organisational models will be discussed in this section.

2.1 The Chosen Models

A number of organisational models were identified and reviewed but only three have been shortlisted because of their three distinct features which will be described later in this section. The three models are:

1. The Burke-Litwin Model (Burke and Litwin, 1992)
2. Kotter 8-Step Change Model (Kotter, June 2012)
3. Mintzberg Model on Organisational Structures (Mintzbergs, June 2012)

Other identified models such as the 7S Model (ESS Consulting, July 2014), Congruence Model (Tushman and Nadler, July 2014) and The Six-Box Model (Weisbord, July 2014) were not down-selected because their features were already exhibited and further developed in the chosen models.

These models were reviewed with various safety professionals to understand the potential value and issues of using an organisational model to establish a safety culture baseline and then influence improvement.

2.1.1 The Burke-Litwin Model

The original thinking underlying the model came from George Litwin and others during the 1960s but has been refined through a series of studies directed by Burke and his colleagues. Recent collaboration has led to the current form of this model shown in figure 1. These type of models are usually developed with academic research as a foundation. However, W. Warner Burke and George H. Litwin wanted to link what they understood from practise to what was known from research and theory. The arrows shown in the diagram represent the two-way interaction between the factors.

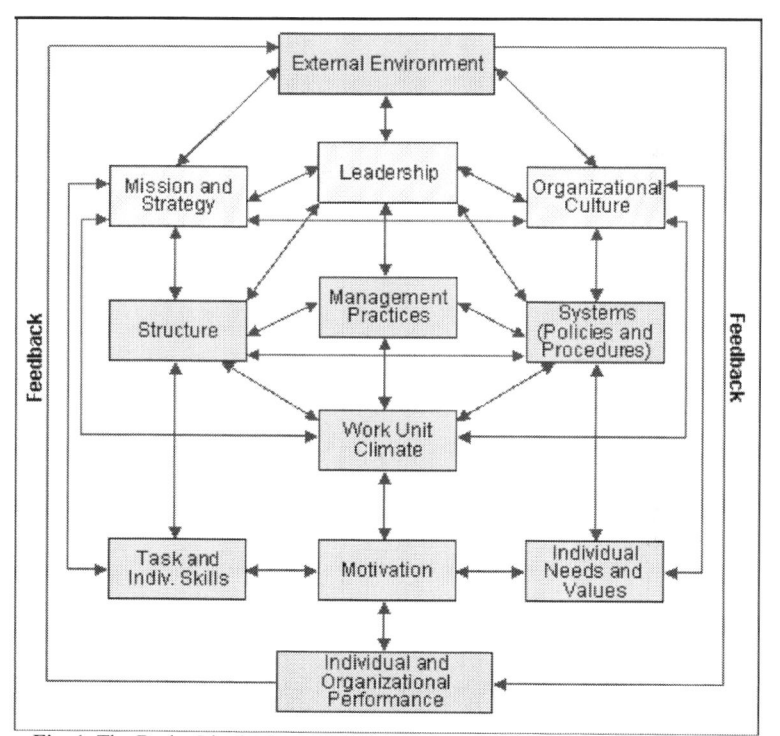

Fig. 1. The Burke-Litwin Causal Model of Organisational Performance and Change (Burke and Litwin, 1992)

The model was constructed to serve as a guide for both organisational performance and planned organisational change. Two types of factors are incorporated within the model, which are Transformational and Transactional factors. These two types of factors collectively include the twelve factors shown in figure 1

The Transformational factors include the External Environment, Mission and Strategy, Leadership, Organisational Culture and Individual and Organisational Performance (i.e. the grey and yellow boxes in figure 1). Transformational "means areas in which alteration is likely caused by interaction with environmental forces and will require entirely new behaviour sets from organisational members" (Burke and Litwin, 1992). The organisation leaders directly initiate a majority of the organisational change. But these leaders can experience influence and direct forces from the organisation's external environment as well.

The Transactional factors include Structure, Management Practices, Systems (Policies and Procedures), Work Unit Climate, Task Requirements and Individual Skills, Motivation, Individual Needs and Values and Individual and Organisational Performance (i.e. the green and lowest grey box in figure 1). Transactional means primary exchange of benefits, in other words, "you do this for me and I'll do that for you" (Burke and Litwin, 1992). This explains the relationship between these factors.

"The transformational factors have more 'weight' than the transactional factors" (Burke and Litwin, 1992) because transformational change, for example, change in leadership, affects the total system whereas transactional change, such as change in structure, may or may not affect the total system. The ranking order by which the greatest impact of change can be influenced by, is the external environment followed by the transformational factors and then the transactional factors.

2.1.2 Kotter 8-Step Change Model

Recognising that a change needs to be made in an organisation and knowing how to practically make that change are two different concepts. The Kotter 8-Step Change Model proposes a process of how a change can be integrated smoothly within an organisation.

There are many theories about how to 'do' change. Many originate with leadership and change management guru, John Kotter. A professor at Harvard Business School and world-renowned change expert, Kotter introduced his eight-step change process in his 1995 book, "Leading Change". Kotter (Kotter, June 2012) eight-step process is shown in figure 2.

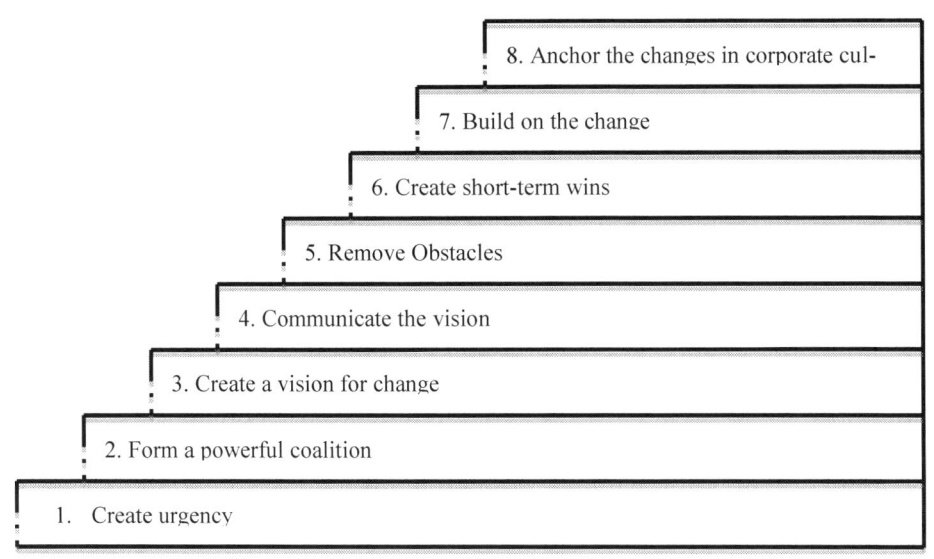

```
                                        8. Anchor the changes in corporate cul-
                                7. Build on the change
                        6. Create short-term wins
                  5. Remove Obstacles
            4. Communicate the vision
      3. Create a vision for change
   2. Form a powerful coalition
1. Create urgency
```

Fig.2. Kotter 8-Step Change Model

2.1.3 Mintzberg Model on Organisational Structures

According to management theorist Henry Mintzberg (Mintzberg, June 2012), an "organisation's structure is formed of five parts: Strategic Apex, Middle Line, Operating Core, Support Staff and Techno Structure". These five parts are shown in figure 3and further explained in Table 1.When these fit together, they combine to create organisations that can perform well. When they do not fit, the organisation is likely to experience severe problems.

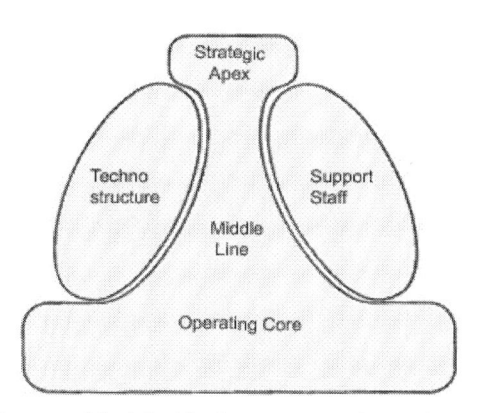

Fig.3. Mintzberg model of the ideal organisation (Mintzberg, June 2012)

Table 1. Explanation of the five parts within the structure of the Mintzberg Model (Mintzberg, June 2012)

Operating core	Those who perform the basic work related directly to the production of products and services
Strategic apex	Charged with ensuring that the organisation serve its mission in an effective way, and also that it serve the needs of those people who control or otherwise have power over the organisation
Middle-line managers	Form a chain joining the strategic apex to the operating core by the use of delegated formal authority
Technostructure	The analysts who serve the organisation by affecting the work of others. They may design it, plan it, change it, or train the people who do it, but they do not do it themselves
Support staff	Composed of specialised units that exist to provide support to the organisation outside the operating work flow

Each of these five parts has a tendency to pull the organisation in a particular direction favourable to them as shown in figure 4.

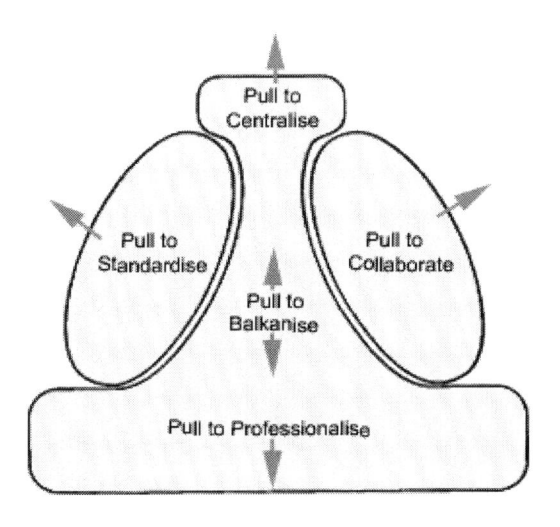

Fig. 4. The effect of various pressures on the model (Mintzberg, June 2012)

2.1.4 Conclusion

The Burke-Litwin Model, Kotter 8-Step Change Model and the Mintzberg Model on Organisational Structures were shortlisted for the safety professional reviews because they each had something different to offer. The Burke-Litwin Model is a causal model that includes important factors needed to be considered within an organisation. The recognition of the importance of the external environment as a factor is a positive quality. The model is a causal model, which means that a change in one or more of the factors can impact the others differently. The Kotter 8-Step change model is more of a process that can be used to slowly integrate the proposed change within the organisation instead of forcing the change. A unique feature of the model is that it does not support a hierarchy approach, but good leadership and a strong coalition is needed for this model to be put into practise. The Minztberg model focuses on how the 'pressures' from the different parts of the organisation can pull the organisation apart and cause unexpected change. Acknowledging 'pressure' is a distinctive feature of this model and no other model researched showed this quality. These differing strengths of the three models could possibly aid safety culture assessment.

2.2 The Safety Professional Interviews

Semi-structured interviews were conducted as it allowed an open discussion instead of a rigid 'question and answer' format. This provides the interviewee with a chance to outline their thoughts as opposed to Likert scales[1]. This also produces a repeatable format which offers a level of consistency between interviews which is important for information comparison. As these semi structured interviews were limited by the amount of time and people, the level of data gathered did not allow for statistical comparison. The purpose of the semi structured interviews with safety professionals was to determine the factors within an organisational model that they consider important to aid safety culture understanding and if any of these factors were present within any of the shortlisted models. The semi structured interviews involved the following professionals:

1. Interview 1 - Ministry of Defence (MOD) Safety Engineer
2. Interview 2 - Atkins Safety Consultant.
3. Interview 3 - Safety and Environmental Management Specialist for Defence Equipment and Support (DE&S).
4. Interview 4 - Aerospace Safety Specialist from DE&S.
5. Interview 5 - An academic safety professional from the University of Aberdeen.

---

[1] Likert scales or rating scales are widely used to scale responses in survey research.

In general, all the interviews concluded that not all of the factors from the organisational models were ideal for assessing safety culture but each had something positive to offer. In relation to the models, the Burke-Litwin model suggests the 'External Environment' factor to be on the top of the model, but the safety professional reviews behind interviewee 1 – interviewee 4 agreed that it would be practical to move this factor to the side of the model and link it to the whole organisation.

Then it was agreed that recognising the 'main influential character' of an organisation was important because this character could easily impose a particular change. 'Influence' was unanimously agreed by all the safety professionals to be contributing factor to change and when analysing 'Influence', figure 5 shows how an influential diagram could be taken into consideration.

Interviewee 1 showed interest in the Burke-Litwin and Minztberg Models. There was an agreement that the 'main influential character' was an important factor in the organisational structure and was not shown on any of the three models. An influential diagram shown in figure 5 was created, to show that 'Management' were the most influential throughout the whole organisation.

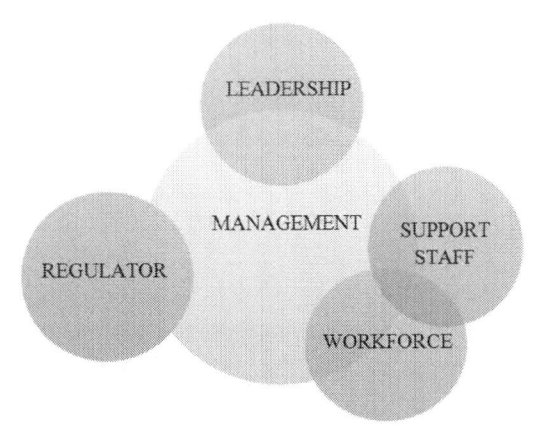

Fig. 5. An Influential Diagram

Interviewee 2 showed interest towards the Minztberg Model (figure 3) as it could be applied to all industries. It was stated that the 'operating core' should focus on the work climate and they should truly understand all the risks involved in their jobs since they are at the forefront of risk. Regarding influence, the techno structure was stated to be the most influential because the organisation needs them as no one else can do their jobs as required. Overall, the Burke-Litwin combined with the Mintzberg model would create an ideal balance and then the Kotter 8-Step change model could be applied. But the Kotter model cannot be applied if the organisation structure is not 'perfect' or practical.

Interviewee 3 was more inclined towards the Burke-Litwin and Kotter 8-Step change model. It was stated that some organisations tend to put "commercial first and safety second". If commercial pressure is applied, any change will be put in place as fast as possible and everyone would comply. Finding one main influencer in the organisation is not possible due to external pressures. This was the only safety professional that emphasised the 'external environment' as the main influential character.

Interviewee 4 dismissed all the models because none of them supported the kind of change experienced by the organisation.

Interviewee 5 was intrigued with the Mintzberg model and the effect of 'pressure' within an organisation. In relation to combining organisational model factors to assess safety culture assessment, the professional stated that scientists choose to have an idealistic view of the organisation instead of a practical view. Scientists assess safety climate at a point in time, but the fact that safety climate is not constant and might change the following week is not considered. The professional reviews behind interview 4 and interview 5 both concluded and agreed that when trying to implement change, it is necessary to thoroughly research on two questions; what is the main purpose of change? What are the long-term effects of this change? This is what is represented behind the factor of 'effective change'.

2.3 Findings

All the safety professionals agreed that the Burke-Litwin model factors could be used to assess safety culture but factors such as "main influential character", "pressure" and "effective change" needed to be considered too. The factors from the organisation model assessment that are going to be used to assess the safety culture tools are:

1. Individual Needs and Values
2. Individual and Organisational Performance
3. Motivation
4. Task and Individual Skills
5. Organisational Culture
6. Structure
7. Mission and Strategy
8. Systems (Policies and Procedures)
9. External Environment
10. Main Influential Character
11. Management
12. Work Climate
13. Effective Change
14. Pressure

3 Safety Culture

"The safety culture of an organisation is the product of individual and group values, attitudes, perceptions, competencies and patterns of behaviour that determine the commitment to, and the style and proficiency of, an organisation's health and safety management" (ACSNI Human Factors Study Group, HSC (1993). A number of other definitions have been established but even though the definitions vary there is a general agreement that safety culture should be a "proactive stance to safety" (Gadd, 2002). The purpose of this section is to identify key factors that aid safety culture assessment.

3.1 Safety Culture Factors

In attempting to assess the Safety Culture of an organisation, a literature review[1] was carried out to identify the factors that will aid safety culture assessment. The factors shown below are those that were most popular during the literature review:

1. Leadership
2. Management
3. Cohesive Culture
4. Rewards
5. Just Culture
6. Safety Archetypes[2]
7. Artefacts[3]
8. Risk
9. Beliefs
10. Safety Values
11. Behaviour
12. Attitudes
13. Training Programmes
14. Communication
15. Working Conditions

---

[1] Literature review included published papers within the last 25 years, mostly recommended by industry consultants and academic lecturers.

[2] Marais and Leveson (2003) state that safety archetypes are system dynamic models that can be used to describe the safety culture of an organisation. In accident analysis, the safety archetypes can be used to identify and highlight change processes and the causal factors that allowed the system to migrate towards an accident state. But they also state that like all models, system archetypes are merely approximations of systems and their behaviour.

[3] Taylor (2010) states that the strength of an organisation's safety culture can be indicated by the presence or absence of artefacts. Artefacts can be formal, documented and physical reminders to all staff of their shared-beliefs, values and behaviours.

3.2 Safety Culture Models

Due to time constraints, the safety culture models and tools discussed in this section are limited. The purpose of this section was to review the two most popular models commonly referred to within Atkins Defence, and assess if any of the chosen models (i.e. The Burke-Litwin Model, Kotter 8-Step Change Model and Mintzberg Model on Organisational Structures) acknowledge either all or more of the factors.

3.1.1 James Reason Safety Culture Model

Professor Reason shows the different components needed to create a positive Safety Culture. According to Reason (1997), the components of a safety culture are included in figure 6.

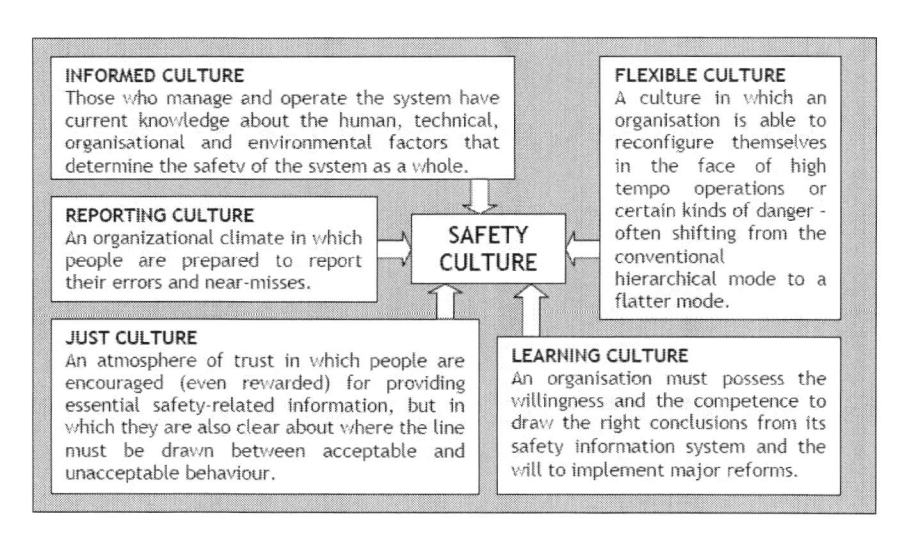

INFORMED CULTURE Those who manage and operate the system have current knowledge about the human, technical, organisational and environmental factors that determine the safety of the system as a whole.		FLEXIBLE CULTURE A culture in which an organisation is able to reconfigure themselves in the face of high tempo operations or certain kinds of danger - often shifting from the conventional hierarchical mode to a flatter mode.
REPORTING CULTURE An organizational climate in which people are prepared to report their errors and near-misses.	SAFETY CULTURE	
JUST CULTURE An atmosphere of trust in which people are encouraged (even rewarded) for providing essential safety-related information, but in which they are also clear about where the line must be drawn between acceptable and unacceptable behaviour.		LEARNING CULTURE An organisation must possess the willingness and the competence to draw the right conclusions from its safety information system and the will to implement major reforms.

Fig. 6. The Components of Safety Culture (Reason, 1997)

3.1.2 Professor Patrick Hudson Safety Culture Model

Hudson (2001) explains that Safety Culture is series of progressive steps as shown in figure 7. Safety cultures can be distinguished along a line from pathological, caring less about safety than about not being caught, through calculative, blindly following all the logically necessary steps, to generative, in which safe behaviour is fully integrated into everything the organisation does. A Safety Culture can only be considered seriously in the later stages of this evolutionary line. Prior to that, up to and including the calculative stage, the term safety culture is best reserved to describe formal and superficial structures rather than an integral part of the overall culture, pervading how the organisation goes about its work. It is obvious that, at the pathological stage, an organisation is not even interested in safety and has to make the first level of acquiring the value system that includes safety as a necessary element. A subsequent stage is one in which safety issues begin to acquire importance, often driven by both internal and external factors as a result of having many incidents. At this first stage of development we can see the values beginning to be acquired, but the beliefs, methods and working practices are still at a primeval stage. At such an early stage, top management believes accidents to be caused by stupidity, inattention and, even, wilfulness on the part of their employees. Many messages may flow from on high, but the majority still reflect the organisation's primary aims, often with 'and be safe' tacked on at the end.

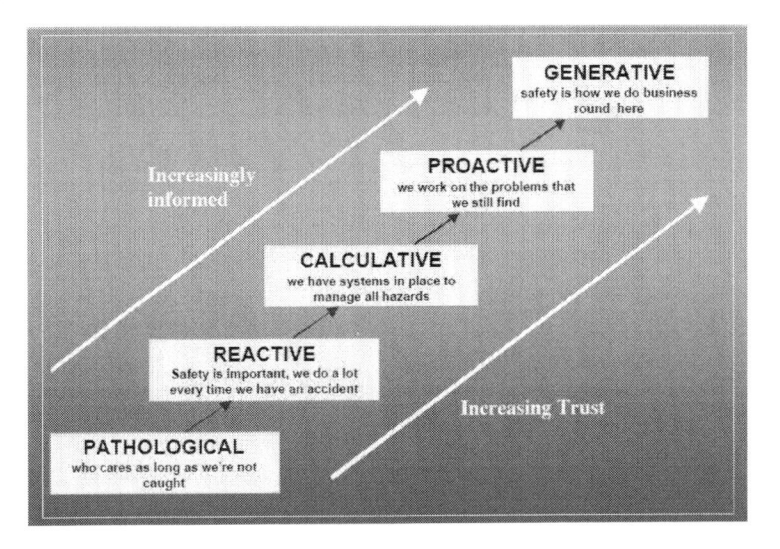

Fig. 7. The Evolutionary Model of Safety Culture (Hudson, 2001)

3.3 Conclusion

The Safety Culture factors consolidated from the research and models are shown in Table 2.

Table 2. A Summary of Factors Discussed in the Section that Aid Safety Culture Assessment
Literature Review: Leadership, Management, Cohesive Culture, Rewards, Just Culture, Safety Archetypes, Artefacts, Risk, Beliefs, Safety Values, Behaviour, Attitudes, Training Programmes, Communication, Working Conditions.
Prof. James Reason Model: Informed Culture, Reporting Culture, Just Culture, Learning Culture and Flexible Culture.
Prof. Patrick Hudson Model[1]: Pathological, Reactive, Calculative, Proactive, And Generative.

As shown in table 2, majority of the factors from the Safety Culture models did not match the literature review. Prof. Hudson's model is a good Safety Culture model as it ideally shows the progressive nature of Safety Culture and can guide organisations to a positive Safety Culture. However, the factors shown in Prof. Reason's model can be adapted to include organisation culture factors to define a combined organisation/safety culture model. For this reason, all the factors concluded from the literature review and Prof. Reason Safety Culture model will be considered during the comparison in the next section.

---

[1] It is noted that the Hudson model does not have variables it has steps. However, it was still added to this table for comparison purposes.

4 Comparison between the Findings of the Organisational Model and the Safety Culture Model

The concept of combining organisational models and safety culture models is nothing new. Flin and Cox (1988) state that "most of the conceptualizations, definitions, and 'measures' developed for safety culture have been derived from the more general notion of organisational culture as used throughout the social and management sciences, and given prominence by organisational theorists". A range of meanings has been attached to safety culture which relate to the concept of organisational culture, three of which were reviewed by the Institution of Occupational Safety and Health (1994). "The first meaning includes those aspects of culture that affect safety (Waring, 1992). The second refers to shared attitudes, values, beliefs and practices concerning safety and the necessity for effective controls. The third relates to the product of individual and group values, attitudes, competencies and patterns of behaviour that determine the commitment to, and the style and proficiency of, an organisation's safety programs (Health and Safety Commission, 1993)".

A model combining the two models should enable easier analysis of the organisation safety culture, predicting the effect of enforcing a particular change and detecting a weakness in the organisation structure. In addition, this 'combined tool' would aid risk management techniques such as Cost Benefit Analysis and in general, the overall process of qualitatively assessing the health and safety performance of the organisation.

4.1 The Comparison

The combined findings from section 2 and 3 are shown in table 3. The table is used to exhibit the common features between the organisational model findings, safety culture literature review and models.

Table 3. A Tick Box Showing Common Factors between the Organisational and Safety Culture Models

Factors	Organisational Model findings	Safety Culture Literature review	Prof. Reason Model
Individual needs and values	✓		
Individual and Organisational Performance	✓		
Motivation	✓		
Task and Individual Skills	✓		
Organisational Culture e.g. Just Culture and Cohesive Culture	✓	✓	✓
Structure	✓		
External Environment	✓		
Main Influential character	✓		
Management	✓	✓	
Work Climate	✓		
Effective Change (Flexible Culture)	✓		✓
Pressure	✓		
Leadership		✓	
Rewards		✓	
Safety Archetypes		✓	
Artefacts	✓	✓	
Risk		✓	
Beliefs		✓	
Safety Values	✓	✓	
Behaviour		✓	
Attitudes		✓	
Training Programme (Learning Culture)		✓	✓
Communication		✓	
Working conditions		✓	
Situation Awareness (Informed Culture)			✓
Decision Making			
Health & Well-being			

Demographics		
Personal Characteristics		
Reporting Culture	✓	✓

For the sake of simpler presentation, a few factors have been grouped under one name since they had similar meaning but just different titles. Table 3 shows a severely low number of common factors between all three columns. There is no reason for this low number and no reason why any of factors considered in Safety Culture cannot be applicable and useful for organisation models too. This shows that there is a potential to develop a common model that can be applicable to any organisation across a wide range of industries.

5 Conclusion

A model that combined both organisation and safety culture model factors was not found. Here lies an opportunity for further research that would enable analysis of the organisation safety culture, predicting the effect of enforcing a particular change and detecting any weaknesses in the organisation structure. In addition, this 'combined model' would aid risk management techniques such as Cost Benefit Analysis and in general, the overall process of assessing the health and safety performance of the organisation.

None of the reviewed literature or safety culture models addressed 'effective change' and 'detecting a weakness' as a factor that would aid safety culture assessment. The factor 'effective change' is important to the safety culture of an organisation because it promotes learning from past changes and encourages continuous improvement. 'Detecting a weakness' within an organisation is important for safety culture assessment because if a weakness cannot be detected then how will the organisation be able to prevent it causing a hazardous event.

My opinion is that factors such as pressure, work climate, motivation are not only important to organisational culture; these factors contribute to the safety culture of an organisation too. As none of the tools reviewed within the safety culture models included all the factors that were understood from a safety culture/organisation point of view, this indicates that a new tool should be developed that addresses this wider scope.

5.1 Recommendations and Suggestions for Future Work

Recommendations from this study would be to continue work on the list of combined factors shown in section 4 in order to develop a tool that could be used across all industries. Further research on a greater selection of organisational models and safety culture models and tools would aid the development of a combined tool.

As semi-structured interviews were carried out for organisational models, a similar process could be carried out with the safety culture models, to get professional opinions on the important factors needed to assess safety culture.

Suggestions for aiding future work would be to research capability maturity models and developing a link/tool that could link organisational models, safety culture models and capability maturity models.

Acknowledgments I would sincerely like to thank Andy German for giving me the opportunity to conduct this research and my Atkins colleagues who were involved in the reviews of this paper. It is much appreciated.

References

A review of Safety Culture and Safety Climate literature for the development of the safety culture inspection toolkit. Research Report 367. Human Engineering for the Health and Safety Executive, 2005.
Safety Culture: A review of the literature. Gadd, S. Sheffield : s.n., 2002, p.1-35.
Burke, W. Warner and Litwin, George H (1992) A Causal Model of Organisational Performance and Change. Vol. 18, p.523-545.
Kotter, John. MindTools. [Online] [Cited: 24 June 2012.] http://www.mindtools.com/pages/article/newPPM_82.htm.
Mintzbergs, Henri. Lindsay Sherwin. [Online] [Cited: 25 June 2012.] http://www.lindsay-sher-win.co.uk/guide_managing_change/html_change_strategy/07_mintzberg.htm.
ESS Consulting. [Online] [Cited: 30 July 2014.] http://www.ess-wa.com/essconsulting/pascale-and-athos-7s-model.
Nadler, David A and Tushman, M L. MindTools. [Online] [Cited: 30 July 2014 http://www.mindtools.com/pages/article/newSTR_95.htm.
Weisbord, Marvin. Proven Models. [Online] [Cited: 30 July 2014.] http://www.provenmodels.com/23/six-boxes/marvin-r.-weisbord.
Marais, Karen and Leveson, Nancy G (2003) Archetypes for Organisational Safety. p.1-15.
Taylor, John Bernard (2010) Safety Culture: Assessing and Changing the Behaviour of Organisations. p.2-38.

Professor James Reason. [Online]. http://www.coloradofirecamp.com/just-culture/definitions-principles.htm.
Professor Patrick Hudson [Online].
http://www.skybrary.aero/bookshelf/books/2417.pdf
Cox, Sue and Flin, Rhona. s.l. : Taylor & Francis (1988) (Work&Stress) Safety Culture: Philosopher's stone or man of straw?, p.189-201.

AUTHOR INDEX

Adrian Allan 1
Katrina Attwood 143
Les Chambers 265
Michele Co 187
Jack W. Davidson 187
M. Elshuber 95
José Miguel Faria 321
Cody Fleming 55
Derek Fowler 117
Nick Golledge 379
S. Gulan 95
Ibrahim Habli 393
Amira Hamilton 285
Paul Hampton 335
Geir Kjetil Hanssen 227
Børge Haugset 227
Benjamin Herd 361
Jason D. Hiser 187
C. Michael Holloway 205
Alexei Iliasov 39
Elizabeth Jacob 409
S. Kandl 95
Tim Kelly 143, 393
John C. Knight 187
Peter Bernard Ladkin 245
Clive Lee 301
Nancy Leveson 55
Michael Luck 361
Kevin Macnish 393
Peter McBurney 361
Chris Megone 393
Simon Miles 361
Thor Myklebust 227
T. Nguyen 95
Anh Nguyen-Tuong 187
Mark Nicholson 393
Mike Parsons 335

Stephen E. Paynter 167
Andrew Rae 393
S. Rieger 95
Benjamin D. Rodes 187
Alexander Romanovsky 39
P. Schrammel 95
R. Sisto 95
Tor Stålhane 227
John Thomas 55
Martin Toland 23
K.R.Wallace 79
Phil Webb 285
Eberechi Weli 1
Chris Wilkinson 55

Made in the USA
Charleston, SC
01 February 2015